GW00372476

L C SMITH

Fundamentals of
Mathematics and Statistics

Fundamentals of Mathematics and Statistics

for Students of Chemistry and Allied Subjects

**C. J. Brookes, I. G. Betteley,
and S. M. Loxston,**
*Department of Mathematics,
University of Aston in Birmingham*

JOHN WILEY & SONS
Chichester · New York · Brisbane · Toronto

Library of Congress Cataloging in Publication Data:

Brookes, C. J.
   Fundamentals of mathematics and statistics
for students of chemistry and allied subjects.

   Includes index.
   1. Mathematics—1961—   2. Mathematical
statistics.   3. Chemistry—Mathematics.
I. Betteley, I. G., joint author.   II. Loxston,
S. M., joint author.   III. Title.
QA37.2.B75   510   78—26110

ISBN 0 471 99733 1   CLOTH
ISBN 0 471 99732 3   PAPER

Typeset by Preface Ltd.,
Salisbury, Wilts.
and printed in Gt Britain
by Page Bros (Norwich) Ltd, Norwich.

Preface

Written primarily for students of Chemistry the book covers much of the mathematics and statistics required in undergraduate Science and Engineering courses at Universities and Colleges. It should also prove useful for engineers and technologists in industry.

Emphasis is placed on the application of mathematics and statistics, each chapter being fully illustrated by means of worked examples. At the end of each chapter there is a large selection of problems with answers, many of which have a chemical bias. The book assumes a knowledge of elementary trigonometry and algebra, and is an extension of the text previously published under the title *Mathematics and Statistics for Chemists*. If preferred, the reader may begin with the statistics and refer back to the mathematics.

Chapter 1 deals with the exponential, logarithmic and hyperbolic functions; it is assumed that the student is familiar with the binomial theorem. Chapter 2 introduces the idea of a complex number and includes a section on De Moivre's theorem. Chapter 3 introduces the differentiation of a single variable together with divers applications including the solution of equations by Newton's approximation technique. It is assumed that the student is familiar with the idea of a limit and also elementary curve sketching. The next two chapters deal with the indefinite and definite integrals and contain for example, sections on areas, volumes and moments of inertia. Taylor's theorem, Maclaurin's theorem and Simpson's rule for numerical integration are also included. A knowledge of partial fractions is assumed. Chapter 6 develops the theory of determinants and matrices, paying particular attention to the solution of equations. Chapter 7 concentrates essentially on vector algebra, including triple products, angular velocity and momentum, and terminates with a short section on the differentiation of vectors. The solution of ordinary differential equations is introduced in Chapter 8 and deals with first order differential equations, second order differential equations, via the Laplace transform, and solution in series using the method of Frobenius. Chapter 9 deals with partial differentiation and contains a section on maxima and minima. Chapter 10 examines the concept of double integration and is readily extended to multiple integrals. The eleventh chapter deals initially with Fourier series and harmonic analysis and introduces the application of Fourier series to the problem of solving certain types of partial differential equations. Group theory is covered in Chapter 12. In addition to the basic mathematical ideas the chapter deals with the symmetry operations for molecules which are so important in molecular theory.

The remaining eight chapters deal with simple statistical theory and assume no prior knowledge of the subject.

Chapter 13 introduces probability and includes permutations and combinations. Statistical terms are defined in the following chapter dealing with general frequency distributions. Methods of simplifying the numerical computation of statistical parameters are explained. In Chapter 15, binomial, Poisson and normal distributions are studied in some detail together with the relationship between them. Measurement and number defective control charts are explained in Chapter 16. Chapter 17 introduces the basic principles of significance testing and confidence limits. The next chapter covers the simple type of analysis of variance problem, considering up to two factors with replicate tests. An illustration of the extensions of the procedures for a larger number of factors is given in the appendix. Chapter 19 deals with simple regression analysis using the method of least squares. The final chapter gives a brief summary of some of the most useful experimental designs and attempts to indicate the general approach to be adopted in order to obtain the maximum efficiency from an experiment.

We are indebted to The Biometrica Trustees for permission to print Tables 1, 4, 5, 6 and 7; to The British Standards Institution for Tables 2, 3 and the table on page 356; also to the Literary Executor of the late Sir Ronald A. Fisher, FRS, Cambridge, to Dr Frank Yates, FRS, Rothampstead, and to Messrs Oliver and Boyd Limited, Edinburgh for permission to reprint sections of Table III from their book, *Statistical Tables for Biological, Agricultural and Medical Research.*

We wish to express our gratitude to the publishers John Wiley and Sons Limited for their cooperation and guidance, and in particular to Dr Howard Jones, Molecular Sciences Editor for his encouragement and helpful suggestions.

C. J. Brookes
I. G. Betteley
S. M. Loxston

Contents

CHAPTER 1

Exponential, Logarithmic and Hyperbolic Functions

1. The exponential function

The *exponential function* exp (x) is defined as

$$\exp(x) = \sum_{r=0}^{\infty} \frac{x^r}{r!}$$

$$= 1 + \frac{x}{1!} + \frac{x^2}{2!} + \frac{x^3}{3!} + \cdots + \frac{x^r}{r!} + \cdots \tag{1}$$

the series being absolutely convergent for all values of x. (See Section 9 of Chapter 8).

From equation (1)

$$\exp(x_1)\exp(x_2) = \left(1 + \frac{x_1}{1!} + \frac{x_1^2}{2!} + \frac{x_1^3}{3!} + \cdots\right)\left(1 + \frac{x_2}{1!} + \frac{x_2^2}{2!} + \frac{x_2^3}{3!} + \cdots\right)$$

$$= 1 + \frac{1}{1!}(x_1 + x_2) + \frac{1}{2!}(x_1^2 + 2x_1 x_2 + x_2^2)$$

$$+ \frac{1}{3!}(x_1^3 + 3x_1^2 x_2 + 3x_1 x_2^2 + x_2^3) + \cdots$$

$$= 1 + \frac{1}{1!}(x_1 + x_2) + \frac{1}{2!}(x_1 + x_2)^2 + \frac{1}{3!}(x_1 + x_2)^3 + \cdots$$

$$= \exp(x_1 + x_2).$$

By repeating the process we obtain

$$\exp(x_1)\exp(x_2) \ldots \exp(x_n) = \exp(x_1 + x_2 + \cdots + x_n) \tag{2}$$

and also

$$\exp(x)\exp(-x) = \exp(0) = 1.$$

Moreover, if we put $x_1 = x_2 = \ldots = x_n = x$ in equation (2), then

$$\{\exp(x)\}^n = \exp(nx)$$

1

and

$$\{\exp(x)\}^{-n} = \frac{1}{\{\exp(x)\}^n} = \frac{1}{\exp(nx)} = \exp(-nx).$$

For $x = 1$,

$$\{\exp(1)\}^n = \exp(n)$$

and hence, writing $\exp(1) = e$

$$\exp(x) \equiv e^x = 1 + \frac{x}{1!} + \frac{x^2}{2!} + \frac{x^3}{3!} + \cdots + \frac{x^r}{r!} + \cdots \qquad (r = 0, 1, 2, \ldots)$$

where e, the *exponential number*, is obtained by putting $x = 1$ in equation (1), i.e.

$$e = 1 + \frac{1}{1!} + \frac{1}{2!} + \frac{1}{3!} + \cdots + \frac{1}{r!} + \cdots$$

$$= 2.71828$$

to five places of decimals.

Similarly,

$$\exp(-x) \equiv e^{-x} = 1 - \frac{x}{1!} + \frac{x^2}{2!} - \frac{x^3}{3!} + \cdots + \frac{(-1)^r x^r}{r!} + \cdots. \qquad (3)$$

The graphs of $y = e^x$ and $y = e^{-x}$ are shown in Fig. 1.1.

Example 1

Show that

$$\lim_{x \to 0} \frac{x\{1 + \exp(-2x)\}}{1 - \exp(-2x)} = 1.$$

From equation (3)

$$\exp(-2x) = 1 - \frac{2x}{1!} + \frac{4x^2}{2!} - \frac{8x^3}{3!} + \cdots$$

$$\therefore \quad 1 + \exp(-2x) = 2\left(1 - x + x^2 - \frac{2x^3}{3} + \cdots\right)$$

and

$$1 - \exp(-2x) = 2x\left(1 - x + \frac{2x^2}{3} - \cdots\right).$$

Hence

$$\frac{x\{1 + \exp(-2x)\}}{1 - \exp(-2x)} = \frac{2x\{1 - x + x^2 - (2x^3/3) + \cdots\}}{2x\{1 - x + (2x^2/3) - \cdots\}}$$

$$= \frac{1 - x + x^2 - \frac{2}{3}x^3 + \cdots}{1 - x + \frac{2}{3}x^2 - \cdots}$$

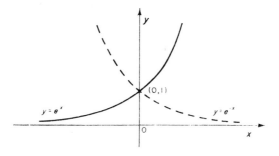

Fig. 1.1

which tends to unity as $x \to 0$, i.e.

$$\lim_{x \to 0} \frac{x\{1 + \exp(-2x)\}}{1 - \exp(-2x)} = 1.$$

Example 2

If x is so small that terms of higher order than x^2 may be neglected, show that

$$e^x\sqrt{(1 + x)} + e^{-x}\sqrt{(1 - x)} \simeq \tfrac{7}{4}x^2 + 2.$$

Since

$$\sqrt{(1 + x)} \equiv (1 + x)^{1/2} = 1 + \tfrac{1}{2}x - \tfrac{1}{8}x^2 + \cdots$$

and

$$\sqrt{(1 - x)} \equiv (1 - x)^{1/2} = 1 - \tfrac{1}{2}x - \tfrac{1}{8}x^2 - \cdots$$

$$\therefore \ e^x\sqrt{(1 + x)} + e^{-x}\sqrt{(1 - x)}$$

$$= \left(1 + \frac{x}{1!} + \frac{x^2}{2!} + \cdots\right)\left(1 + \frac{1}{2}x - \frac{1}{8}x^2 + \cdots\right)$$

$$+ \left(1 - \frac{x}{1!} + \frac{x^2}{2!} + \cdots\right)\left(1 - \frac{1}{2}x - \frac{1}{8}x^2 \cdots\right)$$

$$= 1 + \tfrac{1}{2}x - \tfrac{1}{8}x^2 \qquad 1 - \tfrac{1}{2}x - \tfrac{1}{8}x^2 \quad \text{to } O(x^2)$$

$$+ x + \tfrac{1}{2}x^2 \qquad + \qquad -x + \tfrac{1}{2}x^2$$

$$+ \tfrac{1}{2}x^2 \qquad\qquad + \tfrac{1}{2}x^2$$

$$= 2 + \tfrac{7}{4}x^2.$$

2. The logarithmic function

If $x = e^y$, then y is called the *natural* or *Napierian* logarithm of x and is expressed as

$$y = \log_e x \equiv \ln(x).$$

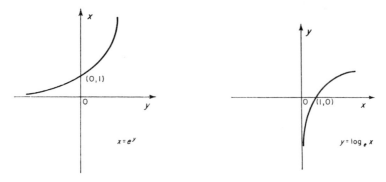

Fig. 1.2

The graph of the function is shown in Fig. 1.2.
From the definition given, we see that

$$\log_e e = 1,$$
$$\log_e e^x = x \log_e e = x,$$
$$\exp(\log_e x) = x.$$
$$\log_e x_1 + \log_e x_2 = \log_e(x_1 x_2)$$

(4)

and

$$\log_e x_1 - \log_e x_2 = \log_e(x_1/x_2).$$

Example 3

In the theory of radiation, the Wien–Planck formula for the intensity of radiation E, of wavelength λ, at a temperature T, is

$$E = \frac{a}{\lambda^5 \{\exp(b/\lambda T) - 1\}}$$

where a and b are constants.

If E_1 and E_2 are the intensities at temperatures T_1 and T_2 respectively for the same wavelength, show that for short wavelengths

$$\log_e\left(\frac{E_1}{E_2}\right) \simeq \frac{b}{\lambda}\left(\frac{1}{T_2} - \frac{1}{T_1}\right).$$

Taking logarithms to base e gives

$$\log_e E = \log_e a - 5 \log_e \lambda - \log_e\left\{\exp\left(\frac{b}{\lambda T}\right) - 1\right\}$$

$$= \log_e a - 5 \log_e \lambda - \log_e\left\{\exp\left(\frac{b}{\lambda T}\right)\right\}\left\{1 - \exp\left(-\frac{b}{\lambda T}\right)\right\}$$

$$= \log_e a - 5 \log_e \lambda - \frac{b}{\lambda T} - \log_e\left\{1 - \exp\left(-\frac{b}{\lambda T}\right)\right\}.$$

But for short wavelengths, $\exp(-b/\lambda T) \to 0$, and therefore

$$\log_e E \simeq \log_e a - 5 \log_e \lambda - \frac{b}{\lambda T} \, .$$

Hence for two different temperatures T_1 and T_2

$$\log_e E_1 \simeq \log_e a - 5 \log_e \lambda - \frac{b}{\lambda T_1} \, ,$$

$$\log_e E_2 \simeq \log_e a - 5 \log_e \lambda - \frac{b}{\lambda T_2} \, .$$

Subtraction now gives

$$\log_e E_1 - \log_e E_2 \equiv \log_e \left(\frac{E_1}{E_2} \right) \simeq \frac{b}{\lambda} \left(\frac{1}{T_2} - \frac{1}{T_1} \right) .$$

3. The logarithmic series

Consider the expansion of the function $(1 + x)^n$ obtained: (i) by means of the binomial theorem, assuming that $|x| < 1$, i.e.

$$(1 + x)^n = 1 + \frac{n}{1!} x + \frac{n(n-1)}{2!} x^2 + \frac{n(n-1)(n-2)}{3!} x^3 + \cdots$$

$$+ \frac{n(n-1)(n-2) \dots (n-r+1)}{r!} x^r + \cdots \, ; \tag{5}$$

and (ii) by re-arranging as an exponential function

$$(1 + x)^n \equiv \exp\{\log_e(1 + x)^n\} \equiv \exp\{n \log_e(1 + x)\}$$

$$= 1 + \frac{n \log_e(1 + x)}{1!} + \frac{\{n \log_e(1 + x)\}^2}{2!} + \cdots$$

$$+ \frac{\{n \log_e(1 + x)\}^r}{r!} + \cdots \, . \tag{6}$$

Equating coefficients of n in equations (5) and (6) now gives

$$\log_e(1 + x) = \frac{x}{1!} - \frac{x^2}{2!} + \frac{2!}{3!} x^3 - \frac{3!}{4!} x^4 + \cdots$$

$$+ \frac{(-1)^{r+1}(r-1)!}{r!} x^r + \cdots$$

i.e.

$$\log_e(1 + x) = x - \frac{1}{2} x^2 + \frac{1}{3} x^3 - \cdots + (-1)^{r+1} \frac{x^r}{r} + \cdots \quad (r = 1, 2, \ldots) \tag{7}$$

which can be shown to be convergent provided $-1 < x \leq +1$.

Replacing x by $-x$ equation (7) becomes

$$\log_e(1-x) = -x - \frac{1}{2}x^2 - \frac{1}{3}x^3 - \cdots - \frac{x^r}{r} - \cdots \quad (r = 1, 2, \ldots) \tag{8}$$

which is convergent for $-1 \leq x < +1$.

On subtraction, equations (7) and (8) give rise to another infinite series which is of particular use in the construction of the Napierian logarithm tables.

For

$$\log_e(1+x) - \log_e(1-x) = \log_e\left(\frac{1+x}{1-x}\right)$$

$$= 2\left(x + \frac{1}{3}x^3 + \frac{1}{5}x^5 + \cdots + \frac{x^{2r-1}}{2r-1} + \cdots\right), \tag{9}$$

for $r = 1, 2, \ldots$, being convergent for $-1 < x < 1$.

Example 4

Determine the value of $\log_e 3$ correct to four places of decimals.

Put $(1+x)/(1-x) = 3$, giving $x = \frac{1}{2}$. Substituting into equation (9) gives

$$\frac{1}{2}\log_e 3 = \frac{1}{2} + \frac{1}{3}\left(\frac{1}{2}\right)^3 + \frac{1}{5}\left(\frac{1}{2}\right)^5 + \frac{1}{7}\left(\frac{1}{2}\right)^7 + \frac{1}{9}\left(\frac{1}{2}\right)^9 + \frac{1}{11}\left(\frac{1}{2}\right)^{11} + \cdots$$

$$= \frac{1}{2} + \frac{1}{24} + \frac{1}{160} + \frac{1}{896} + \frac{1}{4608} + \frac{1}{22,528} + \cdots$$

$$= 0.50000 + 0.04167 + 0.00625 + 0.00112 + 0.00022 + 0.00004 + \cdots$$

$$= 0.54930$$

$\therefore \quad \log_e 3 = 1.0986$ to four places of decimals.

Example 5

Show that the coefficient of x^{-5} in the expansion of

$$\log_e(1+x) + \log_e(2+x) - 2\log_e x$$

is $\frac{33}{5}$.

The given expression is

$$\{\log_e(1+x) - \log_e x\} + \{\log_e(2+x) - \log_e x\}$$

$$= \log_e\left(\frac{1+x}{x}\right) + \log_e\left(\frac{2+x}{x}\right)$$

$$= \log_e\left(1 + \frac{1}{x}\right) + \log_e\left(1 + \frac{2}{x}\right)$$

$$= \left\{\frac{1}{x} - \frac{1}{2}\left(\frac{1}{x}\right)^2 + \cdots + \frac{(-1)^{r+1}}{r}\left(\frac{1}{x}\right)^r + \cdots\right\}$$

$$+ \left\{\frac{2}{x} - \frac{1}{2}\left(\frac{2}{x}\right)^2 + \cdots + \frac{(-1)^{r+1}}{r}\left(\frac{2}{x}\right)^r + \cdots\right\}.$$

Therefore the coefficient of x^{-r} is

$$\frac{(-1)^{r+1}}{r} + \frac{(-1)^{r+1}}{r} 2^r,$$

and hence the coefficient of x^{-5} is

$$\frac{1}{5} + \frac{2^5}{5} = \frac{33}{5}.$$

4. Hyperbolic functions

The function $f(x)$ is said to be an *even* function or *symmetric* if $f(-x) = f(x)$, and an *odd* function or *skew-symmetric* if $f(-x) = -f(x)$.

Thus $\cos x, x^2$ and $x^4 + 3$ are even functions, whereas $\sin x, \tan x$ and x^3 are odd functions.

Although the exponential function e^x is neither an even function nor an odd function, it is convenient in some types of problem to express it as the sum of one even and one odd function.

This is achieved by writing e^x in the form

$$e^x = \tfrac{1}{2}(e^x + e^{-x}) + \tfrac{1}{2}(e^x - e^{-x}).$$

The first of the expressions on the right-hand side of this equation is clearly an even function. It is called the *hyperbolic cosine* of x and is written as

$$\cosh x = \tfrac{1}{2}(e^x + e^{-x}). \tag{10}$$

The second of the expressions however, is an odd function, and is called the *hyperbolic sine* of x. It is written as

$$\sinh x = \tfrac{1}{2}(e^x - e^{-x}). \tag{11}$$

Thus

$$e^x = \cosh x + \sinh x. \tag{12}$$

Replacing x by $-x$ equation (12) gives

$$e^{-x} = \cosh(-x) + \sinh(-x)$$

i.e.

$$e^{-x} = \cosh x - \sinh x, \tag{13}$$

from the definitions given above.

By analogy with the notation used for circular functions, other *hyperbolic functions* are given by

$$\tanh x = \frac{\sinh x}{\cosh x}, \qquad \operatorname{cosech} x = \frac{1}{\sinh x},$$

$$\operatorname{sech} x = \frac{1}{\cosh x}, \qquad \coth x = \frac{1}{\tanh x},$$

each having an exponential form derivable from equations (10) and (11).

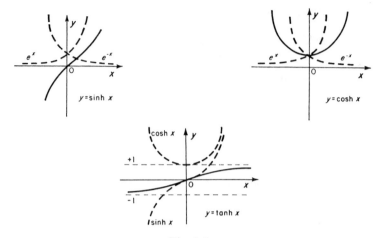

Fig. 1.3

The graphs of the hyperbolic functions can be obtained from those of the exponential functions. Three of these are given in Fig. 1.3.

5. Sum and addition formulae

From equation (11) we have

$$2 \sinh(x + y) = \exp(x + y) - \exp\{-(x + y)\}$$
$$= e^x \cdot e^y - e^{-x} \cdot e^{-y}.$$

Equations (12) and (13) now give

$$2 \sinh(x + y) = (\cosh x + \sinh x)(\cosh y + \sinh y)$$
$$-(\cosh x - \sinh x)(\cosh y - \sinh y)$$
$$= 2(\sinh x \cosh y + \cosh x \sinh y)$$
$$\therefore \sinh(x + y) = \sinh x \cosh y + \cosh x \sinh y. \qquad (14)$$

Similarly

$$\cosh(x + y) = \cosh x \cosh y + \sinh x \sinh y. \qquad (15)$$

Replacing y by $-y$ in equations (14) and (15) gives two more results:

$$\sinh(x - y) = \sinh x \cosh y - \cosh x \sinh y$$

and

$$\cosh(x - y) = \cosh x \cosh y - \sinh x \sinh y.$$

Multiplying together corresponding sides of equations (12) and (13), we obtain a further result

$$e^x \cdot e^{-x} = 1 = (\cosh x + \sinh x)(\cosh x - \sinh x)$$

i.e.

$$\cosh^2 x - \sinh^2 x = 1,$$ (16)

from which may be derived

$$1 - \tanh^2 x = \text{sech}^2 x \quad \text{and} \quad \coth^2 x - 1 = \text{cosech}^2 x.$$

Equations (10) and (16) may now be applied to give other useful formulae. Let y be replaced by x in equations (14) and (15). Then

$$\sinh 2x = 2 \sinh x \cosh x,$$

and

$$\cosh 2x = \cosh^2 x + \sinh^2 x$$
$$= 2 \cosh^2 x - 1$$
$$= 1 + 2 \sinh^2 x,$$

from equation (16).

Re-arranging the last two of these results now gives

$$\sinh^2 x = \tfrac{1}{2}(\cosh 2x - 1) \quad \text{and} \quad \cosh^2 x = \tfrac{1}{2}(\cosh 2x + 1).$$

Example 6

Obtain expansions in ascending powers of x for $\sinh x$ and $\cosh x$.

From the basic definitions

$$2 \sinh x = e^x - e^{-x}$$

$$= \left(1 + \frac{x}{1!} + \frac{x^2}{2!} + \cdots + \frac{x^r}{r!} + \cdots\right)$$

$$- \left(1 - \frac{x}{1!} + \frac{x^2}{2!} - \cdots + \frac{(-1)^r x^r}{r!} + \cdots\right)$$

for $r = 0, 1, 2, \ldots$

$$= 2\left(x + \frac{x^3}{3!} + \frac{x^5}{5!} + \cdots + \frac{x^{2r+1}}{(2r+1)!} + \cdots\right)$$

i.e.

$$\sinh x = x + \frac{x^3}{3!} + \frac{x^5}{5!} + \cdots + \frac{x^{2r+1}}{(2r+1)!} + \cdots \quad (r = 0, 1, 2, \ldots).$$

Similarly,

$$2 \cosh x = e^x + e^{-x}$$

$$= 1 + \frac{x^2}{2!} + \frac{x^4}{4!} + \cdots + \frac{x^{2r}}{(2r)!} + \cdots \quad (r = 0, 1, 2, \ldots).$$

Example 7

Obtain an expression for cosh $3x$ in terms of cosh x only.
  Now

$$\cosh 3x = \cosh(2x + x)$$
$$= \cosh 2x \cosh x + \sinh 2x \sinh x$$
$$= (2 \cosh^2 x - 1)\cosh x + 2 \sinh x \cosh x \sinh x$$
$$= 2 \cosh^3 x - \cosh x + 2 \cosh x(\cosh^2 x - 1)$$

i.e.

$$\cosh 3x = 4 \cosh^3 x - 3 \cosh x.$$

Example 8

Solve the equation $2 \cosh x - \sinh x = 3$.
  In terms of exponentials, the equation may be written

$$(e^x + e^{-x}) - \tfrac{1}{2}(e^x - e^{-x}) = 3$$

i.e.

$$e^x - 6 + 3e^{-x} = 0.$$

Multiplying throughout by e^x now gives

$$e^{2x} - 6e^x + 3 = 0$$

which is a quadratic equation in e^x. Thus

$$e^x = \frac{6 \pm \sqrt{(36 - 12)}}{2} = 3 \pm \sqrt{6}.$$

$$\therefore x = \log_e(3 \pm \sqrt{6}).$$

Example 9

Liquid plastic is forced through a tube of length l and radius r, under a pressure head p.
  If the volume of liquid plastic delivered per second is

$$v = \pi a r^3\left(\frac{1}{x} \cosh x - \frac{2}{x^2} \sinh x + \frac{4}{x^3} \sinh^2 \frac{x}{2}\right)$$

where $x = bpr/(2l)$, a and b being constants, show that for sufficiently small pressures

$$v \simeq \frac{\pi a r^3 x}{4}.$$

We have

$$v = \pi a r^3 \left(\frac{1}{x} \cosh x - \frac{2}{x^2} \sinh x + \frac{2}{x^3} (\cosh x - 1) \right)$$

$$= \pi a r^3 \left\{ \left(\frac{1}{x} + \frac{2}{x^3} \right) \cosh x - \frac{2}{x^2} \sinh x - \frac{2}{x^3} \right\}$$

$$= \pi a r^3 \left\{ \left(\frac{1}{x} + \frac{2}{x^3} \right) \left(1 + \frac{x^2}{2!} + \frac{x^4}{4!} + \cdots \right) - \frac{2}{x^2} \left(x + \frac{x^3}{3!} + \cdots \right) - \frac{2}{x^3} \right\}.$$

Since x is small for small values of p, we may neglect x^2 and higher powers, so that

$$v \simeq \pi a r^3 \left(\frac{x}{2} + \frac{x}{12} - \frac{x}{3} \right)$$

i.e.

$$v \simeq \frac{\pi a r^3 x}{4}.$$

6. Inverse hyperbolic functions

The functions $\sinh^{-1} x$, $\cosh^{-1} x$, $\tanh^{-1} x$ etc., are called *inverse hyperbolic functions*.

If $y = \cosh^{-1} x$, then y is the *number* whose hyperbolic cosine is equal to x, i.e.

$$\cosh y = x, \ \left(not \ y = \frac{1}{\cosh x} \right).$$

Hence

$$\begin{array}{lll} y = \sinh^{-1} x & \text{is equivalent to} & x = \sinh y \\ y = \cosh^{-1} x & \text{is equivalent to} & x = \cosh y \\ y = \tanh^{-1} x & \text{is equivalent to} & x = \tanh y \end{array}$$

and so on.

The graphs of the inverse hyperbolic functions may be obtained from the definitions given above by the same process as that used for inverse circular functions.

This is left as an exercise for the reader.

7. The logarithmic form

Since the hyperbolic functions are defined in terms of exponentials, it is reasonable to suppose that the inverse hyperbolic functions can be expressed in terms of 'inverse exponentials', i.e. in terms of natural logarithms.

That this is in fact so, can be seen from the following examples:

1. $y = \sinh^{-1} x$.

$$\therefore \ x = \sinh y = \tfrac{1}{2}(e^y - e^{-y})$$

$$\therefore \ 2x = e^y - e^{-y}.$$

Multiplying throughout by e^y gives

$$2xe^y = e^{2y} - 1$$

i.e.

$$e^{2y} - 2xe^y - 1 = 0$$

which is a quadratic equation in e^y. The solution is therefore

$$e^y = \frac{2x \pm \sqrt{(4x^2 + 4)}}{2} = x \pm \sqrt{(x^2 + 1)}.$$

Since, however, $e^y > 0$ and $\sqrt{(x^2 + 1)} > x$, the negative sign can be omitted to give

$$e^y = x + \sqrt{(x^2 + 1)}.$$

Hence

$$y = \sinh^{-1} x = \log_e\{x + \sqrt{(x^2 + 1)}\}.$$

2. $y = \tanh^{-1} x$.

$$\therefore \quad x = \tanh y = \frac{\sinh y}{\cosh y} = \frac{e^y - e^{-y}}{e^y + e^{-y}}$$

i.e.

$$x = \frac{e^{2y} - 1}{e^{2y} + 1}$$

$$\therefore \quad x(e^{2y} + 1) = e^{2y} - 1$$

whence

$$e^{2y} = \frac{1 + x}{1 - x}$$

so that

$$y = \tanh^{-1} x = \tfrac{1}{2}\log_e \left(\frac{1 + x}{1 - x}\right), \quad \text{for } |x| < 1.$$

3. $y = \text{sech}^{-1} x$.

$$\therefore \quad x = \text{sech } y = \frac{1}{\cosh y} = \frac{2}{e^y - e^{-y}}$$

i.e.

$$x(e^y - e^{-y}) = 2.$$

Multiplying throughout by e^y now gives

$$xe^{2y} - 2e^y - x = 0.$$

The solution is thus

$$e^y = \frac{2 \pm \sqrt{(4 + 4x^2)}}{2x} = \frac{1 \pm \sqrt{(x^2 + 1)}}{x}.$$

Again, since $\sqrt{(x^2 + 1)} > 1$ we may omit the negative sign to give

$$y = \text{sech}^{-1} x = \log_e \left(\frac{1 + \sqrt{(x^2 + 1)}}{x} \right).$$

The logarithmic form for the remaining inverse functions may be obtained in the same way.

Example 10

Show that

$$2 \tanh^{-1} \left(\frac{x^2}{x^2 + 2} \right) = \log_e (x^2 + 1).$$

Let

$$y = \tanh^{-1} \left(\frac{x^2}{x^2 + 2} \right),$$

then

$$\frac{x^2}{x^2 + 2} = \tanh y = \frac{e^y - e^{-y}}{e^y + e^{-y}} = \frac{e^{2y} - 1}{e^{2y} + 1}$$

i.e.

$$x^2 (e^{2y} + 1) = (x^2 + 2)(e^{2y} - 1)$$

giving

$$e^{2y} = x^2 + 1$$

$$\therefore \ 2y = 2 \tanh^{-1} \left(\frac{x^2}{x^2 + 2} \right) = \log_e (x^2 + 1).$$

Example 11

Show that the curve represented by the equation

$$\cosh^{-1} (4 + y - y^2) = \tanh^{-1} \left(1 - \frac{1}{(x + 1)^4} \right)^{1/2}$$

is the circle $x^2 + y^2 + 2x - y - 3 = 0$.

From the given equation

$$\tanh\{\cosh^{-1} (4 + y - y^2)\} = \left(1 - \frac{1}{(x + 1)^4} \right)^{1/2}$$

$$\therefore \ \tanh^2\{\cosh^{-1} (4 + y - y^2)\}$$

$$\equiv 1 - \text{sech}^2\{\cosh^{-1} (4 + y - y^2)\} = 1 - \frac{1}{(x + 1)^4}.$$

Hence

$$\text{sech}^2\{\cosh^{-1}(4+y-y^2)\} = \frac{1}{(x+1)^4}.$$

∴ Inverting

$$\cosh^2\{\cosh^{-1}(4+y-y^2)\} = (x+1)^4$$

$$\therefore \quad \cosh\{\cosh^{-1}(4+y-y^2)\} = (x+1)^2, \text{ since } \cosh x > 0$$

i.e.

$$4+y-y^2 = (x+1)^2$$

giving

$$x^2 + y^2 + 2x - y - 3 = 0.$$

PROBLEMS

1. Show that

(i) $e^{1/5} = 1.2214$, (ii) $\cosh(\tfrac{1}{4}) = 1.0314$, (iii) $\sinh(\tfrac{1}{4}) = 0.2526$,

and evaluate $\cosh^{-1}(5)$ to four decimal places.

2. Simplify the following expressions:

(i) $\dfrac{\cosh x + 1}{\sinh x}$, (ii) $\dfrac{\cosh 3x + \sinh 3x}{\cosh x - \sinh x}$.

3. If x is small, show that

$$(1 + x + x^2)^{1/x} \simeq (1 + \tfrac{1}{2}x - \tfrac{13}{24}x^2)e \quad \text{to } O(x^2).$$

4. Prove that

$$\lim_{x \to 0} \left(\frac{e^x - 1}{x} \right) = 1.$$

5. Given that

$$\frac{x}{n+x} < \log_e\left(1 + \frac{x}{n}\right) < \frac{x}{n},$$

prove that

$$\lim_{n \to \infty} \left(1 + \frac{x}{n}\right)^n = \exp(x).$$

6. Show that

$$2\sin^2\theta \sinh^2\phi + 2\cos^2\theta \cosh^2\phi = \cosh 2\phi + \cos 2\theta.$$

7. Show that

$$\cosh^3 x + \sinh^3 x - \cosh^2 x \sinh x - \sinh^2 x \cosh x = e^{-x}.$$

8. Obtain expressions for $\sinh 7x$, $\cosh 7x$ and $\tanh 7x$ in terms of $\sinh x$, $\cosh x$ and $\tanh x$ respectively.

9. If $x = \log_e \tan(\theta + \pi/4)$, show that:

    (i) $\sin 2\theta = \tanh x$
    (ii) $\cos 2\theta = \text{sech } x$
    (iii) $\tan 2\theta = \sinh x$
    (iv) $\tan \theta = \tanh (x/2)$.

10. If $3 \sinh x + 2 \cosh x = 6$, determine the value of x.

11. Solve the equation

$$\sinh(\log_e x) + \cosh(\log_e 2x) = 1.$$

12. Prove that $|\tanh x| < 1$, and show that

$$\tanh x = x - \frac{x^3}{3} + \frac{2x^5}{15} + \cdots.$$

Hence evaluate $\tanh(\tfrac{1}{3})$.

13. Given that

$$\sin x = \sum_{n=1}^{\infty} \frac{(-1)^{n+1} x^{2n-1}}{(2n-1)!},$$

show that

$$\lim_{x \to 0} \left(\frac{\sin x - 2 \sinh x + x \cosh x}{x^5} \right) = \frac{1}{30}.$$

14. Show that

    (i) $\tanh^{-1}\left(\dfrac{x^2 - 1}{x^2 + 1}\right) = \log_e x$

    (ii) $\cosh^{-1}(x + 1) = \pm\sinh^{-1}\sqrt{\{x(x + 2)\}}$

    (iii) $\text{cosech}^{-1}(1/x) = \log_e \{x + \sqrt{(x^2 + 1)}\}$.

15. The mass of a crystal at time $t = 0$ is 25 mg, and one hour later its mass is 31 mg. If its increase in mass per hour is equal to two-thirds of its increase in mass for the previous hour, determine the maximum possible mass of the crystal.

16. In the irreversible reaction $A + B \to C$, the volume of C produced at time $t = 0$ is 16.00 cm^3. If there are 18.00 cm^3 of C present one minute later, and subsequently the increase in the amount of C present in the nth minute is $(1/n)$th of the increase produced during the previous minute, determine the maximum volume of C that can be produced.

17. In the experiment of Clément and Desormes for the measurement of γ, using a gas enclosed within a carboy,

$$\gamma = \frac{\log_e P_1 - \log_e P_0}{\log_e P_1 - \log_e P_2}$$

where P_1 is the initial gas pressure, P_2 is the final gas pressure and P_0 is the atmospheric pressure.

If $p_1 = P_1 - P_0$ and $p_2 = P_2 - P_0$, show that for small values of p_1 and p_2

$$\gamma \simeq \frac{p_1}{p_1 - p_2}.$$

18. The osmotic pressure P of a solution at temperature T is given by

$$P = \log_e \left(\frac{1}{1-x}\right)^{RT/V_0}$$

where V_0 is the molecular volume of the solvent under normal pressure and

$$x = \frac{n_1}{n_1 + n_2}$$

n_1 and n_2 being the number of gram molecules of solute and solvent respectively in the solution.

Given that for very dilute solutions n_1 is very much less that n_2, obtain the van't Hoff equation

$$P = \frac{RT}{V} n_1$$

where $V = n_2 V_0$, is the volume of the solution.

CHAPTER 2

Complex Numbers

Consider the algebraic equation $x^2 = c$, c being a given real number. If $c > 0$ we can always find a real number x satisfying this equation. If however, $c < 0$, there is no real solution and further progress would normally be regarded as impossible.

In order to extend our ideas of algebra and, at the same time, overcome difficulties of this kind, it is necessary to introduce the concept of a *complex number*. If a and b are real numbers and $i^2 = -1$ then $a + ib$ is called a complex number, a being the *real part* and b the *imaginary part*.

1. Laws of algebra

Suppose that $a + ib = 0$ and that we may employ ordinary algebraic techniques and write $a = -ib$. Then

$$a^2 = i^2 b^2 = -b^2,$$

which is inadmissable unless both a and b are zero. We conclude therefore, that $a + ib = 0$ if, and only if, $a = b = 0$.

Two complex numbers $a + ib$ and $c + id$ are defined to be equal if, and only if, their real parts are equal and their imaginary parts are equal, i.e. $a = c$ and $b = d$.

The laws of addition, subtraction and multiplication are defined to be analogous to those of the real numbers, i^2 being replaced by -1, real and imaginary parts retaining their separate identities.

Thus

$$(a + ib) \pm (c + id) = (a \pm c) + i(b \pm d)$$

and

$$(a + ib)(c + id) = ac + ibc + iad + i^2 bd = (ac - bd) + i(bc + ad).$$

Division is defined to be the inverse of multiplication, so that if

$$z = \frac{a + ib}{c + id}$$

where $z = x + iy$, then x and y must be found such that

$$(x + iy)(c + id) = a + ib,$$

giving

$$xc - yd = a, \quad yc + xd = b.$$

18

Hence

$$x = \frac{ac + bd}{c^2 + d^2}, \quad y = \frac{bc - ad}{c^2 + d^2}$$

provided $c^2 + d^2 \neq 0$.

Now

$$(c + id)(c - id) = c^2 + d^2$$

and

$$(a + ib)(c - id) = ac + bd + i(bc - ad)$$

verifying that the same values of x and y would have resulted from the conventional process of rationalization. That is, multiplying numerator and denominator by $c - id$, the *conjugate complex* of $c + id$.

Thus

$$\frac{a + ib}{c + id} = \frac{a + ib}{c + id} \times \frac{c - id}{c - id} = \left(\frac{ac + bd}{c^2 + d^2}\right) + i\left(\frac{bc - ad}{c^2 + d^2}\right).$$

Example 1

(i) $(1 + 3i)(3 - 2i) = 3 - 2i + 9i - 6i^2 = 9 + 7i$

(ii) $\dfrac{4 - 3i}{1 + 2i} = \dfrac{4 - 3i}{1 + 2i} \times \dfrac{1 - 2i}{1 - 2i} = \dfrac{4 - 8i - 3i + 6i^2}{1 + 4} = -\dfrac{2}{5} - i\dfrac{11}{5}$

(iii) The solution of the equation $x^2 - 4x + 13 = 0$, is

$$x = \frac{4 \pm \sqrt{(16 - 52)}}{2} = 2 \pm \sqrt{(-9)} = 2 \pm 3i.$$

2, The Argand diagram

Since a complex number $z = x + iy$ is essentially an ordered pair of real numbers (x, y) it is convenient to have a geometrical representation and to regard z as a point in the *complex plane* or *Argand diagram*. This is achieved by referring the point to rectangular axes Ox, Oy — the x-axis being called the *real axis* and the y-axis the *imaginary axis* (see Fig. 2.1).

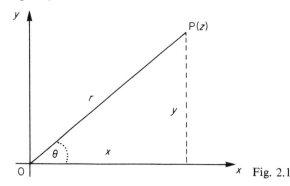

Fig. 2.1

Introducing polar coordinates (r, θ), leads to an alternative, or polar, form of the complex number, and we may write

$$z = x + iy = r(\cos \theta + i \sin \theta)$$

The *positive* length OP is called the *modulus* of z and is written

$$r = OP = +\sqrt{(x^2 + y^2)} = |z|.$$

The angle θ, measured in radians, between OP and the positive direction of the x-axis, is called the *argument* (or *amplitude*) of z and is written

$$\theta = \arg(z) = \tan^{-1}(y/x).$$

We observe that the $\arg(z)$ is multi-valued and choose the principal value to be that for which $-\pi < \theta \leqslant +\pi$.

Example 2

Express $\sqrt{(3)} + i$, $-\sqrt{(3)} + i$, $-2 - 2\sqrt{(3)}i$, $1 - i$ in polar form.
    Now,

$$|\sqrt{(3)} + i| = 2; \quad |-\sqrt{(3)} + i| = 2; \quad |-2 - 2\sqrt{(3)}i| = 4; \quad |1 - i| = \sqrt{2};$$

$$\arg\{\sqrt{(3)} + i\} = \tan^{-1}\{1/\sqrt{(3)}\} = \pi/6; \quad \arg\{-\sqrt{(3)} + i\} = \tan^{-1}\{-1/\sqrt{(3)}\} = 5\pi/6;$$

$$\arg\{-2 - 2\sqrt{(3)}i\} = \tan^{-1}(+\sqrt{3}) = -2\pi/3; \quad \arg(1 - i) = \tan^{-1}(-1) = -\pi/4,$$

the angles in each case corresponding to the appropriate quadrant in which the respective points lie.

$$\therefore \quad \sqrt{(3)} + i = 2\left(\cos\frac{\pi}{6} + i \sin\frac{\pi}{6}\right); \quad -\sqrt{(3)} + i = 2\left(\cos\frac{5\pi}{6} + i \sin\frac{5\pi}{6}\right),$$

$$-2 - 2\sqrt{(3)}i = 4\left(\cos\frac{2\pi}{3} - i \sin\frac{2\pi}{3}\right); \quad 1 - i = \sqrt{(2)}\left(\cos\frac{\pi}{4} - i \sin\frac{\pi}{4}\right).$$

Example 3

Find the locus of z if $\arg(z + 4i) = \pi/3$.
    Let $z = x + iy$. Then

$$\arg(z + 4i) = \arg\{x + i(y + 4)\}$$

$$= \tan^{-1}\left(\frac{y + 4}{x}\right)$$

$$\therefore \quad (y + 4)/x = \tan(\pi/3) = \sqrt{3}$$

showing that the locus is the straight line $y + 4 = \sqrt{(3)}x$.

3. Properties of the modulus and argument

Let

$$z_1 = r_1(\cos \theta_1 + i \sin \theta_1) \quad \text{and} \quad z_2 = r_2(\cos \theta_2 + i \sin \theta_2).$$

Then

$$z_1 z_2 = r_1 r_2 (\cos\theta_1 + i\sin\theta_1)(\cos\theta_2 + i\sin\theta_2)$$
$$= r_1 r_2 \{(\cos\theta_1 \cos\theta_2 - \sin\theta_1 \sin\theta_2) + i(\sin\theta_1 \cos\theta_2 + \cos\theta_1 \sin\theta_2)\}$$
$$= r_1 r_2 \{\cos(\theta_1 + \theta_2) + i\sin(\theta_1 + \theta_2)\}$$

and, it follows that

$$|z_1 z_2| = r_1 r_2 = |z_1||z_2|, \quad \arg(z_1 z_2) = \theta_1 + \theta_2 = \arg(z_1) + \arg(z_2)$$

the latter being chosen in accordance with the definition of principal values outlined above.

Similarly

$$\frac{z_1}{z_2} = \frac{r_1(\cos\theta_1 + i\sin\theta_1)}{r_2(\cos\theta_2 + i\sin\theta_2)} = \frac{r_1(\cos\theta_1 + i\sin\theta_1)(\cos\theta_2 - i\sin\theta_2)}{r_2(\cos\theta_2 + i\sin\theta_2)(\cos\theta_2 - i\sin\theta_2)}$$

$$= \frac{r_1}{r_2}\left(\frac{(\cos\theta_1 \cos\theta_2 + \sin\theta_1 \sin\theta_2) + i(\sin\theta_1 \cos\theta_2 - \cos\theta_1 \sin\theta_2)}{\cos^2\theta_2 + \sin^2\theta_2}\right)$$

$$= \frac{r_1}{r_2}\{\cos(\theta_1 - \theta_2) + i\sin(\theta_1 - \theta_2)\}.$$

Thus

$$\left|\frac{z_1}{z_2}\right| = \frac{r_1}{r_2} = \frac{|z_1|}{|z_2|}, \quad \arg\left(\frac{z_1}{z_2}\right) = \theta_1 - \theta_2 = \arg(z_1) - \arg(z_2).$$

Extending the procedure to the general case we find

$$\left|\frac{z_1 z_2 \cdots z_n}{\omega_1 \omega_2 \cdots \omega_n}\right| = \frac{|z_1||z_2|\cdots|z_n|}{|\omega_1||\omega_2|\cdots|\omega_n|}$$

and

$$\arg\left(\frac{z_1 z_2 \cdots z_n}{\omega_1 \omega_2 \cdots \omega_n}\right) = \arg(z_1) + \arg(z_2) + \cdots + \arg(z_n) - \arg(\omega_1) - \arg(\omega_2)$$
$$- \cdots - \arg(\omega_n).$$

In particular,

$$|z^n| = |z|^n, \quad \arg(z^n) = n\arg(z).$$

Example 4

Simplify

$$z = \frac{\{1 + \sqrt(3)i\}^2 \{2 + 2i\}^3}{\{\sqrt(3) + i\}^4}.$$

We write

$$|z| = \frac{|1 + \sqrt{(3)}i|^2 \ |2 + 2i|^3}{|\sqrt{(3)} + i|^4} = \frac{(2)^2 (2\sqrt{2})^2}{(2)^4} = 2$$

and

$$\arg(z) = \arg\{1 + \sqrt{(3)}i\}^2 + \arg(2 + 2i)^3 - \arg\{\sqrt{(3)} + i\}^4$$
$$= 2 \arg\{1 + \sqrt{(3)}i\} + 3 \arg(2 + 2i) - 4 \arg\{\sqrt{(3)} + i\}$$
$$= 2(\pi/3) + 3(\pi/4) - 4(\pi/6)$$
$$= 3\pi/4$$

$$\therefore \quad z = 2\left(\cos\frac{3\pi}{4} + i \sin\frac{3\pi}{4}\right) = -\sqrt{(2)}(1 - i).$$

4. Equation of a circle

Let $z = x + iy$ and $z_1 = x_1 + iy_1$, so that $z - z_1 = (x - x_1) + i(y - y_1)$. Then

$$|z - z_1| = \sqrt{\{(x - x_1)^2 + (y - y_1)^2\}}$$
$$= \text{distance between the points } z, z_1$$

from which it follows that the equation

$$|z - z_1| = a$$

represents a circle of radius a and centre (x_1, y_1). (See Fig. 2.2.)

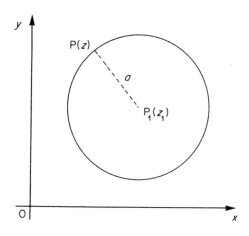

Fig. 2.2

Example 5

Sketch the regions for which

$$\pi/6 \leqslant \arg z \leqslant 2\pi/3 \quad \text{and} \quad |z| \leqslant 3.$$

Since $|z| = 3$ is the equation of a circle of radius 3 units and centre the origin, the interior of the circle may be represented by the inequality $|z| < 3$. Similarly, the interior of the circle *and* the circle itself may be represented by $|z| \leqslant 3$. The specified region is illustrated in Fig. 2.3.

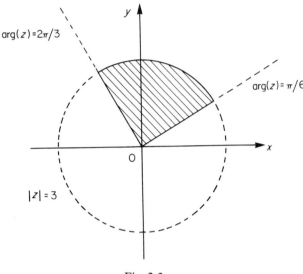

Fig. 2.3

Example 6

Find the locus of z if $|z - 2i| = 3|z + 2i|$.

Let $z = x + iy$. Then

$$|x + i(y - 2)| = 3|x + i(y + 2)|$$

$$\therefore \{x^2 + (y - 2)^2\}^{1/2} = 3\{x^2 + (y + 2)^2\}^{1/2}$$

$$\therefore x^2 + (y - 2)^2 = 9x^2 + 9(y + 2)^2$$

$$\therefore 8x^2 + 8y^2 + 40y + 32 = 0$$

giving

$$x^2 + (y + \tfrac{5}{2})^2 = \tfrac{9}{4}$$

i.e. a circle centre $(0, -5/2)$ and radius $3/2$.

5. De Moivre's theorem

We consider the evaluation of $(\cos\theta + i\sin\theta)^n$ for different forms of n.

Case (i) : n a positive integer Referring to Section 3 we see that

$$(\cos\theta_1 + i\sin\theta_1)(\cos\theta_2 + i\sin\theta_2) = \cos(\theta_1 + \theta_2) + i\sin(\theta_1 + \theta_2).$$

Similarly,

$$(\cos\theta_1 + i\sin\theta_1)(\cos\theta_2 + i\sin\theta_2)(\cos\theta_3 + i\sin\theta_3) = \cos(\theta_1 + \theta_2 + \theta_3)$$
$$+ i\sin(\theta_1 + \theta_2 + \theta_3)$$

and in general

$$(\cos\theta_1 + i\sin\theta_1)(\cos\theta_2 + i\sin\theta_2)\cdots(\cos\theta_n + i\sin\theta_n)$$
$$= \cos(\theta_1 + \theta_2 + \cdots + \theta_n) + i\sin(\theta_1 + \theta_2 + \cdots + \theta_n).$$

Putting $\theta_1 = \theta_2 = \cdots = \theta_n = \theta$, this equation becomes

$$(\cos\theta + i\sin\theta)^n = \cos n\theta + i\sin n\theta$$

Case (ii) : n a negative integer Let $n = -m$ where m is a positive integer. Then

$$(\cos\theta + i\sin\theta)^n = (\cos\theta + i\sin\theta)^{-m}$$

$$= \frac{1}{(\cos\theta + i\sin\theta)^m} = \frac{1}{\cos m\theta + i\sin m\theta}$$

$$= \cos m\theta - i\sin m\theta, \quad \text{by rationalization}$$

$$= \cos(-m\theta) + i\sin(-m\theta)$$

$$= \cos n\theta + i\sin n\theta.$$

Case (iii) : n fractional Let $n = p/q$, p and q being positive integers. Now

$$\left(\cos\frac{p\theta}{q} + i\sin\frac{p\theta}{q}\right)^q = \cos p\theta + i\sin p\theta = (\cos\theta + i\sin\theta)^p \tag{1}$$

and

$$\cos\frac{p\theta}{q} + i\sin\frac{p\theta}{q} = (\cos\theta + i\sin\theta)^{p/q},$$

which is *one* of the several solutions.

To determine the general solution, put $p\theta/q = \phi$ in equation (1), so that

$$(\cos\phi + i\sin\phi)^q = (\cos\theta + i\sin\theta)^p$$

$$\therefore \quad \cos q\phi + i\sin q\phi = \cos p\theta + i\sin p\theta.$$

Equating real parts, $\cos q\phi = \cos p\theta$, and $q\phi = p\theta + 2k\pi$ for $k = 0, \pm1, \pm2, \ldots$. Therefore

$$(\cos\theta + i\sin\theta)^{p/q} = \cos\left(\frac{p\theta}{q} + \frac{2k\pi}{q}\right) + i\sin\left(\frac{p\theta}{q} + \frac{2k\pi}{q}\right). \tag{2}$$

Suppose that k takes q consecutive integral values $k = N$ to $k = N + q - 1$. Then

$$\frac{2N\pi}{q} \leqslant \frac{2k\pi}{q} \leqslant \frac{2(N + q - 1)\pi}{q},$$

a range of less than 2π. Consequently the values of (2) are distinct for each value of k within the specified range. Any other value of k would provide duplicate solutions. It should be noted however, that k can take *any* q integral values which generate q distinct solutions.

Example 7

To evaluate $(1 - i)^{16}$ we express $1 - i$ in modulus–argument form and then apply de Moivre's theorem. Thus,

$$1 - i = \sqrt{(2)}\left(\cos\frac{\pi}{4} - i\sin\frac{\pi}{4}\right)$$

$$\therefore \ (1 - i)^{16} = 2^8\left(\cos\frac{\pi}{4} - i\sin\frac{\pi}{4}\right)^{16} = 2^8(\cos 4\pi - i\sin 4\pi) = 2^8$$

Example 8

Determine the nth roots of unity.
    We write,

$$z^n = 1 = \cos(0) + i\sin(0) = \cos(2k\pi) + i\sin(2k\pi),$$

for *all* integral values of k.
    Applying de Moivre's theorem we have

$$z = \cos\frac{2k\pi}{n} + i\sin\frac{2k\pi}{n}, \quad \text{for } k = 0, 1, 2, \ldots, n - 1.$$

Thus the *cube roots* of unity may be obtained by putting $n = 3$ and $k = 0, 1, 2$, viz.,

$$z_1 = \cos(0) + i\sin(0), \quad z_2 = \cos\frac{2\pi}{3} + i\sin\frac{2\pi}{3}, \quad z_3 = \cos\frac{4\pi}{3} + i\sin\frac{4\pi}{3}$$

$$= 1, \qquad\qquad\qquad = -\frac{1}{2} + \frac{i\sqrt{3}}{2}, \qquad\qquad = -\frac{1}{2} - \frac{i\sqrt{3}}{2}.$$

Example 9

Find the five values of $\{\sqrt{(3)} + i\}^{1/5}$.
    In terms of modulus and argument

$$\sqrt{(3)} + i = 2(\cos\tfrac{1}{6}\pi + i\sin\tfrac{1}{6}\pi) = 2\{\cos(2k + \tfrac{1}{6})\pi + i\sin(2k + \tfrac{1}{6})\pi\}$$

for all integral values of k. Thus

$$\{\sqrt{(3)} + i\}^{1/5} = 2^{1/5}\left(\cos\frac{(2k + \tfrac{1}{6})\pi}{5} + i\sin\frac{(2k + \tfrac{1}{6})\pi}{5}\right), \quad \text{for } k = 0, 1, 2, 3, 4.$$

Example 10

Express $\cos 5\theta$ in terms of multiples of $\cos\theta$.

We write,

$$\cos 5\theta + i \sin 5\theta = (\cos \theta + i \sin \theta)^5$$

$$= \cos^5 \theta + 5 \cos^4 \theta(i \sin \theta) + 10 \cos^3 \theta(i \sin \theta)^2$$
$$+ 10 \cos^2 \theta(i \sin \theta)^3 + 5 \cos \theta(i \sin \theta)^4 + (i \sin \theta)^5$$
$$= \cos^5 \theta + 5i \cos^4 \theta \sin \theta - 10 \cos^3 \theta \sin^2 \theta - 10i \cos^2 \theta \sin^3 \theta$$
$$+ 5 \cos \theta \sin^4 \theta + i \sin^5 \theta.$$

Equating real parts:

$$\cos 5\theta = \cos^5 \theta - 10 \cos^3 \theta \sin^2 \theta + 5 \cos \theta \sin^4 \theta$$

$$= \cos^5 \theta - 10 \cos^3 \theta(1 - \cos^2 \theta) + 5 \cos \theta(1 - \cos^2 \theta)^2$$

$$= 16 \cos^5 \theta - 20 \cos^3 \theta + 5 \cos \theta.$$

6. Expansion of $\cos^n \theta$, $\sin^n \theta$ (n integral)

Let

$$z = \cos \theta + i \sin \theta, \quad \text{so that} \quad z^n = \cos n\theta + i \sin n\theta,$$

$$\frac{1}{z} = \cos \theta - i \sin \theta \quad \text{and} \quad \frac{1}{z^n} = \cos n\theta - i \sin n\theta.$$

Then

$$z + \frac{1}{z} = 2 \cos \theta, \qquad z - \frac{1}{z} = 2i \sin \theta,$$

$$z^n + \frac{1}{z^n} = 2 \cos n\theta, \quad z^n - \frac{1}{z^n} = 2i \sin n\theta,$$

and application of the binomial theorem to $\{z \pm (1/z)\}^n$ will generate the required expansions, as in the following examples.

Example 11

Express $\cos^5 \theta$ in terms of multiple angles.
   We write,

$$2^5 \cos^5 \theta = \left(z + \frac{1}{z}\right)^5$$

$$= z^5 + 5z^3 + 10z + \frac{10}{z} + \frac{5}{z^3} + \frac{1}{z^5}$$

$$= \left(z^5 + \frac{1}{z^5}\right) + 5\left(z^3 + \frac{1}{z^3}\right) + 10\left(z + \frac{1}{z}\right)$$

$$= 2 \cos 5\theta + 5(2 \cos 3\theta) + 10(2 \cos \theta)$$

$$\therefore \quad \cos^5 \theta = \tfrac{1}{16}(\cos 5\theta + 5 \cos 3\theta + 10 \cos \theta).$$

Example 12

Express $\sin^7 \theta$ in terms of multiple angles.

In this case we write

$$(2i)^7 \sin^7 \theta = -i2^7 \sin^7 \theta$$

$$= \left(z - \frac{1}{z}\right)^7$$

$$= z^7 - 7z^5 + 21z^3 - 35z + \frac{35}{z} - \frac{21}{z^3} + \frac{7}{z^5} - \frac{1}{z^7}$$

$$= \left(z^7 - \frac{1}{z^7}\right) - 7\left(z^5 - \frac{1}{z^5}\right) + 21\left(z^3 - \frac{1}{z^3}\right) - 35\left(z - \frac{1}{z}\right)$$

$$= 2i \sin 7\theta - 7(2i \sin 5\theta) + 21(2i \sin 3\theta) - 35(2i \sin \theta)$$

$\therefore \qquad \sin^7 \theta = \frac{1}{64}(-\sin 7\theta + 7 \sin 5\theta - 21 \sin 3\theta + 35 \sin \theta).$

7. Euler's relation

In Chapter 1 we introduced the *real* exponential function

$$e^x = \sum_{r=0}^{\infty} \frac{x^r}{r!} = 1 + \frac{x}{1!} + \frac{x^2}{2!} + \frac{x^3}{3!} + \cdots + \frac{x^r}{r!} + \cdots.$$

Extending our ideas to the *complex* domain, we define

$$e^z = \sum_{r=0}^{\infty} \frac{z^r}{r!} = 1 + \frac{z}{1!} + \frac{z^2}{2!} + \frac{z^3}{3!} + \cdots + \frac{z^r}{r!} + \cdots \qquad (3)$$

and preserve the properties of the real domain by postulating that $e^{z_1 + z_2} = e^{z_1} e^{z_2}$.

Replacing the variable by iy in equation (3)

$$e^{iy} = 1 + \frac{iy}{1!} + \frac{i^2 y^2}{2!} + \frac{i^3 y^3}{3!} + \cdots + \frac{i^r y^r}{r!} + \cdots$$

$$= \left(1 - \frac{y^2}{2!} + \frac{y^4}{4!} - \cdots\right) + i\left(y - \frac{y^3}{3!} + \frac{y^5}{5!} - \cdots\right)$$

$$= \cos y + i \sin y, \qquad (4)$$

(see Chapter 5) which is known as Euler's relation.

With $z = x + iy$, it follows that

$$e^z = e^{x+iy} = e^x \cdot e^{iy} = e^x(\cos y + i \sin y).$$

Replacing y by $-y$, equation (4) gives

$$e^{-iy} = \cos y - i \sin y$$

and combining this equation with equation (4) we find

$$\cos y = \tfrac{1}{2}(e^{iy} + e^{-iy}), \qquad \sin y = \frac{1}{2i}(e^{iy} - e^{-iy}).$$

Comparison with the exponential forms of the hyperbolic functions (Chapter 1) shows that

$$\cos y = \cosh iy, \quad i \sin y = \sinh iy,$$

and

$$\cosh y = \cos iy, \quad i \sinh y = \sin iy.$$

Example 13

Find the real and imaginary parts of

(i) $\cos z$, (ii) $\log_e z$, (iii) $\cos^{-1} 2$.

(i) $\quad \cos z = \cos(x + iy)$

$$= \cos x \cos iy - \sin x \sin iy$$

$$= \cos x \cosh y - i \sin x \sinh y.$$

(ii) $\quad \log_e z = \log_e r(\cos \theta + i \sin \theta) = \log_e re^{i\theta}$

$$= \log_e r + \log_e \exp\{i(\theta \pm 2k\pi)\}, \quad k \text{ integral}$$

$$= \log_e r + i(\theta \pm 2k\pi).$$

(iii) Let $z = x + iy = \cos^{-1} 2$, i.e. $\cos(x + iy) = 2$. Using (i) and equating real and imaginary parts

$$\cos x \cosh y = 2, \quad \sin x \sinh y = 0.$$

From the second equation, either $\sinh y = 0$ or $\sin x = 0$. If $\sinh y = 0$ then $y = 0$ and $\cosh y = 1$, resulting in $\cos x = 2$ which is inadmissable. Thus $\sin x = 0$ and $x = 2k\pi$ (k integral), the value $2k\pi$ being necessary — rather than $k\pi$ — to ensure that $\cos x > 0$. With $x = 2k\pi$, $\cos x = +1$ and $\cosh y = 2$, i.e. $y = \cosh^{-1} 2$
$$= \log_e(2 \pm \sqrt{3}) = \pm \log_e(2 + \sqrt{3}).$$
Hence

$$\cos^{-1} 2 = 2k\pi \pm i \log_e(2 + \sqrt{3}), \quad k \text{ integral}.$$

Example 14

Sum the series

$$\cos \theta + \cos 3\theta + \cdots + \cos(2n - 1)\theta.$$

Let

$$C = \cos \theta + \cos 3\theta + \cdots + \cos(2n - 1)\theta$$

$$S = \sin \theta + \sin 3\theta + \cdots + \sin(2n - 1)\theta,$$

then

$$C + iS = (\cos \theta + i \sin \theta) + (\cos 3\theta + i \sin 3\theta) + \cdots + \{\cos(2n - 1)\theta$$

$$+ i \sin(2n - 1)\theta\}$$

$$= e^{i\theta} + e^{i3\theta} + \cdots + e^{i(2n-1)\theta},$$

which is a geometric progression with common ratio $e^{i2\theta}$.

$$\therefore \; C + iS = \frac{e^{i\theta}(1 - e^{i2n\theta})}{1 - e^{i2\theta}}$$

$$= \frac{e^{i\theta} \cdot e^{in\theta}(e^{-in\theta} - e^{in\theta})}{e^{i\theta}(e^{-i\theta} - e^{i\theta})}$$

$$= \frac{(\cos n\theta + i \sin n\theta)(-2 \sin n\theta)}{-2 \sin \theta}.$$

Equating real parts

$$C = \cos \theta + \cos 3\theta + \cdots + \cos(2n - 1)\theta = \frac{\cos n\theta \, \sin n\theta}{\sin \theta}$$

PROBLEMS

1. Express in the form $a + ib$:

  (i) $(2 - 5i)(3 + i)$,   (ii) $\dfrac{1 + 2i}{4 - i}$,   (iii) $\dfrac{(1 + i)(2 - 3i)}{(i - 1)(2 - 2i)}$.

2. Solve the following equations:

  (i) $z^2 = -49$,   (ii) $z^2 + 4z + 8 = 0$,   (iii) $z^3 + z + 2 = 0$.

3. Express the following in polar form:

  (i) $-3i$,   (ii) $1 + i$,   (iii) $-1 + \sqrt{(3)}i$,   (vi) $-\sqrt{(3)} - i$,   (v) $1 - \sqrt{(3)}i$.

4. Simplify the expression

$$z = \frac{\{1 - \sqrt{(3)}i\}^2 \{\sqrt{(3)} + i\}^4}{(2 + 2i)^3}.$$

5. Determine the locus of the point $P \equiv P(z)$, if

  $\arg(z - 3i) = 3\pi/4$.

6. Find the locus of z if

  $2|z + 3i| - |z - i| = 0$.

7. Simplify

  (i) $(\sin \theta - i \cos \theta)^5$,   (ii) $(\cos \theta - i \sin \theta)^{-2}$,   (iii) $\{2(\cos \tfrac{1}{4}\pi + i \sin \tfrac{1}{4}\pi)\}^6$.

8. Use de Moivre's theorem to evaluate

  (i) $(\sqrt{(3)} - i)^{12}$,   (ii) $(-\sqrt{(3)} + i)^{18}$.

9. Determine the values of

  (i) $i^{1/6}$,   (ii) $(2 - 2i)^{1/3}$.

10. Express $\sin 6\theta$, $\cos 6\theta$ in terms of $\sin \theta$, $\cos \theta$ and deduce an expression for $\tan 6\theta$ in terms of $\tan \theta$ only.

11. Express $\sin^8 \theta$ in terms of multiple angles.

12. Find the real and imaginary parts of

(i) $\sin z$, (ii) $\log_e(5 + 12i)$, (iii) $\sin^{-1} 3$.

13. Prove that

$$|z_1| - |z_2| \leqslant |z_1 + z_2| \leqslant |z_1| + |z_2|.$$

14. If z is a complex number, show that

(i) $\sin^{-1} z = -i \log_e\{iz \pm \sqrt{(1 - z^2)}\}$, (ii) $\tan^{-1} z = \dfrac{i}{2} \log_e\left(\dfrac{i + z}{i - z}\right)$.

15. In the mathematical model of the Morse curve for diatomic molecules, the wavefunctions ψ_1, ψ_2 corresponding to zero and finite potential energy levels respectively, are given by

$$\psi_1 = a_1 e^{inx} + b_1 e^{-inx}$$

$$\psi_2 = a_2 e^{mx} + b_2 e^{-mx},$$

where n and m are functions of the mass and energy of the particle, and a_1, a_2, b_1, b_2 are constants. If $\psi_1 = \psi_2$ and $\psi_1' = \psi_2'$ at $x = \alpha$ (dashes denoting differentiation with respect to x — see Chapter 3), show that

$$m \tan(n\alpha) + n = 0.$$

16. In the determination of linear combinations of atomic orbitals, suitable for use in the theory of wave mechanics, an equation of the form

$$z^2 + z + 1 = 0$$

arises. Show that the solutions may be expressed in the form

$$z_1 = e^{2i\pi/3}, \quad z_2 = e^{-2i\pi/3}.$$

17. In the theory of quantum chemistry, it can be shown that the probability p, of transmission of a particle through a potential energy barrier, is given by $p = (z\bar{z})^{-1}$, where

$$z = \tfrac{1}{2} e^{i\alpha}(a \cosh \beta - ib \sinh \beta),$$

a, b, α, β being functions of the mass and energy of the particle.
   Show that for large values of β

$$p \simeq \frac{4e^{-2\beta}}{a^2 + b^2}.$$

18. The complex potential for a two-dimensional fluid flow is given by

$$w = \frac{1}{z - a} - \frac{1}{z + a}$$

where a is a constant. If $w = \phi + i\psi$, show that the streamlines corresponding to $\psi = 1$ are given by

$$(x^2 - y^2 - a^2)^2 + 4x^2 y^2 - 4axy = 0.$$

19. In the flow of fluid around a circular cylinder of radius a, due to sources of equal strength m at points on opposite sides of the cylinder, the complex velocity of a particle at a point $P(a, \theta)$ is given by

$$u - iv = \frac{m(e^{4i\theta} - 1)}{a e^{i\theta}(1 - 9e^{2i\theta})(e^{2i\theta} - 9)}.$$

Show that the magnitude of the fluid velocity at P is

$$\frac{m \sin 2\theta}{a(41 - 9 \cos 2\theta)}.$$

CHAPTER 3

Differentiation and Applications

1. The derivative of a function

Let $f(x)$ be a continuous and single-valued function of the variable x, i.e. to every value of x there corresponds one and only one value of $f(x)$.

Then, if $y = f(x)$, an increment δx in x will produce a corresponding increment δy in y, so that

$$y = f(x) \tag{1}$$

and

$$y + \delta y = f(x + \delta x). \tag{2}$$

Subtracting equation (1) from equation (2) now gives

$$\delta y = f(x + \delta x) - f(x)$$

so that

$$\frac{\delta y}{\delta x} = \frac{f(x + \delta x) - f(x)}{\delta x} \tag{3}$$

which represents the slope of the straight line through PP′ (see Fig. 3.1).

The limiting value of the quotient $\delta y/\delta x$ as δx tends to zero is called the *derivative* of y with respect to x and is written

$$\frac{dy}{dx} = \lim_{\delta x \to 0} \left(\frac{\delta y}{\delta x} \right) = \lim_{\delta x \to 0} \frac{f(x + \delta x) - f(x)}{\delta x}, \tag{4}$$

the process of obtaining dy/dx being called *differentiation*.

Referring to Fig. 3.1 we note that as $\delta x \to 0$, the point P′ → P. In the limiting case the chord PP′ thus becomes the tangent to the curve $y = f(x)$ at the point P. The gradient at P is therefore

$$\frac{dy}{dx} = \tan \psi$$

where ψ is the angle between the tangent and the positive direction of the x-axis.

Alternative notations for dy/dx are:

$$y', \quad y_1, \quad f'(x), \quad \frac{df}{dx}.$$

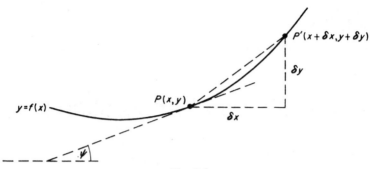

Fig. 3.1

In general dy/dx will still be a function of x and may be differentiated with respect to x to give the *second derivative* of y or $f(x)$ with respect to x, which is written as $(d/dx)(dy/dx)$ or d^2y/dx^2.

Alternatively, we may write,

$$y'', \quad y_2, \quad f''(x), \quad \frac{d^2f}{dx^2}.$$

Similarly, higher derivatives such as d^3y/dx^3, d^4y/dx^4, etc., may exist and are obtained as above.

Returning now to equation (3) we see that

$$\frac{\delta x}{\delta y} = \frac{\delta x}{f(x+\delta x) - f(x)} = \frac{1}{\{f(x+\delta x) - f(x)\}/\delta x}.$$

Thus since $\delta y \to 0$ as $\delta x \to 0$

$$\frac{dx}{dy} = \lim_{\delta x \to 0} \frac{1}{\{f(x+\delta x) - f(x)\}/\delta x}$$

$$= \frac{\lim_{\delta x \to 0}(1)}{\lim_{\delta x \to 0} \{f(x+\delta x) - f(x)\}/\delta x}$$

$$= \frac{1}{dy/dx}.$$

It is important to note, however, that in general

$$\frac{d^2y}{dx^2} \neq \frac{1}{d^2x/dy^2}, \quad \text{(see p. 215)}.$$

2. Derivative of a sum

Let y be the sum of several continuous and single-valued functions of x, say $f_1(x)$, $f_2(x), \ldots$, i.e.

$$y = f_1(x) + f_2(x) + \cdots.$$

If an increment δx in x produces an increment δy in y, then

$$y + \delta y = f_1(x + \delta x) + f_2(x + \delta x) + \cdots,$$

giving, according to equation (4),

$$\frac{dy}{dx} = \lim_{\delta x \to 0} \frac{1}{\delta x} \{f_1(x + \delta x) + f_2(x + \delta x) + \cdots - f_1(x) - f_2(x) - \cdots\}$$

$$= \lim_{\delta x \to 0} \frac{f_1(x + \delta x) - f_1(x)}{\delta x} + \lim_{\delta x \to 0} \frac{f_2(x + \delta x) - f_2(x)}{\delta x} + \cdots$$

i.e.

$$\frac{dy}{dx} = \frac{df_1}{dx} + \frac{df_2}{dx} + \cdots.$$

The derivative of the sum of several functions is thus equal to the sum of their respective derivatives.

3. Derivative of a product

Let $y = f(x)g(x)$, where $f(x)$ and $g(x)$ are continuous and single-valued functions of x.

Then,

$$y + \delta y = f(x + \delta x)g(x + \delta x)$$

giving

$$\delta y = f(x + \delta x)g(x + \delta x) - f(x)g(x)$$

$$= \{f(x + \delta x) - f(x)\}g(x + \delta x) + \{g(x + \delta x) - g(x)\}f(x)$$

$$\therefore \quad \frac{dy}{dx} = \lim_{\delta x \to 0} \left(\frac{\delta y}{\delta x}\right) = \lim_{\delta x \to 0} \frac{f(x + \delta x) - f(x)}{\delta x} \cdot g(x + \delta x)$$

$$+ \lim_{\delta x \to 0} \frac{g(x + \delta x) - g(x)}{\delta x} \cdot f(x).$$

Since $g(x + \delta x) \to g(x)$ as $\delta x \to 0$

$$\therefore \quad \frac{dy}{dx} = \frac{df(x)}{dx} \cdot g(x) + \frac{dg(x)}{dx} \cdot f(x).$$

4. Derivative of a quotient

Let

$$y = \frac{f(x)}{g(x)},$$

$f(x)$ and $g(x)$ being continuous and single-valued functions of x.

Then

$$y + \delta y = \frac{f(x + \delta x)}{g(x + \delta x)}$$

and

$$\delta y = \frac{f(x + \delta x)}{g(x + \delta x)} - \frac{f(x)}{g(x)}$$

$$= \frac{f(x + \delta x)g(x) - g(x + \delta x)f(x)}{g(x + \delta x)g(x)}$$

$$= \frac{\{f(x + \delta x) - f(x)\}g(x) - \{g(x + \delta x) - g(x)\}f(x)}{g(x + \delta x)g(x)}$$

$$\therefore \frac{dy}{dx} = \lim_{\delta x \to 0} \left(\frac{\delta y}{\delta x}\right)$$

$$= \lim_{\delta x \to 0} \left(\frac{\dfrac{f(x + \delta x) - f(x)}{\delta x} \cdot g(x) - \dfrac{g(x + \delta x) - g(x)}{\delta x} \cdot f(x)}{g(x + \delta x)g(x)}\right)$$

$$= \frac{\dfrac{df(x)}{dx} \cdot g(x) - \dfrac{dg(x)}{dx} \cdot f(x)}{\{g(x)\}^2}$$

5. Function of a function

Suppose that $y = F(u)$ where u is a function $f(x)$ of x.
Then

$$\frac{dy}{dx} = \frac{dF(u)}{dx} = \lim_{\substack{\delta x \to 0 \\ \delta u \to 0}} \frac{F(u + \delta u) - F(u)}{\delta x}$$

$$= \lim_{\substack{\delta x \to 0 \\ \delta u \to 0}} \frac{F(u + \delta u) - F(u)}{\delta u} \cdot \frac{\delta u}{\delta x} = \frac{dF(u)}{du} \cdot \frac{du}{dx}.$$

But $u = f(x)$ so that $du/dx = f'(x)$.

$$\therefore \frac{dy}{dx} = F'(u)f'(x).$$

6. Derivatives of known functions

(i) $f(x) = c$ (c = constant).

Since c is a constant, the increment δx in x will produce no change in c, so that the derivative of c with respect to x is zero.

(ii) $f(x) = \sin x$.

$$f(x + \delta x) = \sin(x + \delta x)$$

$$\therefore \frac{d}{dx}(\sin x) = \lim_{\delta x \to 0} \frac{\sin(x + \delta x) - \sin x}{\delta x}$$

$$= \lim_{\delta x \to 0} \frac{2\cos(x + \frac{1}{2}\delta x)\sin(\frac{1}{2}\delta x)}{\delta x}$$

$$= \lim_{\delta x \to 0} \cos(x + \frac{1}{2}\delta x) \cdot \frac{\sin(\frac{1}{2}\delta x)}{(\frac{1}{2}\delta x)} \; .$$

But

$$\lim_{\delta x \to 0} \frac{\sin(\frac{1}{2}\delta x)}{(\frac{1}{2}\delta x)} = 1, \quad \therefore \quad \frac{d(\sin x)}{dx} = \cos x.$$

Similarly

$$\frac{d}{dx}(\cos x) = -\sin x.$$

(iii) $f(x) = \tan x$.

$$\therefore \frac{d}{dx}(\tan x) = \frac{d}{dx}\left(\frac{\sin x}{\cos x}\right)$$

$$= \frac{\cos x \cos x - \sin x(-\sin x)}{\cos^2 x}$$

i.e.

$$\frac{d(\tan x)}{dx} = \frac{\cos^2 x + \sin^2 x}{\cos^2 x} = \sec^2 x.$$

Similarly, it can be shown that

$$\frac{d}{dx}(\operatorname{cosec} x) = -\operatorname{cosec} x \cot x,$$

$$\frac{d}{dx}(\sec x) = \sec x \tan x,$$

and

$$\frac{d}{dx}(\cot x) = -\operatorname{cosec}^2 x.$$

(iv) $f(x) = x^n$, where n is
(a) *a positive integer*:

$$f(x + \delta x) = (x + \delta x)^n$$

$$\therefore \frac{d}{dx}(x^n)$$

$$= \lim_{\delta x \to 0} \frac{(x + \delta x)^n - x^n}{\delta x}$$

$$= \lim_{\delta x \to 0} \frac{\left(x^n + \frac{n}{1!}x^{n-1}(\delta x) + \frac{n(n-1)}{2!}x^{n-2}(\delta x)^2 + \cdots + (\delta x)^n\right) - x^n}{\delta x}$$

$$= \lim_{\delta x \to 0} \left(\frac{n}{1!}x^{n-1} + \frac{n(n-1)}{2!}x^{n-2}(\delta x) + \cdots + (\delta x)^{n-1}\right)$$

i.e.

$$\frac{d}{dx}(x^n) = nx^{n-1}.$$

(b) *a negative integer*:

Let $n = -k$ where k is a positive integer. Then

$$f(x) = x^{-k} = \frac{1}{x^k}.$$

Now

$$\frac{d}{dx}\left(\frac{1}{x^k}\right) = \frac{0 \cdot x^k - (1)(kx^{k-1})}{x^{2k}} = -kx^{-(k+1)}$$

$$\therefore \frac{d}{dx}(x^{-k}) = \frac{d}{dx}\left(\frac{1}{x^k}\right) = -kx^{-(k+1)}.$$

(c) *fractional*:

Let $n = k_1/k_2$ where k_1 and k_2 are positive integers. Then

$$f(x) = x^{k_1/k_2} = y \text{ (say)}$$

$$\therefore y^{k_2} = x^{k_1}.$$

But

$$\frac{d}{dx}(y^{k_2}) = \frac{d}{dy}(y^{k_2})\frac{dy}{dx} = k_2 y^{k_2 - 1}\frac{dy}{dx}$$

$$\therefore k_2 y^{k_2 - 1}\frac{dy}{dx} = k_1 x^{k_1 - 1}$$

where

$$y^{k_2 - 1} = (x^{k_1/k_2})^{k_2 - 1} = x^{k_1(k_2 - 1)/k_2}$$

$$\therefore \frac{dy}{dx} = \frac{k_1}{k_2} \cdot x^{k_1 - 1} \cdot x^{-k_1(k_2 - 1)/k_2} = \frac{k_1}{k_2}x^{k_1/k_2 - 1}.$$

Thus

$$\frac{d}{dx}(x^5) = 5x^4; \quad \frac{d}{dx}(x^{-7}) = -7x^{-8}; \quad \frac{d}{dx}(x^{3/4}) = \frac{3}{4}x^{-1/4};$$

and so on.

Example 1

Determine dy/dx from first principles if $y = \cot x$.

Let an increment δx in x produce an increment δy in y. Then

$$y = \cot x.$$

$$y + \delta y = \cot(x + \delta x)$$

$$\therefore \quad \delta y = \cot(x + \delta x) - \cot x = \frac{\cos(x + \delta x)}{\sin(x + \delta x)} - \frac{\cos x}{\sin x}$$

$$= \frac{\cos(x + \delta x)\sin x - \cos x \sin(x + \delta x)}{\sin(x + \delta x)\sin x} = \frac{-\sin \delta x}{\sin(x + \delta x)\sin x}$$

$$\therefore \quad \frac{dy}{dx} = \lim_{\delta x \to 0}\left(\frac{\delta y}{\delta x}\right) = \lim_{\delta x \to 0}\frac{-1}{\sin(x + \delta x)\sin x} \cdot \frac{\sin \delta x}{\delta x}$$

$$= \frac{-1}{\sin^2 x} = -\operatorname{cosec}^2 x.$$

Example 2

Find the derivative with respect to x of $\sin 3x$, $\cos 4x$ and $\tan 2x$ respectively.

Let

$$y = \sin 3x = \sin u \quad \text{where} \quad u = 3x.$$

Then

$$\frac{dy}{dx} = \frac{d(\sin u)}{dx} = \frac{d(\sin u)}{du}\frac{du}{dx} = \cos u \times 3$$

i.e.

$$\frac{dy}{dx} = 3\cos 3x.$$

Similarly

$$\frac{d}{dx}(\cos 4x) = (-\sin 4x)(4) = -4\sin 4x$$

and

$$\frac{d}{dx}(\tan 2x) = (\sec^2 2x)(2) = 2\sec^2 2x.$$

Example 3

Find dy/dx if $y = (5x^3 - 7x^2 + x - 1)^4$.

Let $u = 5x^3 - 7x^2 + x - 1$, so that $du/dx = 15x^2 - 14x + 1$. Then

$$\frac{dy}{dx} = \frac{d}{dx}(u^4) = \frac{d}{du}(u^4)\frac{du}{dx}$$

$$= 4u^3(15x^2 - 14x + 1)$$

$$= 4(5x^3 - 7x^2 + x - 1)^3(15x^2 - 14x + 1).$$

(v) $f(x) = \log_e x$.

$$f(x + \delta x) = \log_e(x + \delta x)$$

$$\therefore \quad \frac{d(\log_e x)}{dx} = \lim_{\delta x \to 0}\frac{\log_e(x + \delta x) - \log_e x}{\delta x}$$

$$= \lim_{\delta x \to 0}\frac{\log_e(1 + \delta x/x)}{\delta x}.$$

Putting $\delta x/x = 1/n$, we have

$$\frac{d(\log_e x)}{dx} = \lim_{n \to \infty}\frac{n\log_e(1 + 1/n)}{x}.$$

Since $\log_e$ is a continuous function, the $\log_e$ and limit may be interchanged to give

$$\frac{d(\log_e x)}{dx} = \frac{1}{x}\log_e\left\{\lim_{n \to \infty}\left(1 + \frac{1}{n}\right)^n\right\} = \frac{1}{x}\log_e(e)$$

(see Chapter 1).

Hence,

$$\frac{d}{dx}(\log_e x) = \frac{1}{x}.$$

(vi) $f(x) = e^x$.

The derivative of e^x with respect to x may be derived from (v). For, if $y = e^x$ then $\log_e y = x$ and

$$\frac{d}{dx}(\log_e y) = \frac{d}{dy}(\log_e y)\frac{dy}{dx} = \frac{d}{dx}(x)$$

i.e.

$$\frac{1}{y}\frac{dy}{dx} = 1$$

$$\therefore \quad \frac{d}{dx}(e^x) = e^x.$$

Example 4

Find the derivative with respect to x of the functions

$$\log_e(x^2 + 3x + 1), \quad \exp(5 \sin x).$$

(a) Let $y = \log_e u$, where $u = x^2 + 3x + 1$. Then

$$\frac{du}{dx} = 2x + 3,$$

and

$$\frac{dy}{dx} = \frac{d}{dx}(\log_e u) = \frac{d}{du}(\log_e u)\frac{du}{dx}$$

$$= \frac{1}{u}(2x + 3) = \frac{2x + 3}{x^2 + 3x + 1}.$$

(b) Let $y = e^u$ where $u = 5 \sin x$ and $du/dx = 5 \cos x$. Then

$$\frac{dy}{dx} = \frac{d}{du}(e^u)\frac{du}{dx} = e^u \cdot 5 \cos x$$

i.e.

$$\frac{dy}{dx} = 5 \cos x \cdot \exp(5 \sin x).$$

Example 5

Show that $y = e^{-x}(\cos 2x + \sin 2x)$ satisfies the differential equation

$$\frac{d^2y}{dx^2} + 2\frac{dy}{dx} + 5y = 0.$$

Differentiating with respect to x

$$\frac{dy}{dx} = -e^{-x}(\cos 2x + \sin 2x) + e^{-x}(-2 \sin 2x + 2 \cos 2x)$$

$$= e^{-x}(\cos 2x - 3 \sin 2x).$$

$$\therefore \quad \frac{dy^2}{dx^2} = -e^{-x}(\cos 2x - 3 \sin 2x) + e^{-x}(-2 \sin 2x - 6 \cos 2x)$$

$$= e^{-x}(\sin 2x - 7 \cos 2x).$$

Hence

$$\frac{d^2y}{dx^2} + 2\frac{dy}{dx} + 5y = e^{-x}(\sin 2x - 7 \cos 2x) + 2e^{-x}(\cos 2x - 3 \sin 2x)$$

$$+ 5e^{-x}(\cos 2x + \sin 2x)$$

$$= 0.$$

Example 6

Find dy/dx if $y = (\sec x)/x$.

Section 4 gives

$$\frac{dy}{dx} = \frac{x\dfrac{d}{dx}(\sec x) - \sec x\dfrac{d}{dx}(x)}{x^2}$$

$$= \frac{x\sec x\tan x - \sec x}{x^2}$$

$$= \frac{\sec x}{x^2}(x\tan x - 1).$$

Example 7

In a vertical cylinder of water containing a suspension of particles of gamboge, if there are x particles per unit volume at a height y, then

$$ky = \frac{1}{1 - bx} + \log_e\left(\frac{x}{1 - bx}\right) + \text{constant}$$

where k is a constant, b corresponding to the b in Van der Waal's equation.
Show that

$$\frac{dy}{dx} = \frac{1}{kx(1 - bx)^2}$$

and that d^2y/dx^2 is zero when $x = 1/(3b)$.

Differentiating with respect to x

$$k\frac{dy}{dx} = \frac{b}{(1 - bx)^2} + \frac{1 - bx}{x}\cdot\frac{(1)(1 - bx) - x(-b)}{(1 - bx)^2}$$

$$= \frac{b}{(1 - bx)^2} + \frac{1}{x(1 - bx)}$$

$$= \frac{1}{x(1 - bx)^2}$$

$$\therefore\; k\frac{dy^2}{dx^2} = \frac{0\cdot x(1 - bx)^2 - (1)\{(1 - bx)^2 + x\cdot 2(1 - bx)(-b)\}}{x^2(1 - bx)^4}$$

$$= \frac{(3bx - 1)}{x^2(1 - bx)^3}, \qquad x \neq \frac{1}{b}.$$

which is zero when $x = 1/(3b)$.

7. Differentiation of hyperbolic functions

(i) $f(x) = \sinh x$.

$$\frac{d}{dx}(\sinh x) = \frac{d}{dx}\left(\frac{e^x - e^{-x}}{2}\right) = \frac{1}{2}\frac{d}{dx}(e^x - e^{-x})$$

$$= \tfrac{1}{2}(e^x + e^{-x})$$

i.e.

$$\frac{d}{dx}(\sinh x) = \cosh x.$$

(ii) $f(x) = \cosh x.$

$$\frac{d}{dx}(\cosh x) = \frac{d}{dx}\left(\frac{e^x + e^{-x}}{2}\right) = \frac{1}{2}\frac{d}{dx}(e^x + e^{-x})$$

$$= \tfrac{1}{2}(e^x - e^{-x})$$

i.e.

$$\frac{d}{dx}(\cosh x) = \sinh x.$$

(iii) $f(x) = \coth x.$

$$\frac{d}{dx}(\coth x) = \frac{d}{dx}\left(\frac{\cosh x}{\sinh x}\right)$$

$$= \frac{\sinh x \cdot \sinh x - \cosh x \cdot \cosh x}{\sinh^2 x}$$

$$= \frac{\sinh^2 x - \cosh^2 x}{\sinh^2 x} = \frac{-1}{\sinh^2 x}$$

i.e.

$$\frac{d}{dx}(\coth x) = -\operatorname{cosech}^2 x.$$

Table 3.1 compares the derivatives of circular and hyperbolic functions.

TABLE 3.1

Circular function	Derivative w.r.t. x	Hyperbolic function	Derivative w.r.t. x
$\sin x$	$\cos x$	$\sinh x$	$\cosh x$
$\cos x$	$-\sin x$	$\cosh x$	$\sinh x$
$\tan x$	$\sec^2 x$	$\tanh x$	$\operatorname{sech}^2 x$
$\operatorname{cosec} x$	$-\operatorname{cosec} x \cot x$	$\operatorname{cosech} x$	$-\operatorname{cosech} x \coth x$
$\sec x$	$\sec x \tan x$	$\operatorname{sech} x$	$-\operatorname{sech} x \tanh x$
$\cot x$	$-\operatorname{cosec}^2 x$	$\coth x$	$-\operatorname{cosech}^2 x$

Example 8

If

$$y = \frac{\sinh 2x + \cosh 2x}{\sinh x + \cosh x}$$

obtain dy/dx.

Differentiating with respect to x gives

$$\frac{dy}{dx} = \frac{\begin{array}{c}(2\cosh 2x + 2\sinh 2x)(\sinh x + \cosh x)\\ -(\sinh 2x + \cosh 2x)(\cosh x + \sin x)\end{array}}{(\sinh x + \cosh x)^2}$$

$$= \frac{\sinh 2x + \cosh 2x}{\sinh x + \cosh x} = y.$$

Example 9

If

$$y = \log_e\left(\frac{\coth x + 1}{\coth x - 1}\right)$$

show that $dy/dx = 2$.

Re-arranging the given expression, we obtain

$$y = \log_e(\coth x + 1) - \log_e(\coth x - 1)$$

$$\therefore \frac{dy}{dx} = \frac{-\text{cosech}^2 x}{\coth x + 1} + \frac{\text{cosech}^2 x}{\coth x - 1}$$

$$= \frac{-\text{cosech}^2 x(\coth x - 1) + \text{cosech}^2 x(\coth x + 1)}{(\coth x + 1)(\coth x - 1)}$$

$$= \frac{2\,\text{cosech}^2 x}{\coth^2 x - 1} = 2$$

since $\coth^2 x - 1 = \text{cosech}^2 x$.

8. Inverse functions

(i) $f(x) = \sin^{-1} x$.

Let $y = \sin^{-1} x$, then $x = \sin y$. Differentiating with respect to y gives

$$\frac{dx}{dy} = \cos y = \pm\sqrt{(1 - \sin^2 y)} = \pm\sqrt{(1 - x^2)}.$$

$$\therefore \frac{dy}{dx} = \frac{1}{dx/dy} = \frac{1}{\sqrt{(1 - x^2)}}$$

if we take the principal value of $\sin^{-1} x$.

(ii) $f(x) = \tan^{-1} x$.

Let $y = \tan^{-1} x$, then $x = \tan y$ and

$$\frac{dx}{dy} = \sec^2 y = 1 + \tan^2 y = 1 + x^2$$

i.e.

$$\frac{dy}{dx} = \frac{1}{1 + x^2}.$$

(iii) $f(x) = \operatorname{cosec}^{-1} x$.

Let $y = \operatorname{cosec}^{-1} x$, then $x = \operatorname{cosec} y$ and

$$\frac{dx}{dy} = -\operatorname{cosec} y \cot y = \pm \operatorname{cosec} y \sqrt{(\operatorname{cosec}^2 y - 1)}$$

$$= \pm x \sqrt{(x^2 - 1)}.$$

$$\therefore \frac{dy}{dx} = \frac{-1}{x\sqrt{(x^2 - 1)}}$$

taking the principal value of $\operatorname{cosec}^{-1} x$.

(iv) $f(x) = \sinh^{-1} x$.

Let $y = \sinh^{-1} x$, then $x = \sinh y$ and

$$\frac{dx}{dy} = \cosh y = +\sqrt{(1 + \sinh^2 y)}, \quad \text{since } \cosh y > 0$$

$$= \sqrt{(1 + x^2)}.$$

$$\therefore \frac{dy}{dx} = \frac{1}{\sqrt{(1 + x^2)}}.$$

(v) $f(x) = \operatorname{sech}^{-1} x$.

Let $y = \operatorname{sech}^{-1} x$, then $x = \operatorname{sech} y$ and

$$\frac{dx}{dy} = -\operatorname{sech} y \tanh y = \pm\operatorname{sech} y \sqrt{(1 - \operatorname{sech}^2 y)}$$

$$= \pm x\sqrt{(1 - x^2)}.$$

$$\therefore \frac{dy}{dx} = \frac{\pm 1}{x\sqrt{(1 - x^2)}},$$

both signs being admissible, since there are two equal and opposite values of y for each value of x.

Table 3.2 compares the derivatives of inverse circular and inverse hyperbolic functions.

TABLE 3.2

Inverse circular	Derivative w.r.t. x (principal value)	Inverse hyperbolic	Derivative w.r.t. x
$\sin^{-1} x$	$\dfrac{1}{\sqrt{(1-x^2)}}$	$\sinh^{-1} x$	$\dfrac{1}{\sqrt{(x^2+1)}}$
$\cos^{-1} x$	$\dfrac{-1}{\sqrt{(1-x^2)}}$	$\cosh^{-1} x$	$\dfrac{\pm 1}{\sqrt{(x^2+1)}}$
$\tan^{-1} x$	$\dfrac{1}{1+x^2}$	$\tanh^{-1} x$	$\dfrac{1}{1-x^2},\quad x^2 < 1$
$\operatorname{cosec}^{-1} x$	$\dfrac{-1}{x\sqrt{(x^2-1)}}$	$\operatorname{cosech}^{-1} x$	$\dfrac{\pm 1}{x\sqrt{(x^2+1)}}$
$\sec^{-1} x$	$\dfrac{1}{x\sqrt{(x^2-1)}}$	$\operatorname{sech}^{-1} x$	$\dfrac{\pm 1}{x\sqrt{(x^2+1)}}$
$\cot^{-1} x$	$\dfrac{-1}{1+x^2}$	$\coth^{-1} x$	$\dfrac{-1}{x^2-1},\quad x^2 > 1$

Example 10

Determine dy/dx if $y = (x^2 - 1)\sin^{-1} 3x$.

Differentiating with respect to x gives

$$\frac{dy}{dx} = 2x \sin^{-1} 3x + (x^2 - 1)\frac{3}{\sqrt{(1 - 9x^2)}}.$$

Example 11

If

$$y = \tan^{-1}\left(\frac{x}{x+1}\right) + \tan^{-1}\left(\frac{x+1}{x}\right)$$

show that $dy/dx = 0$.

Differentiating with respect to x, we obtain

$$\frac{dy}{dx} = \frac{1}{\left\{1 + \left(\dfrac{x}{x+1}\right)^2\right\}} \cdot \frac{(1)(x+1) - (x)(1)}{(x+1)^2}$$

$$+ \frac{1}{\left\{1 + \left(\dfrac{x+1}{x}\right)^2\right\}} \cdot \frac{(1)(x) - (x+1)(1)}{x^2}$$

$$= \frac{(x+1)^2}{\{(x+1)^2 + x^2\}} \cdot \frac{1}{(x+1)^2} + \frac{x^2}{\{x^2 + (x+1)^2\}} \cdot \frac{(-1)}{x^2}$$

$$= 0.$$

9. Implicit differentiation

So far we have only considered cases in which y has been expressed explicitly as a function of x. When this is not so, it is usually easier to obtain dy/dx by differentiating the expression containing y 'as it stands', rather than first attempting to derive y explicitly as a function of x.

This will necessitate the determination of the derivative of terms containing combinations of y and its derivatives.

Referring to Section 5, dealing with a function of a function, we see that

$$\frac{d}{dx}(y^2) = \frac{d}{dy}(y^2)\frac{dy}{dx} = 2y\frac{dy}{dx},$$

$$\frac{d}{dx}\left(\frac{dy}{dx}\right)^2 = 2\left(\frac{dy}{dx}\right)\frac{d}{dx}\left(\frac{dy}{dx}\right) = 2\frac{dy}{dx}\cdot\frac{d^2y}{dx^2},$$

$$\frac{d}{dx}\left(y^3\frac{dy}{dx}\right) = \left(3y^2\frac{dy}{dx}\right)\left(\frac{dy}{dx}\right) + y^3\left(\frac{d^2y}{dx^2}\right)$$

$$= 3y^2\left(\frac{dy}{dx}\right)^2 + y^3\frac{d^2y}{dx^2},$$

and so on.

Example 12

Determine dy/dx and d^2y/dx^2 at the point $(1, 0)$ on the curve

$$3x^2y - xy^2 + 2xy + x - y - 1 = 0.$$

Differentiating *implicitly* with respect to x, we obtain

$$\left(6xy + 3x^2\frac{dy}{dx}\right) - \left(y^2 + x\,2y\frac{dy}{dx}\right) + 2\left(y + x\frac{dy}{dx}\right) + 1 - \frac{dy}{dx} = 0$$

i.e.

$$\frac{dy}{dx}(3x^2 - 2xy + 2x - 1) + 6xy - y^2 + 2y + 1 = 0 \tag{5}$$

giving $dy/dx = -\frac{1}{4}$ at the point $(1, 0)$.

Differentiating equation (5) implicitly with respect to x gives

$$\frac{d^2y}{dx^2}(3x^2 - 2xy + 2x - 1) + \frac{dy}{dx}\left(6x - 2y - 2x\frac{dy}{dx} + 2\right)$$

$$+ 6y + 6x\frac{dy}{dx} - 2y\frac{dy}{dx} + 2\frac{dy}{dx} = 0.$$

Substituting in the values $x = 1$, $y = 0$ and $dy/dx = -\frac{1}{4}$ now gives $d^2y/dx^2 = \frac{33}{32}$.

Example 13

Show that $y = \sqrt{(x^2 + 1)}\sinh^{-1} x$ satisfies the differential equation

$$(x^2 + 1)\frac{d^2 y}{dx^2} + x\frac{dy}{dx} - y = 2x.$$

Given

$$y = (x^2 + 1)^{1/2} \sinh^{-1} x$$

$$\therefore \frac{dy}{dx} = \tfrac{1}{2}(x^2 + 1)^{-1/2} (2x)\sinh^{-1} x + (x^2 + 1)^{1/2} \frac{1}{\sqrt{(x^2 + 1)}}$$

$$\therefore (x^2 + 1)^{1/2} \frac{dy}{dx} = x \sinh^{-1} x + (x^2 + 1)^{1/2}.$$

Differentiating implicitly with respect to x

$$\tfrac{1}{2}(x^2 + 1)^{-1/2} (2x)\frac{dy}{dx} + (x^2 + 1)^{1/2} \frac{d^2 y}{dx^2}$$

$$= \sinh^{-1} x + \frac{x}{\sqrt{(x^2 + 1)}} + \tfrac{1}{2}(x^2 + 1)^{-1/2} (2x)$$

i.e.

$$x\frac{dy}{dx} + (x^2 + 1)\frac{d^2 y}{dx^2} = \sqrt{(x^2 + 1)}\sinh^{-1} x + x + x$$

$$= y + 2x$$

$$\therefore (x^2 + 1)\frac{d^2 y}{dx^2} + x\frac{dy}{dx} - y = 2x.$$

Example 14

The distance of a particle from the origin at time t is x, where

$$xe^t = a \sinh t + b \cosh t$$

a and b being constants.

Show that the motion of the particle satisfies the differential equation

$$\frac{d^2 x}{dt^2} + 2\frac{dx}{dt} = 0.$$

Differentiating implicitly with respect to t

$$\therefore \frac{dx}{dt}e^t + xe^t = a \cosh t + b \sinh t$$

$$\therefore \frac{d^2 x}{dt^2}e^t + \frac{dx}{dt}e^t + \frac{dx}{dt}e^t + xe^t = a \sinh t + b \cosh t = xe^t$$

$$\therefore \quad e^t\left(\frac{d^2x}{dt^2} + 2\frac{dx}{dt}\right) = 0$$

i.e.

$$\frac{d^2x}{dt^2} + 2\frac{dx}{dt} = 0 \quad \text{since } e^t \neq 0.$$

10. Parametric differentiation

Suppose that $y = f(x)$, i.e. y is a function of the single variable x.

If x is now expressed as a function of t, i.e. $x = X(t)$ say, then y may also be expressed as a function of t, i.e. $y = Y(t)$ say.

The equations

$$x = X(t), \quad y = Y(t)$$

are called *parametric equations* for the function $y = f(x)$, t being the *parameter*.
Since

$$y = Y(t)$$

$$\therefore \quad \frac{dy}{dx} = \frac{d}{dx}Y(t) = \frac{d}{dt}Y(t)\frac{dt}{dx}$$

$$= \frac{dy}{dt}\frac{dt}{dx}$$

i.e.

$$\frac{dy}{dx} = \frac{dy/dt}{dx/dt}.$$

Example 15

Determine dy/dx and d^2y/dx^2 if $x = 3t^2$ and $y = 6t$.

Now

$$\frac{dx}{dt} = 6t, \quad \frac{dy}{dt} = 6$$

$$\therefore \quad \frac{dy}{dx} = \frac{dy/dt}{dx/dt} = \frac{1}{t}.$$

$$\therefore \quad \frac{d^2y}{dx^2} = \frac{d}{dx}\left(\frac{1}{t}\right) = \frac{d}{dt}\left(\frac{1}{t}\right)\frac{dt}{dx}$$

$$= -\frac{1}{t^2}\cdot\frac{1}{6t} = -\frac{1}{6t^3}.$$

Example 16

Determine dy/dx and d^2y/dx^2 if $x = \sin 2\theta$ and $y = \cos 2\theta$.

We have

$$\frac{dx}{d\theta} = 2 \cos 2\theta, \quad \frac{dy}{d\theta} = -2 \sin 2\theta$$

$$\therefore \quad \frac{dy}{dx} = \frac{dy/d\theta}{dx/d\theta} = -\frac{2 \sin 2\theta}{2 \cos 2\theta} = -\tan 2\theta.$$

$$\therefore \quad \frac{d^2y}{dx^2} = \frac{d}{dx}(-\tan 2\theta) = -\frac{d}{d\theta}(\tan 2\theta)\frac{d\theta}{dx}$$

$$= -2 \sec^2 2\theta \; \frac{1}{2 \cos 2\theta}$$

$$= -\sec^3 2\theta.$$

Example 17

Determine dy/dx and d^2y/dx^2 if

$$x = 2 \cos \theta + \cos 2\theta \quad \text{and} \quad y = 2 \sin \theta + \sin 2\theta.$$

In this case

$$\frac{dx}{d\theta} = -2 \sin \theta - 2 \sin 2\theta, \quad \frac{dy}{d\theta} = 2 \cos \theta + 2 \cos 2\theta$$

$$= -2(\sin \theta + \sin 2\theta) \qquad\qquad = 2(\cos \theta + \cos 2\theta)$$

$$= -4 \sin \frac{3\theta}{2} \cos \frac{\theta}{2} \qquad\qquad = 4 \cos \frac{3\theta}{2} \cos \frac{\theta}{2}.$$

$$\therefore \quad \frac{dy}{dx} = \frac{dy/d\theta}{dx/d\theta} = \frac{-4 \cos(3\theta/2)\cos(\theta/2)}{4 \sin(3\theta/2)\cos(\theta/2)} = -\cot(3\theta/2).$$

Differentiating again with respect to x, we obtain

$$\frac{d^2y}{dx^2} = \frac{d}{dx}\left(-\cot\frac{3\theta}{2}\right) = -\frac{d}{d\theta}\left(\cot\frac{3\theta}{2}\right)\frac{d\theta}{dx}$$

$$= \frac{3}{2} \operatorname{cosec}^2 \frac{3\theta}{2} \cdot \frac{1}{-4 \sin(3\theta/2)\cos(\theta/2)}$$

$$= -\tfrac{3}{8} \operatorname{cosec}^3(3\theta/2)\sec(\theta/2).$$

11. Maxima and minima

Consider the point D at $x = d$, on the curve $y = f(x)$ shown in Fig. 3.2.

For values of x *just* less than $x = d$, it can be seen from the diagram that the gradient of the given curve, and hence $f'(x)$, is negative, whereas for values of x *just* greater than $x = d$, the gradient and hence $f'(x)$, is positive.

At the point $x = d$ itself, the gradient of the curve is zero, so that $f'(x)$ is also zero.

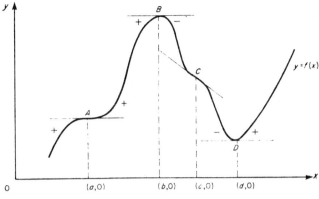

Fig. 3.2

The function $f(x)$ is said to possess a *relative minimum* or simply a *minimum* at the point D.

Alternatively, since *in the neighbourhood* of $x = d$, the gradient of the curve moves from a negative value, through zero, to a positive value as x increases, i.e. $f'(x)$ increases as x increases, the rate of change of $f'(x)$ with respect to x is positive, i.e. $f''(x) > 0$.

Thus at the point D, the function satisfies the conditions

$$f'(d) = 0, \quad f''(d) > 0.$$

Consider next the point B at which $x = b$. By a similar argument to that above, it can be seen that for x *just* less than b, $f'(x)$ is positive; for x *just* greater than b, $f'(x)$ is negative; and for $x = b$, $f'(x)$ is again zero.

In this case the function $f(x)$ is said to possess a *relative maximum* or simply a *maximum* at the point B.

Thus, in the neighbourhood of $x = b$, the value of $f'(x)$ decreases as x increases, so that $f''(x) < 0$, and the function satisfies the conditions

$$f'(b) = 0, \quad f''(b) < 0.$$

It can be seen that as we pass through the points A and C the curve changes from 'concave-up' to 'concave-down' or vice versa. Such points, at which $f''(x) = 0$, are called *points of inflexion*. It should be noted that for the point $x = a$ to be a point of inflexion, $f'(a)$ need not necessarily be zero.

The values of the function $f(x)$ at A, B and D, where $f'(x) = 0$, are referred to as *stationary values* of the function.

Thus in general, for the function $f(x)$ to possess a stationary value at $x = a$, then $f'(a) = 0$.

In particular, for the point to be a maximum we must also have at

$$x = a - \epsilon, \quad f'(x) > 0$$

$$x = a, \qquad f'(x) = 0$$

$$x = a + \epsilon, \quad f'(x) < 0$$

or

$$f''(\alpha) < 0$$

where ϵ is a small positive number.

For a minimum, at

$$x = \alpha - \epsilon, \quad f'(x) < 0$$
$$x = \alpha \qquad \quad f'(x) = 0$$
$$x = \alpha + \epsilon, \quad f'(x) > 0$$

or

$$f''(\alpha) > 0.$$

If $f''(\alpha) = 0$, then the point may be a maximum, minimum or point of inflexion. For a point of inflexion there must be no change in the sign of $f'(x)$ as x increases from $\alpha - \epsilon$ through α to $\alpha + \epsilon$.

Example 18

Determine the maxima, minima and points of inflexion of the function

$$y = \frac{x + 2}{(x + 1)(x - 2)}$$

and sketch the curve.

(i) When $x = 0, y = -1$.
 When $y = 0, x = -2$.

(ii) As $x \to -1, y \to \infty$ giving an asymptote at $x = -1$.
 As $x \to 2, y \to \infty$ giving an asymptote at $x = 2$.
 As $x \to \pm\infty, y \to 0$.

(iii) Since

$$y = \frac{x + 2}{x^2 - x - 2}$$

$$\therefore \frac{dy}{dx} = \frac{(1)(x^2 - x - 2) - (x + 2)(2x - 1)}{(x^2 - x - 2)^2} = -\frac{x(x + 4)}{(x^2 - x - 2)^2}$$

Thus stationary values occur at $x = 0$ and $x = -4$.

For

$$x = -\epsilon, \quad f'(x) > 0$$
$$x = 0, \qquad f'(x) = 0$$
$$x = +\epsilon, \quad f'(x) < 0, \text{ giving a maximum at } (0, -1).$$

For

$$x = -4 - \epsilon, \quad f'(x) < 0$$

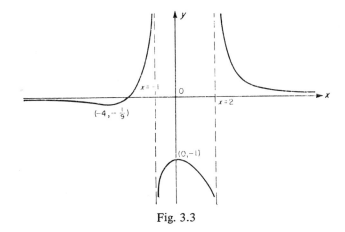

Fig. 3.3

$x = -4,$ $\qquad f'(x) = 0$

$x = -4 + \epsilon,$ $\quad f'(x) > 0$, giving a minimum at $(-4, -\frac{1}{9})$.

The graph of the function is shown in Fig. 3.3.

Example 19

Determine the nature of the stationary values of the function

$$y = \frac{(x-2)^3}{(x-1)(x-3)}$$

and sketch the curve.

(i) When $x = 0, y = -\frac{8}{3}$.
    When $y = 0, x = 2$.

(ii) As $x \to 1, y \to \infty$ giving an asymptote at $x = 1$.
    As $x \to 3, y \to \infty$ giving an asymptote at $x = 3$.

Also

$$y = \frac{(x-2)^3}{x^2 - 4x + 3} = \frac{(x-2)^3}{(x-2)^2 - 1}$$

$$= x - 2 + \frac{x - 2}{(x-2)^2 - 1},$$

showing that $y \to x - 2$ for large values of x.

(iii)

$$\frac{dy}{dx} = \frac{3(x-2)^2(x^2 - 4x + 3) - (x-2)^3(2x-4)}{(x^2 - 4x + 3)^2}$$

$$= \frac{(x-2)^2(x^2-4x+1)}{(x^2-4x+3)^2}$$

$$= \frac{(x-2)^2\{x-(2+\sqrt{3})\}\{x-(2-\sqrt{3})\}}{(x^2-4x+3)^2}.$$

Thus stationary values of $f(x)$ occur at $x = 2$ and $x = 2 \pm \sqrt{3}$.

For

$$x = 2 - \epsilon, \quad f'(x) < 0$$

$$x = 2, \qquad f'(x) = 0$$

$$x = 2 + \epsilon, \quad f'(x) < 0, \text{ giving a point of inflexion at } (2, 0).$$

For

$$x = 2 + \sqrt{3} - \epsilon, \quad f'(x) < 0$$

$$x = 2 + \sqrt{3}, \qquad f'(x) = 0$$

$$x = 2 + \sqrt{3} + \epsilon, \quad f'(x) > 0, \text{ giving a minimum at } \left(2 + \sqrt{3}, \frac{3\sqrt{3}}{2}\right).$$

For

$$x = 2 - \sqrt{3} - \epsilon, \quad f'(x) > 0$$

$$x = 2 - \sqrt{3}, \qquad f'(x) = 0$$

$$x = 2 - \sqrt{3} + \epsilon, \quad f'(x) < 0, \text{ giving a maximum at } \left(2 - \sqrt{3}, -\frac{3\sqrt{3}}{2}\right).$$

The curve is as shown in Fig. 3.4.

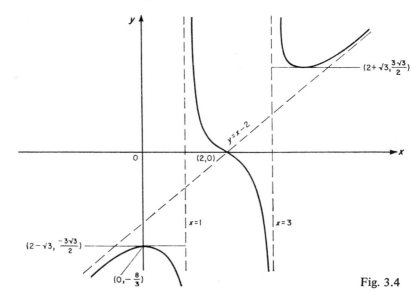

Fig. 3.4

Example 20

Show that the maximum volume of a right circular cone which can be inscribed inside a sphere of radius R, is equal to $\frac{8}{27}$ of the volume of the sphere.

Let the cross section of the configuration be as shown in Fig. 3.5, h being the height of the cone.

Then the volume of the cone is

$$V = \tfrac{1}{3}\pi h^3 \tan^2 \alpha \qquad (6)$$

Now

$$AE \cdot EC = BE \cdot ED$$

i.e.

$$h^2 \tan^2 \alpha = h(2R - h)$$

so that

$$\tan^2 \alpha = \frac{2R - h}{h}$$

and equation (6) becomes

$$V = \tfrac{1}{3}\pi h^2 (2R - h) = \tfrac{1}{3}\pi(2Rh^2 - h^3).$$

Differentiating with respect to h gives

$$\frac{dV}{dh} = \tfrac{1}{3}\pi(4Rh - 3h^2)$$

and

$$\frac{d^2 V}{dh^2} = \tfrac{1}{3}\pi(4R - 6h).$$

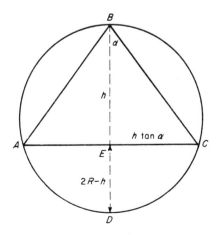

Fig. 3.5

For a maximum or minimum

$$\frac{dV}{dh} = 0, \quad \text{giving} \quad h = 0, \quad \text{or} \quad h = \frac{4R}{3}.$$

For

$$h = 0, \quad \frac{d^2 V}{dh^2} = \frac{4\pi R}{3} > 0$$

giving a minimum value.

For

$$h = \frac{4R}{3}, \quad \frac{d^2 V}{dh^2} = \frac{-4\pi R}{3} < 0$$

giving a maximum value.

Hence the maximum volume of the inscribed cone is

$$V_{max} = \frac{\pi}{3} \left\{ 2R \left(\frac{16R^2}{9} \right) - \left(\frac{64R^3}{27} \right) \right\} = \frac{32\pi R^3}{81}$$

$$= \frac{8}{27} \times \text{volume of the sphere.}$$

Example 21

In the theory of ionic processes in solution, k which is dependent upon the equilibrium constant, is given by the expression

$$\log_e k = -\frac{a + b \exp(T/\theta)}{T}$$

where T is the temperature, θ a temperature characteristic of the solvent, and a and b are constants.

Show that the maximum value of k occurs when T satisfies the transcendental equation

$$\frac{a\theta}{b} = (T - \theta)\exp(T/\theta).$$

Differentiating with respect to T gives

$$\frac{1}{k}\frac{dk}{dT} = -\frac{(b/\theta)\exp(T/\theta) \cdot T - \{a + b \exp(T/\theta)\}}{T^2}$$

i.e.

$$\frac{dk}{dT} = -\frac{k}{T^2} \left\{ b \left(\frac{T}{\theta} - 1 \right) \exp\left(\frac{T}{\theta} \right) - a \right\}$$

For stationary values, $dk/dT = 0$ giving

$$b\left(\frac{T}{\theta} - 1\right)\exp\left(\frac{T}{\theta}\right) = a$$

i.e.

$$(T - \theta)\exp(T/\theta) = a\theta/b.$$

An equation of this type from which T cannot be determined explicitly is called *transcendental*.

Let the stationary value occur when $T = T'$ say, then for

$$T = T' - \epsilon, \quad \frac{dk}{dT} > 0$$

$$T = T', \quad \frac{dk}{dT} = 0$$

$$T = T' + \epsilon, \quad \frac{dk}{dT} < 0,$$

thereby indicating a maximum.

12. Newton's method (An approximation technique for the solution of equations.)

Let the equation $f(x) = 0$ have a real root at $x = \alpha$, represented by the point A in Fig. 3.6.

In general α will not be an integer and will thus be difficult to determine.

Suppose that it is possible to find by some means a value $x = \alpha_1$ such that $f(\alpha_1)$ is almost zero, i.e. $x = \alpha_1$ is a good approximation to $x = \alpha$. Let $x = \alpha_1$ be represented by the point N in Fig. 3.6.

If the tangent to the curve at P is now drawn to meet the x-axis in T, then we see that AT < AN so that the value of x at T, i.e. α_2 say, is a closer approximation to $x = \alpha$ than is $x = \alpha_1$.

Referring to Fig. 3.6

$$AT = AN - TN$$

$$= \alpha_1 - \frac{PN}{\tan \psi}$$

$$= \alpha_1 - \frac{f(\alpha_1)}{f'(\alpha_1)}$$

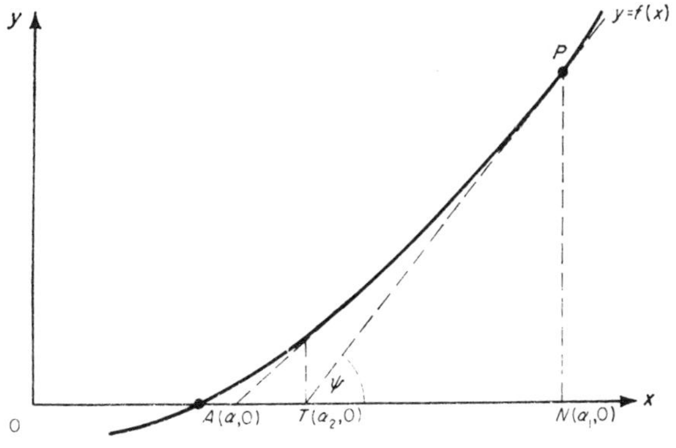

Fig. 3.6

c

i.e. if $x = \alpha_1$ is a good approximation to a root of the equation $f(x) = 0$ then

$$\alpha_2 = \alpha_1 - \frac{f(\alpha_1)}{f'(\alpha_1)} \tag{7}$$

is a better approximation.

The process may now be repeated using $x = \alpha_2$ as a second approximation.

Continuing in this way the value of α may be estimated with considerable accuracy. Care must be taken however, in choosing the first approximation $x = \alpha_1$ since difficulties may arise.

Referring to Fig. 3.7, it can be seen that the method explained above would, in this case, lead to a second approximation α_2 which is in fact a worse approximation than α_1.

Thus whether or not α_2 is a better approximation than α_1 depends upon the rate at which the slope of the curve changes as x increases, i.e. upon $f''(x)$.

If for example $x = \alpha_1'$ had been chosen as the first approximation, i.e. where $f(\alpha_1') < 0$, then we *would* have obtained a better approximation.

Difficulties of this nature may thus be avoided by choosing the first approximation α_1 such that $f(\alpha_1)$ and $f''(\alpha_1)$ are either both positive or both negative.

The analytical proof of Newton's method of approximation is to be found in Chapter 5.

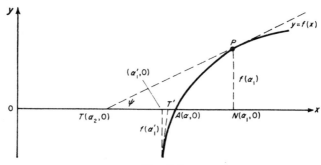

Fig. 3.7

Example 22

Find the root of the equation

$$x^3 + 3x - 1 = 0$$

which lies between $x = 0$ and $x = 1$, correct to three places of decimals.

Let

$$f(x) = x^3 + 3x - 1$$

so that

$$f'(x) = 3x^2 + 3, \quad f''(x) = 6x.$$

Then $f(0) = -1$ and $f(1) = 3$, showing that there is a root between $x = 0$ and $x = 1$.

Try as a first approximation $\alpha_1 = \frac{1}{2}$, then

$$f(\tfrac{1}{2}) = \tfrac{5}{8}, \quad f'(\tfrac{1}{2}) = \tfrac{15}{4}, \quad f''(\tfrac{1}{2}) = 3$$

and equation (7) gives

$$\alpha_2 = \tfrac{1}{2} - \frac{(\tfrac{5}{8})}{(\tfrac{15}{4})} = \tfrac{1}{3}.$$

Taking $\alpha_2 = \frac{1}{3}$ as a second approximation

$$f(\tfrac{1}{3}) = \tfrac{1}{27} \quad \text{and} \quad f'(\tfrac{1}{3}) = \tfrac{10}{3}$$

so that

$$\alpha_3 = \tfrac{1}{3} - \frac{(\tfrac{1}{27})}{(\tfrac{10}{3})} = \tfrac{1}{3} - \tfrac{1}{90} = 0.3222.$$

Similarly

$$\alpha_4 = 0.3222 - \frac{f(0.3222)}{f'(0.3222)} = 0.3222 - \frac{0.0001}{3.3117}$$

$$= 0.322.$$

Hence $x = 0.322$ to three decimal places.

Example 23

Determine the value of the positive root of the equation

$$2 \cos x = \cosh x$$

correct to two decimal places.

Let

$$f(x) = 2 \cos x - \cosh x$$

giving

$$f'(x) = -2 \sin x - \sinh x$$

$$f''(x) = -2 \cos x - \cosh x.$$

Then

$$f(0) = 2 - 1 = 1, \quad f(1) = 2(0.5403) - 1.5431 = -0.4625$$

showing that there is a root between $x = 0$ and $x = 1$. Try $\alpha_1 = 0.5$, then

$$f(0.5) = 0.6276, \quad f'(0.5) = -1.4799, \quad f''(0.5) = -2.8828.$$

Since $f(0.5)$ and $f''(0.5)$ are of opposite sign we try a different value for α_1. To do this we note that $f''(x)$ is always negative for values of x between 0 and 1, so that α_1 must be chosen such that $f(\alpha_1) < 0$.

Let $\alpha_1 = 0.9$, then

$$f(0.9) = -0.1899, \quad f'(0.9) = -2.5931, \quad f''(0.9) = -2.6763.$$

Equation (7) now gives

$$\alpha_2 = 0.9 - \frac{f(0.9)}{f'(0.9)} = 0.9 - \frac{(-0.1899)}{(-2.5931)}$$

$$= 0.827.$$

Taking $\alpha_2 = 0.827$

$$\alpha_3 = 0.827 - \frac{f(0.827)}{f'(0.827)} = 0.827 - \frac{(-0.007)}{(-2.395)}$$

$$= 0.827 - 0.003$$

$$= 0.824.$$

Since $f(0.824)/f'(0.824)$ is negligible compared with 0.824, the value of the positive root is 0.82 correct to two decimal places.

Example 24

The base of a large storage tank of volume 250 ft^3, is a square of side x ft. If the total area of the base and sides is 400 ft^2, use Newton's method of approximation to determine the value of x correct to one decimal place.

Let the height of the tank be y ft. Then the volume is

$$x^2 y = 250, \quad \text{i.e.} \quad y = \frac{250}{x^2}.$$

The total area of base and sides is

$$400 = x^2 + 4xy = x^2 + \frac{1000}{x}.$$

We now apply Newton's method to the equation

$$f(x) = x^2 + \frac{1000}{x} - 400.$$

$$\therefore f'(x) = 2x - \frac{1000}{x^2}, \quad f''(x) = 2 + \frac{2000}{x^3}.$$

Let $\alpha_1 = 20$, then

$$f(20) = 50, \quad f'(20) = 37.5, \quad f''(20) = 2.25.$$

Equation (7) now gives

$$\alpha_2 = 20 - \frac{f(20)}{f'(20)} = 20 - \frac{50}{37.5} = 20 - 1.334$$

$$= 18.666.$$

Similarly

$$\alpha_3 = 18.666 - \frac{f(18.666)}{f'(18.666)} = 18.666 - \frac{1.980}{34.462} = 18.666 - 0.057$$

$$= 18.609$$

Hence $x = 18.6$ ft, to one decimal place.

PROBLEMS

1. Differentiate with respect to x

(i) $(x^2 - 7x + 2)(x - 7)$, (ii) $\left(\dfrac{x}{x+1}\right)^{1/2}$,

(iii) $\dfrac{(x+1)(x+2)}{(x+3)(x+4)}$, (iv) $\dfrac{x^3 - 1}{\sqrt{(x+1)}}$.

2. Differentiate with respect to x

(i) $\sin^4 \sqrt{(x-1)}$, (ii) $\dfrac{\cos x}{1 + \cot x}$, (iii) $\cos^{-1}(3x - 2)$,

(iv) $\operatorname{cosec}^{-1}(\tan x/2)$.

3. Differentiate with respect to x

(i) $\sinh^{-1}(x^2 + 2)$, (ii) $\tanh^{-1}\left(\cos \dfrac{x}{2}\right)$, (iii) $\dfrac{\cosh 4x - \sinh 4x}{\sinh 2x + \cosh 2x}$.

If $\tan y = e^{2x}$, show that $dy/dx = \operatorname{sech} 2x$.

4. Differentiate with respect to x

(i) $\dfrac{3x - 7}{\sqrt{(2x^2 + 3x - 11)}}$, (ii) $x^{\sin x}$, (iii) $\tan^{-1}\left(\dfrac{x^2 + 1}{x^2 - 1}\right)$,

(iv) $\log_e(\sec 3x + \tan 3x)$.

5. If $y = \sin x + \tfrac{1}{3} \sin 3x + \tfrac{1}{5} \sin 5x$, show that

$$2\frac{dy}{dx} = \frac{\sin 6x}{\sin x},$$

provided $x \neq n\pi (n = 0, 1, 2, \ldots)$.

6. Show that $x^2 + y^2 - \sin x = 2$, satisfies the differential equation

$$2y \frac{d^2 y}{dx^2} + 2\left(\frac{dy}{dx}\right)^2 + x^2 + y^2 = 0.$$

7. If $y = e^{3x} \cos 4x$, show that

$$\frac{d^n y}{dx^n} = 5^n e^{3x} \cos(4x + n\alpha)$$

where $\tan \alpha = \tfrac{4}{3}$.

8. If $y \sqrt{(1 - x^2)} = \sin^{-1} x$, show that

$$(1 - x^2)\frac{dy}{dx} = 1 + xy.$$

9. Show that $y = \tan^{-1}(\cosh x)$ satisfies the differential equation

$$\sec^2 y \frac{d^2 y}{dx^2} + \sinh 2x \frac{dy}{dx} - \cosh x = 0.$$

10. If $x = f(y)$, show that

$$\frac{d^2 x}{dy^2}\left(\frac{dy}{dx}\right)^2 + \frac{dx}{dy}\left(\frac{d^2 y}{dx^2}\right) = 0.$$

Hence deduce, that if $y = \tan^{-1}(\sinh x)$, then

$$\frac{d^2 y}{dx^2} + \left(\frac{dy}{dx}\right)^2 \tan y = 0.$$

11. Determine dy/dx and $d^2 y/dx^2$ at the point $(0, 1)$ on the curve

$$x^4 y - 2x^3 y + xy^2 - 3xy + 7x - y + 1 = 0.$$

12. Determine dy/dx and $d^2 y/dx^2$ if

(i) $x = 2at$, $\quad y = at^2$,
(ii) $x = a \cos \theta$, $\quad y = b \sin \theta$,
(iii) $x = 3 \sin \theta - 2 \sin^3 \theta$; $\quad y = 3 \cos \theta - 2 \cos^3 \theta$,
(iv) $x = \log \cot \frac{1}{2}\theta - \cos \theta$, $\quad y = \sin \theta$.

13. Determine the nature of the stationary values of the function

$$y = \frac{x^2 + 3}{x(x - 3)}$$

and sketch the curve.

14. Sketch the curves

(i) $y = \dfrac{x^2 - 3x + 2}{x^2 + 3x + 2}$, $\quad$ (ii) $y = \dfrac{x}{2x^2 - 3x + 1}$, $\quad$ (iii) $y^2 = \dfrac{x^3}{x^2 - 4}$.

15. Find the height and radius of the cylinder of maximum volume which can be inscribed inside a sphere of radius a.

16. Determine to three places of decimals, the root of the equation

$$x^3 - 4x + 1 = 0$$

which lies between $x = 1$ and $x = 2$.

17. Sketch the curves $y = \log_e x^3$ and $y = x$, and hence show that the equation $\log_e x^3 = x$ has a root between $x = 4.0$ and $x = 5.0$.
Use Newton's method to determine its value correct to two decimal places.

18. If the relation between the pressure p of aqueous vapour, and its temperature T, is

$$p = \exp(a + ba^T + c\beta^T)$$

where a, b, c, α and β are constants, show that

$$\frac{dp}{dT} = p(b\alpha^T \log_e \alpha + c\beta^T \log_e \beta).$$

19. If the number of ions produced at time t by Röntgen rays, in an experiment, is n where

$$\left(\frac{a+n}{a-n}\right) = \exp(2at)$$

a being constant, show that the rate of production of ions is

$$\frac{dn}{dt} = a^2 - n^2.$$

20. If the relation between the pressure p of aqueous vapour and its temperature T, is

$$p = ab^{T/(\gamma+T)}$$

where a, b, and γ are constants, show that

$$\frac{dp}{dT} = p\,\frac{\gamma \log_e b}{(\gamma + T)^2}.$$

21. A source of heat of intensity I_1 is placed at a point A_1 and another source of heat of intensity I_2 is placed at a point A_2 distance a from A_1.

If the intensity of heat at a point distance x from A_1 on the line $A_1 A_2$ is

$$I = \frac{I_1}{x^2} + \frac{I_2}{(a-x)^2},$$

show that the maximum value of I is

$$\frac{1}{a^2}\,(I_1^{1/3} + I_2^{1/3})^3.$$

CHAPTER 4

Methods of Integration

In Chapter 3, we were concerned with the problem of determining the derivative of a given function. We now consider the problem of deriving the function corresponding to a given derivative.

If $f(x)$ and $F(x)$ are two functions of x only, such that

$$f(x) = \frac{dF(x)}{dx},$$

then we say that the integral of $f(x)$ with respect to x is $F(x)$, and write

$$\int f(x)\,dx = F(x),$$

the process of obtaining $F(x)$ being called *integration*.

As the derivative of a constant C with respect to x is zero, the more general result of integrating $f(x)$ is

$$\int f(x)\,dx = F(x) + C,$$

which is called the *indefinite integral* of $f(x)$ with respect to x. The function $f(x)$ is referred to as the *integrand*.

For example, since the derivative of $x^{n+1}/(n+1)$ with respect to x is x^n, therefore from the above definition

$$\int x^n\,dx = \frac{x^{n+1}}{n+1} + C, \quad \text{provided } n \neq -1$$

and similarly, since

$$\frac{d}{dx}\left(\frac{e^{ax}}{a}\right) = e^{ax},$$

therefore

$$\int e^{ax}\,dx = \frac{e^{ax}}{a} + C, \quad \text{where } a \text{ is constant.}$$

By referring to the known derivatives of various functions, we may build up a list of 'standard forms', and by manipulating the integrand into one of these forms, we can evaluate a large number of integrals.

Three simple rules which will be required in the ensuing examples are as follows.
(i) If k is a constant then

$$\int kf(x)\ dx = k \int f(x)\ dx.$$

(ii) If $f(x), g(x)$ and $h(x)$ are functions of x only, then

$$\int \{f(x) + g(x) + h(x)\}\ dx = \int f(x)\ dx + \int g(x)\ dx + \int h(x)\ dx.$$

Thus,

$$\int (x^2 + 2x - 3)\ dx = \int x^2\ dx + 2\int x\ dx - 3\int dx$$

$$= \frac{x^3}{3} + x^2 - 3x + C.$$

(iii) If, in the integrand, x is replaced by $ax + b$ where a, b are constants, then the value of the new integral is equal to the value of the original integral (with x replaced by $ax + b$) divided by a, i.e.

$$\int x^n\ dx = \frac{x^{n+1}}{n+1} + C, \quad \therefore \int (ax + b)^n\ dx = \frac{(ax+b)^{n+1}}{n+1} \times \frac{1}{a} + C.$$

All three results are easily verified by differentiating both sides of each equation with respect to x.

Consider the following cases:

1. Rational fractions

The type of integrand concerned takes the form

$$\frac{g(x)}{h(x)},$$

where $g(x)$ and $h(x)$ are both polynomials in x. In the ensuing theory it is important to note that, if the numerator $g(x)$ is of equivalent or higher degree than that of the denominator $h(x)$, then a division must be performed until the numerator is less in degree than that of the denominator.

We deal with three separate cases.

(a) Denominator of first degree

The most simple case of all is clearly when $g(x) = 1$ and $h(x) = x$. Since

$$\frac{d}{dx}(\log_e x) = \frac{1}{x}, \quad \therefore \int \frac{1}{x}\ dx = \log_e x + C.$$

Hence,

$$\int \frac{1}{3x-2}\, dx = \frac{1}{3}\log_e(3x-2) + C.$$

Example 1

$$\int \frac{10x-3}{5x+1}\, dx = \int\left(2 - \frac{5}{5x+1}\right) dx$$

$$= 2x - \log_e(5x+1) + C.$$

Example 2

$$\int \frac{8x^2 - 30x + 11}{4x-1}\, dx = \int\left(2x - \frac{28x-11}{4x-1}\right) dx$$

$$= \int\left(2x - 7 + \frac{4}{4x-1}\right) dx$$

$$= x^2 - 7x + \log_e(4x-1) + C.$$

Example 3

The variation of the saturation vapour pressure p of water, with temperature T, is given by

$$T^2 \frac{dp}{dT} = p(a + bT),$$

where a and b are constants. Determine the value of p.

Re-arranging the given expression, we write

$$\int \frac{dp}{p} = \int \frac{a + bT}{T^2}\, dT$$

i.e.

$$\log_e p = a\int \frac{dT}{T^2} + b \int \frac{dT}{T}$$

$$= -\frac{a}{T} + b \log_e T + \text{constant}.$$

(b) Denominator of second degree

There are two cases to be considered: (i) denominator possessing factors; and (ii) denominator possessing no factors.

For the first case, we express the integrand in terms of partial fractions, thus reducing the integrand to type (*a*). Case (i):

Example 4

$$I = \int \frac{9x + 11}{3x^2 + 5x - 2} \, dx.$$

The integrand is

$$\frac{9x + 11}{(3x - 1)(x + 2)} = \frac{6}{3x - 1} + \frac{1}{x + 2}.$$

Hence,

$$I = \int \frac{6}{3x - 1} \, dx + \int \frac{1}{x + 2} \, dx.$$

i.e.

$$I = 2 \log_e(3x - 1) + \log_e(x + 2) + C.$$

Example 5

$$I = \int \frac{2x + 7}{(2x + 1)^2} \, dx.$$

The integrand is

$$\frac{2x + 7}{(2x + 1)^2} = \frac{1}{2x + 1} + \frac{6}{(2x + 1)^2}.$$

Hence,

$$I = \int \frac{1}{2x + 1} \, dx + \int \frac{6}{(2x + 1)^2} \, dx$$

i.e.

$$I = \tfrac{1}{2} \log_e(2x + 1) - \frac{3}{2x + 1} + C.$$

Example 6

$$I = \int \frac{x^2 + 7}{x^2 + 3x} \, dx.$$

The integrand is

$$\frac{x^2 + 7}{x(x + 3)} = 1 + \frac{7 - 3x}{x(x + 3)} = 1 + \frac{7}{3x} - \frac{16}{3(x + 3)}.$$

Hence,

$$I = \int dx + \frac{7}{3} \int \frac{1}{x} \, dx - \frac{16}{3} \int \frac{1}{x + 3} \, dx$$

$$= x + \tfrac{7}{3} \log_e x - \tfrac{16}{3} \log_e(x + 3) + C.$$

Case (ii): These integrals may be evaluated by rewriting the denominator as the sum or difference of two squares, and then using the following standard forms. Since,

$$\frac{d}{dx}\left\{\tan^{-1}\left(\frac{x}{a}\right)\right\} = \frac{a}{a^2 + x^2}, \quad \therefore \int \frac{1}{a^2 + x^2}\, dx = \frac{1}{a}\tan^{-1}\left(\frac{x}{a}\right) + C,$$

$$\frac{d}{dx}\left\{\tanh^{-1}\left(\frac{x}{a}\right)\right\} = \frac{a}{a^2 - x^2}, \quad \therefore \int \frac{1}{a^2 - x^2}\, dx = \frac{1}{a}\tanh^{-1}\left(\frac{x}{a}\right) + C,$$

and

$$\frac{d}{dx}\left\{\coth^{-1}\left(\frac{x}{a}\right)\right\} = \frac{-a}{x^2 - a^2}, \quad \therefore \int \frac{1}{x^2 - a^2}\, dx = -\frac{1}{a}\coth^{-1}\left(\frac{x}{a}\right) + C,$$

the second result being used when $x^2 < a^2$ and the third when $x^2 > a^2$, a being constant.

Referring to Chapter 1, we see that alternative forms for the second and third integrals are

$$\frac{1}{2a}\log_e\left(\frac{a+x}{a-x}\right) + C, \quad \frac{1}{2a}\log_e\left(\frac{x-a}{x+a}\right) + C.$$

Example 7

$$I = \int \frac{1}{x^2 + 6x + 13}\, dx = \int \frac{1}{(x+3)^2 + 4}\, dx = \frac{1}{2}\tan^{-1}\left(\frac{x+3}{2}\right) + C.$$

Example 8

$$I = \int \frac{1}{6 - x - x^2}\, dx = \int \frac{1}{\frac{25}{4} - (x + \frac{1}{2})^2}\, dx = \frac{2}{5}\tanh^{-1}\left(\frac{x + \frac{1}{2}}{\frac{5}{2}}\right) + C.$$

i.e.

$$I = \frac{2}{5}\tanh^{-1}\left(\frac{2x+1}{5}\right) + C.$$

For the cases when the numerator is other than unity, we use the following result:

$$\int \frac{2ax + b}{ax^2 + bx + c}\, dx = \log_e(ax^2 + bx + c) + \text{constant},$$

which is easily verified by differentiating both sides of the equation with respect to x; a, b and c being constants.

Example 9

$$I = \int \frac{x + 3}{x^2 + 9}\, dx.$$

The integrand is

$$\frac{x+3}{x^2+9} = \frac{1}{2} \cdot \frac{2x}{x^2+9} + \frac{3}{x^2+9}$$

$$\therefore I = \frac{1}{2} \int \frac{2x}{x^2+9} \, dx + 3 \int \frac{1}{x^2+9} \, dx$$

i.e.

$$I = \frac{1}{2} \log_e(x^2+9) + \tan^{-1}\left(\frac{x}{3}\right) + C.$$

Example 10

$$I = \int \frac{2x+13}{x^2+14x+48} \, dx.$$

The integrand is

$$\frac{2x+13}{x^2+14x+48} = \frac{2x+14}{x^2+14x+48} - \frac{1}{x^2+14x+48}.$$

Hence,

$$I = \int \frac{2x+14}{x^2+14x+48} \, dx - \int \frac{1}{(x+7)^2-1} \, dx$$

$$= \log_e(x^2+14x+48) + \coth^{-1}(x+7) + C.$$

Example 11

$$I = \int \frac{x^2+2x-12}{5-4x-x^2} \, dx.$$

The integrand is

$$\frac{x^2+2x-12}{5-4x-x^2} = -1 - \frac{2x+7}{5-4x-x^2}$$

$$= -1 - \frac{2x+4}{5-4x-x^2} - \frac{3}{5-4x-x^2}.$$

Hence,

$$I = -\int dx - \int \frac{2x+4}{5-4x-x^2} \, dx - 3 \int \frac{1}{9-(x+2)^2} \, dx$$

$$= -x + \log_e(5-4x-x^2) - \tanh^{-1}\left(\frac{x+2}{3}\right) + C.$$

(c) Denominator of higher degree

We consider here only denominators which possess rational factors, and again make use of partial fractions to reduce the integral to type (*a*) or (*b*).

Example 12

$$I = \int \frac{25x^2 + 16x - 65}{5x^3 + 11x^2 - 13x - 3} \, dx.$$

The integrand is

$$\frac{25x^2 + 16x - 65}{(x + 3)(x - 1)(5x + 1)} = \frac{2}{x + 3} - \frac{1}{x - 1} + \frac{20}{5x + 1}.$$

Hence,

$$I = 2 \int \frac{1}{x + 3} \, dx - \int \frac{1}{x - 1} \, dx + 4 \int \frac{5}{5x + 1} \, dx$$

$$= 2 \log_e(x + 3) - \log_e(x - 1) + 4 \log_e(5x + 1) + C.$$

Example 13

$$I = \int \frac{8x^2 - 45x + 45}{(2x - 1)(4 - x)^2} \, dx.$$

The integrand is

$$\frac{2}{2x - 1} - \frac{3}{4 - x} - \frac{1}{(4 - x)^2}$$

Hence,

$$I = 2 \int \frac{1}{2x - 1} \, dx - 3 \int \frac{1}{4 - x} \, dx - \int \frac{1}{(4 - x)^2} \, dx$$

$$= \log_e(2x - 1) + 3 \log_e(4 - x) - \frac{1}{4 - x} + C.$$

Example 14

$$I = \int \frac{x^3 - 2x^2 - 23x - 32}{(x^2 - 1)(x^2 + 4x + 9)} \, dx.$$

The integrand is

$$\frac{x^3 - 2x^2 - 23x - 32}{(x + 1)(x - 1)(x^2 + 4x + 9)} = \frac{1}{x + 1} - \frac{2}{x - 1} + \frac{2x + 5}{x^2 + 4x + 9}.$$

Hence,

$$I = \int \frac{1}{x + 1} \, dx - 2 \int \frac{1}{x - 1} \, dx + \int \frac{2x + 4}{x^2 + 4x + 9} \, dx + \int \frac{1}{(x + 2)^2 + 5} \, dx$$

$$= \log_e(x + 1) - 2 \log_e(x - 1) + \log_e(x^2 + 4x + 9)$$

$$+ \frac{1}{\sqrt{5}} \tan^{-1}\left(\frac{x + 2}{\sqrt{5}}\right) + C.$$

Example 15

In the irreversible ter-molecular reaction $A + B + C \to D$, the initial concentrations of A, B, C are 3, 4 and 6 gram molecules per litre respectively.

If x gram molecules per litre of D are produced after a time t minutes, then

$$kt = \int \frac{1}{(3-x)(4-x)(6-x)} \, dx$$

where k is the velocity constant for the reaction. Show that x is given by

$$(1-x/3)^2(1-x/6) = (1-x/4)^3 \exp(-6kt).$$

Expressing the integrand in partial fractions gives

$$kt = \int \left(\frac{\frac{1}{3}}{3-x} - \frac{\frac{1}{2}}{4-x} + \frac{\frac{1}{6}}{6-x} \right) dx.$$

$$\therefore \ kt = -\tfrac{1}{3}\log_e(3-x) + \tfrac{1}{2}\log_e(4-x) - \tfrac{1}{6}\log_e(6-x) + C.$$

At $t = 0$, $x = 0$,

$$\therefore \ C = \tfrac{1}{3}\log_e 3 - \tfrac{1}{2}\log_e 4 + \tfrac{1}{6}\log_e 6$$

giving

$$kt = -\tfrac{1}{3}\log_e(1-x/3) + \tfrac{1}{2}\log_e(1-x/4) - \tfrac{1}{6}\log_e(1-x/6)$$

$$\therefore \ 6kt = \log_e \frac{(1-x/4)^3}{(1-x/3)^2(1-x/6)},$$

giving the result.

2. Irrational fractions

An illustration of the most simple case to occur is

$$\int \frac{x^3 + 4x^2 - 3x + 1}{\sqrt{x}} \, dx = \int (x^{5/2} + 4x^{3/2} - 3x^{1/2} + x^{-1/2}) \, dx$$

$$= \frac{2x^{7/2}}{7} + \frac{8x^{5/2}}{5} - 2x^{3/2} + 2x^{1/2} + C.$$

The more general type of integral in this category may be evaluated by the use of one of the following three standard forms.

Since,

$$\frac{d}{dx}\left\{ \sin^{-1}\left(\frac{x}{a}\right) \right\} = \frac{1}{\sqrt{(a^2 - x^2)}}, \quad \therefore \int \frac{1}{\sqrt{(a^2 - x^2)}} \, dx = \sin^{-1}\left(\frac{x}{a}\right) + C,$$

$$\frac{d}{dx}\left\{ \sinh^{-1}\left(\frac{x}{a}\right) \right\} = \frac{1}{\sqrt{(a^2 + x^2)}}, \quad \therefore \int \frac{1}{\sqrt{(a^2 + x^2)}} \, dx = \sinh^{-1}\left(\frac{x}{a}\right) + C,$$

and

$$\frac{d}{dx}\left\{ \cosh^{-1}\left(\frac{x}{a}\right) \right\} = \frac{1}{\sqrt{(x^2 - a^2)}}, \quad \therefore \int \frac{1}{\sqrt{(x^2 - a^2)}} \, dx = \cosh^{-1}\left(\frac{x}{a}\right) + C,$$

the first integral having the alternative form $-\cos^{-1}(x/a) + C$. The second and third integrals have logarithmic equivalents, as shown in Chapter 1.

Example 16

$$\int \frac{1}{\sqrt{(3 + 6x - 9x^2)}}\, dx = \int \frac{1}{\sqrt{\{4 - (3x - 1)^2\}}}\, dx$$

$$= \frac{1}{3} \sin^{-1}\left(\frac{3x - 1}{2}\right) + C.$$

Example 17

$$\int \frac{1}{\sqrt{(x^2 - 4x + 29)}}\, dx = \int \frac{1}{\sqrt{\{(x - 2)^2 + 25\}}}\, dx$$

$$= \sinh^{-1}\left(\frac{x - 2}{5}\right) + C.$$

Example 18

$$\int \frac{1}{\sqrt{(x^2 + 14x + 33)}}\, dx = \int \frac{1}{\sqrt{\{(x + 7)^2 - 16\}}}\, dx$$

$$= \cosh^{-1}\left(\frac{x + 7}{4}\right) + C.$$

For the cases when the numerator is a linear function of x, we use the result:

$$\int \frac{2ax + b}{\sqrt{(ax^2 + bx + c)}}\, dx = 2\sqrt{(ax^2 + bx + c)} + \text{constant},$$

which, as before, is easily verified by differentiation.

Example 19

$$I = \int \frac{2x - 3}{\sqrt{(x^2 - 2x + 10)}}\, dx.$$

The integrand is

$$\frac{2x - 2}{\sqrt{(x^2 - 2x + 10)}} - \frac{1}{\sqrt{(x^2 - 2x + 10)}}.$$

Hence,

$$I = \int \frac{2x - 2}{\sqrt{(x^2 - 2x + 10)}}\, dx - \int \frac{1}{\sqrt{\{(x - 1)^2 + 9\}}}\, dx$$

$$= 2\sqrt{(x^2 - 2x + 10)} - \sinh^{-1}\left(\frac{x - 1}{3}\right) + C.$$

Example 20

$$I = \int \frac{x - 2}{\sqrt{(x^2 - 6x + 5)}}\, dx.$$

The integrand is

$$\frac{\frac{1}{2}(2x-6)}{\sqrt{(x^2-6x+5)}} + \frac{1}{\sqrt{(x^2-6x+5)}}.$$

Hence,

$$I = \frac{1}{2}\int\frac{2x-6}{\sqrt{(x^2-6x+5)}}\,dx + \int\frac{1}{\sqrt{\{(x-3)^2-4\}}}\,dx$$

i.e.

$$I = \sqrt{(x^2-6x+5)} + \cosh^{-1}\left(\frac{x-3}{2}\right) + C.$$

Example 21

$$I = \int\frac{3x+17}{\sqrt{(x^2+10x+34)}}\,dx.$$

The integrand is

$$\frac{\frac{3}{2}(2x+10)}{\sqrt{(x^2+10x+34)}} + \frac{2}{\sqrt{(x^2+10x+34)}}.$$

Hence,

$$I = \frac{3}{2}\int\frac{2x+10}{\sqrt{(x^2+10x+34)}}\,dx + 2\int\frac{1}{\sqrt{\{(x+5)^2+9\}}}\,dx$$

$$= 3\sqrt{(x^2+10x+34)} + 2\sinh^{-1}\left(\frac{x+5}{3}\right) + C.$$

Example 22

$$I = \int\sqrt{\left(\frac{2x-3}{x+1}\right)}\,dx.$$

The integrand is

$$\sqrt{\left(\frac{2x-3}{x+1}\right)} = \frac{2x-3}{\sqrt{(x+1)}\sqrt{(2x-3)}} = \frac{2x-3}{\sqrt{(2x^2-x-3)}}$$

$$= \frac{\frac{1}{2}(4x-1)}{\sqrt{(2x^2-x-3)}} - \frac{\frac{5}{2}}{\sqrt{(2x^2-x-3)}}.$$

Hence,

$$I = \frac{1}{2}\int\frac{4x-1}{\sqrt{(2x^2-x-3)}}\,dx - \frac{5}{2\sqrt{2}}\int\frac{1}{\sqrt{(x^2-x/2-3/2)}}\,dx$$

$$= \sqrt{(2x^2-x-3)} - \frac{5}{2\sqrt{2}}\int\frac{1}{\sqrt{\{(x-\frac{1}{4})^2-\frac{25}{16}\}}}\,dx$$

$$= \sqrt{(2x^2-x-3)} - \frac{5}{2\sqrt{2}}\cosh^{-1}\left(\frac{4x-1}{5}\right) + C.$$

3. Circular functions

From the knowledge of derivatives already obtained we see that

$$\int \cos x \, dx = \sin x + C; \qquad \int \sin x \, dx = -\cos x + C.$$

$$\int \sec^2 x \, dx = \tan x + C; \qquad \int \operatorname{cosec} x \cot x \, dx = -\operatorname{cosec} x + C.$$

$$\int \sec x \tan x \, dx = \sec x + C; \qquad \int \operatorname{cosec}^2 x \, dx = -\cot x + C.$$

These results, together with various trigonometric formulae, are sufficient to enable the following types of integral to be evaluated.

Example 23

$$\int \sin^2 x \, dx = \tfrac{1}{2} \int (1 - \cos 2x) \, dx = \tfrac{1}{2}(x - \tfrac{1}{2} \sin 2x) + C.$$

Example 24

$$\int \tan x \, dx = \int \frac{\sin x}{\cos x} \, dx = - \int \frac{1}{\cos x} \, d(\cos x)$$

$$= -\log_e \cos x + C.$$

Example 25

$$\int \cos^3 x \, dx = \int \cos^2 x \cos x \, dx = \int (1 - \sin^2 x) \, d(\sin x)$$

$$= \sin x - \tfrac{1}{3} \sin^3 x + C.$$

Example 26

$$I = \int \tan^4 x \, dx.$$

The integrand is

$$\tan^4 x = \tan^2 x(\sec^2 x - 1) = \tan^2 x \sec^2 x - (\sec^2 x - 1).$$

Hence,

$$I = \int \tan^2 x \, d(\tan x) - \int \sec^2 x \, dx + \int dx$$

$$= \tfrac{1}{3} \tan^3 x - \tan x + x + C.$$

Example 27

$$\int \frac{2\cos^2 x}{\sin^2 2x}\,dx = \int \frac{1 + \cos 2x}{\sin^2 2x}\,dx$$

$$= \int (\operatorname{cosec}^2 2x + \operatorname{cosec} 2x \cot 2x)\,dx$$

$$= -\tfrac{1}{2}\cot 2x - \tfrac{1}{2}\operatorname{cosec} 2x + C.$$

Example 28

$$\int \frac{\sin^3 x}{\cos^2 x}\,dx = \int \frac{\sin^2 x}{\cos^2 x}\,d(-\cos x) = -\int \frac{(1 - \cos^2 x)}{\cos^2 x}\,d(\cos x)$$

$$= \int d(\cos x) - \int \frac{1}{\cos^2 x}\,d(\cos x)$$

$$= \cos x + \frac{1}{\cos x} + C.$$

Example 29

$$I = \int \cos^4 x\,dx.$$

Now

$$\cos^4 x = \{\tfrac{1}{2}(1 + \cos 2x)\}^2 = \tfrac{1}{4}(1 + 2\cos 2x + \cos^2 2x)$$

and

$$\cos^2 2x = \tfrac{1}{2}(1 + \cos 4x).$$

Hence,

$$I = \int (\tfrac{1}{4} + \tfrac{1}{2}\cos 2x + \tfrac{1}{8} + \tfrac{1}{8}\cos 4x)\,dx$$

$$= \tfrac{3}{8}x + \tfrac{1}{4}\sin 2x + \tfrac{1}{32}\sin 4x + C.$$

We now deal with integrands in which the numerator is unity and the denominator takes the form $a\cos x + b\sin x + c$, where a, b, and c are constants.

These integrals are easily converted to 'standard forms' by expressing the denominator in terms of half-angles, so that

$$a\cos x \equiv a\{\cos^2 (x/2) - \sin^2 (x/2)\},$$

$$b\sin x \equiv 2b\,\sin(x/2)\cos(x/2),$$

$$c \equiv c\{\sin^2 (x/2) + \cos^2 (x/2)\},$$

and then dividing numerator and denominator by $\cos^2 (x/2)$.

Example 30

$$\int \frac{1}{\cos x}\,dx = \int \frac{1}{\cos^2 (x/2) - \sin^2 (x/2)}\,dx = \int \frac{\sec^2 (x/2)}{1 - \tan^2 (x/2)}\,dx.$$

Putting $u = \tan(x/2)$ so that $du = \frac{1}{2} \sec^2(x/2)dx$, the integral becomes

$$\int \frac{2}{1 - u^2} du = 2 \tanh^{-1} u + C$$

$$= 2 \tanh^{-1}\{\tan(x/2)\} + C.$$

Example 31

$$I = \int \frac{1}{2 + 3 \cos x - \sin x} dx.$$

The denominator of the integrand is:

$$2 + 3 \cos x - \sin x = 2\{\sin^2(x/2) + \cos^2(x/2)\}$$
$$+ 3\{\cos^2(x/2) - \sin^2(x/2)\} - 2 \sin(x/2)\cos(x/2)$$
$$= 5 \cos^2(x/2) - 2 \sin(x/2)\cos(x/2) - \sin^2(x/2)$$

$$\therefore \quad \frac{1}{2 + 3 \cos x - \sin x} = \frac{\sec^2(x/2)}{5 - 2 \tan(x/2) - \tan^2(x/2)}.$$

Hence, putting $u = \tan(x/2)$ so that $du = \frac{1}{2} \sec^2(x/2) dx$, the integral becomes,

$$I = \int \frac{2 \, du}{5 - 2u - u^2} = 2 \int \frac{1}{6 - (u + 1)^2} du$$

$$= \frac{2}{\sqrt{6}} \tanh^{-1}\left(\frac{u + 1}{\sqrt{6}}\right) + C$$

i.e.

$$I = \frac{2}{\sqrt{6}} \tanh^{-1}\left(\frac{\tan(x/2) + 1}{\sqrt{6}}\right) + C.$$

For integrands of the type

$$\frac{a \sin x + b \cos x}{p \sin x + q \cos x} \equiv H(x) \quad \text{say,}$$

where a, b, p and q are constants, let

$$a \sin x + b \cos x \equiv \lambda(p \sin x + q \cos x) + \mu(p \cos x - q \sin x)$$

where λ, μ are constants to be determined. Then,

$$\int H(x) \, dx = \int \lambda \, dx + \mu \int \frac{p \cos x - q \sin x}{p \sin x + q \cos x} dx$$

$$= \lambda x + \mu \log_e(p \sin x + q \cos x) + C.$$

Example 32

$$I = \int \frac{5 \sin x - 2 \cos x}{2 \sin x + \cos x} dx.$$

Let

$$5 \sin x - 2 \cos x \equiv \lambda(2 \sin x + \cos x) + \mu(2 \cos x - \sin x).$$

Equating the coefficients of $\sin x$ and $\cos x$ respectively, gives

$$5 = 2\lambda - \mu, \quad -2 = \lambda + 2\mu$$

whence,

$$\lambda = \tfrac{8}{5} \quad \text{and} \quad \mu = -\tfrac{9}{5}$$

$$\therefore I = \frac{8}{5} \int dx - \frac{9}{5} \int \frac{2 \cos x - \sin x}{2 \sin x + \cos x} dx$$

i.e.

$$I = \frac{8x}{5} - \frac{9}{5} \log_e(2 \sin x + \cos x) + C.$$

Example 33

If a spherical particle of radius r is fixed in a liquid which is moving horizontally with uniform velocity, the pressure at any point (r, θ) of the particle is p where

$$p = 2 \int \{a - (2a + 3b)\sin^2 \theta + 4b \sin^4 \theta\} \, d\theta,$$

a and b being constants.

Show that

$$p = (a - b \sin^2 \theta)\sin 2\theta + \text{constant}.$$

We have

$$p = 2a \int d\theta - (4a + 6b) \int \sin^2 \theta \, d\theta + 8b \int \sin^4 \theta \, d\theta$$

$$= 2a\theta - (4a + 6b)I_1 + 8bI_2 + \text{constant}, \quad \text{say}.$$

Now,

$$I_1 = \int \sin^2 \theta \, d\theta = \int \tfrac{1}{2}(1 - \cos 2\theta) \, d\theta = \tfrac{1}{2}\theta - \tfrac{1}{4} \sin 2\theta + \text{constant}$$

and

$$I_2 = \int \sin^4 \theta \, d\theta = \int \{\tfrac{1}{2}(1 - \cos 2\theta)\}^2 \, d\theta$$

$$= \frac{1}{4} \int (1 - 2 \cos 2\theta + \cos^2 2\theta) \, d\theta$$

$$= \frac{\theta}{4} - \frac{\sin 2\theta}{4} + \frac{1}{8} \int (1 + \cos 4\theta) \, d\theta$$

$$= \frac{\theta}{4} - \frac{\sin 2\theta}{4} + \frac{\theta}{8} + \frac{\sin 4\theta}{32} + \text{constant}.$$

Hence,

$$p = 2a\theta - (4a + 6b)\left(\frac{\theta}{2} - \frac{\sin 2\theta}{4}\right) + 8b\left(\frac{3\theta}{8} - \frac{\sin 2\theta}{4} + \frac{\sin 4\theta}{32}\right) + \text{constant}$$

$$= \left(a - \frac{b}{2}\right)\sin 2\theta + \frac{b}{4}\sin 4\theta + \text{constant}$$

$$= \left(a - \frac{b}{2}\right)\sin 2\theta + \frac{b}{2}\sin 2\theta \cos 2\theta + \text{constant}$$

$$= \left(a - \frac{b}{2}\right)\sin 2\theta + \frac{b}{2}\sin 2\theta(1 - 2\sin^2 \theta) + \text{constant}$$

i.e.

$$p = (a - b\sin^2 \theta)\sin 2\theta + \text{constant}.$$

4. Change of variable

Certain types of integral may be reduced to a 'standard form' by means of a simple substitution.

For example, by using a suitable substitution, certain irrational functions may be transformed into rational functions.

The general method is best illustrated by the following examples.

Example 34

$$I = \int \sqrt{(a^2 - x^2)}\, dx.$$

Let $x = a\sin\theta$, so that $dx = a\cos\theta\, d\theta$. Then

$$I = \int a\cos\theta \cdot a\cos\theta\, d\theta$$

$$= \int a^2 \cos^2\theta\, d\theta$$

$$= \tfrac{1}{2}a^2 \int (1 + \cos 2\theta)\, d\theta$$

$$= \tfrac{1}{2}a^2 (\theta + \tfrac{1}{2}\sin 2\theta) + C.$$

Hence, as

$$\sin 2\theta = 2\sin\theta\cos\theta = 2 \cdot \frac{x}{a} \cdot \sqrt{(1 - x^2/a^2)}$$

$$= \frac{2x}{a^2}\sqrt{(a^2 - x^2)},$$

$$I = \tfrac{1}{2}a^2 \sin^{-1}(x/a) + \tfrac{1}{2}x\sqrt{(a^2 - x^2)} + C.$$

Example 35

$$I = \int \sqrt{\left(\frac{3+x}{3-x}\right)}\, dx.$$

Let $x = 3 \cos 2\theta$, so that $dx = -6 \sin 2\theta\, d\theta$. Then,

$$I = \int \sqrt{\left(\frac{1 + \cos 2\theta}{1 - \cos 2\theta}\right)} (-6 \sin 2\theta)\, d\theta$$

$$= -6 \int \frac{\cos \theta}{\sin \theta} \sin 2\theta\, d\theta$$

$$= -12 \int \cos^2 \theta\, d\theta = -6(\theta + \tfrac{1}{2} \sin 2\theta) + C,$$

from Example 34.

Now

$$\sin 2\theta = \sqrt{(1 - \cos^2 2\theta)} = \sqrt{(1 - x^2/9)} = \tfrac{1}{3} \sqrt{(9 - x^2)}$$

and hence

$$I = -3 \cos^{-1}(x/3) - \sqrt{(9 - x^2)} + C.$$

Note that this result could also have been obtained by the method used in Example 20.

Example 36

$$I = \int \frac{1}{\sqrt{\{(a - x)(x - b)\}}}\, dx.$$

Let

$$x = a \cos^2 \theta + b \sin^2 \theta$$

$$\therefore\ a - x = a(1 - \cos^2 \theta) - b \sin^2 \theta$$

$$= (a - b)\sin^2 \theta$$

and

$$x - b = a \cos^2 \theta + b(\sin^2 \theta - 1)$$

$$= (a - b)\cos^2 \theta.$$

Also,

$$dx = -2(a - b)\cos \theta \sin \theta\, d\theta.$$

Hence,

$$I = \int \frac{-2(a - b)\cos \theta \sin \theta\, d\theta}{(a - b)\sin \theta \cos \theta}$$

$$= -2 \int d\theta = -2\theta + C.$$

But,

$$2x = a(1 + \cos 2\theta) + b(1 - \cos 2\theta)$$

$$\therefore \ 2x - (a + b) = (a - b)\cos 2\theta.$$

Thus, in terms of the original variable, the integral becomes,

$$I = C - \cos^{-1}\left(\frac{2x - (a + b)}{a - b}\right) = C_1 + \sin^{-1}\left(\frac{2x - (a + b)}{a - b}\right)$$

where $C_1 = C - \pi/2$.

Example 37

$$I = \int \frac{1}{x\sqrt{(5x^2 - 6x + 1)}} \, dx.$$

Let $x = 1/u$, so that $dx = -du/u^2$. Then the integrand becomes,

$$-\frac{1/u^2}{u^{-1}\sqrt{(5/u^2 - 6/u + 1)}} = -\frac{1}{\sqrt{(5 - 6u + u^2)}}.$$

Hence,

$$I = \int \frac{-1}{\sqrt{(u^2 - 6u + 5)}} \, du = -\int \frac{1}{\sqrt{\{(u - 3)^2 - 4\}}} \, du$$

$$= -\cosh^{-1}\left(\frac{u - 3}{2}\right) + C$$

$$= C - \cosh^{-1}\left(\frac{1 - 3x}{2x}\right).$$

5. Product of two functions

Suppose now, that the integrand takes the form of a product of two functions of x, $f(x)$ and $g(x)$ say, and consider the product $F(x)g(x)$, where $F(x)$ is some other function of x.

Then,

$$\frac{d}{dx}\{F(x)g(x)\} = \frac{dF(x)}{dx}g(x) + F(x)\frac{dg(x)}{dx}$$

and

$$\therefore \ \int \frac{dF(x)}{dx} g(x) \, dx = F(x)g(x) - \int F(x)\frac{dg(x)}{dx} \, dx.$$

If we now choose $dF(x)/dx = f(x)$, then the above equation becomes,

$$\int f(x)g(x) \, dx = F(x)g(x) - \int F(x)\frac{dg(x)}{dx} \, dx,$$

where

$$F(x) = \int f(x) \, dx.$$

Hence, in order to integrate the product of two functions: (i) form the product—second function times integral of first function; and (ii) subtract the integral of the product—integral of first function times derivative of second function.

Since the integrand can be written as $f(x)g(x)$ or $g(x)f(x)$, we may choose the 'first' function to be $g(x)$ or $f(x)$, whichever is the most convenient.

The process (i) and (ii) is referred to as 'integration by parts', and is most frequently used when $f(x)$ is a geometrical, exponential, logarithmic or binomial function, and $g(x)$ is one of the three remaining functions.

Example 38

$$\int x \sin x \, dx = x(-\cos x) - \int (-\cos x) \, dx$$

$$= -x \cos x + \sin x + C.$$

Example 39

$$\int x \log_e(1 + x^2) \, dx = \frac{x^2}{2} \log_e(1 + x^2) - \int \left(\frac{x^2}{2}\right) \frac{2x}{1 + x^2} \, dx$$

$$= \frac{x^2}{2} \log_e(1 + x^2) - \int \frac{x^3}{1 + x^2} \, dx.$$

Now

$$\int \frac{x^3}{1 + x^2} \, dx = \int \left(x - \frac{x}{1 + x^2}\right) dx$$

$$= \tfrac{1}{2}x^2 - \tfrac{1}{2} \log_e(1 + x^2) + C.$$

$$\therefore \int x \log_e(1 + x^2) \, dx = \tfrac{1}{2}(1 + x^2)\log_e(1 + x^2) - \tfrac{1}{2}x^2 - C.$$

Example 40

$$\int \sinh^{-1} x \, dx = \int 1 \cdot \sinh^{-1} x \, dx$$

$$= x \sinh^{-1} x - \int x \cdot \frac{1}{\sqrt{(1 + x^2)}} \, dx$$

$$= x \sinh^{-1} x - \sqrt{(1 + x^2)} + C$$

Example 41

$$I = \int e^x(x+3)\log_e(x+2) \, dx.$$

Now,

$$\int e^x(x+3) \, dx = e^x(x+3) - \int e^x \, dx$$

$$= e^x(x+3) - e^x + \text{constant},$$

$$= e^x(x+2) + \text{constant}.$$

Hence,

$$I = e^x(x+2)\log_e(x+2) - \int e^x(x+2) \cdot \frac{1}{x+2} \, dx + \text{constant}$$

$$= e^x(x+2)\log_e(x+2) - \int e^x \, dx + \text{constant}$$

$$= e^x(x+2)\log_e(x+2) - e^x + C.$$

Example 42

$$I = \int e^{2x} \sin 5x \, dx$$

$$\therefore \quad I = \tfrac{1}{2}e^{2x} \sin 5x - \tfrac{5}{2}\int e^{2x} \cos 5x \, dx$$

$$= \tfrac{1}{2}e^{2x} \sin 5x - \tfrac{5}{2}\left(\tfrac{1}{2}e^{2x} \cos 5x + \tfrac{5}{2}\int e^{2x} \sin 5x \, dx\right)$$

i.e.

$$I = \tfrac{1}{2}e^{2x} \sin 5x - \tfrac{5}{4}e^{2x} \cos 5x - \tfrac{25}{4}I + \text{constant}$$

$$\therefore \quad \tfrac{29}{4}I = \tfrac{1}{2}e^{2x} \sin 5x - \tfrac{5}{4}e^{2x} \cos 5x + \text{constant},$$

$$\therefore \quad I = \frac{e^{2x}}{29}(2 \sin 5x - 5 \cos 5x) + C.$$

Example 43

In the theory of electrochemical corrosion, concerning the attack by a liquid on a metal in the form of a wire, the amount m corroded in time t is given by

$$\frac{dt}{dm} = \lambda \log_e \sqrt{\left(1 + \frac{\epsilon m}{M - m}\right)},$$

where M is the original mass of the wire and λ and ϵ are constants. Determine t in terms of m.

Integrating the given equation with respect to m gives

$$t = \frac{\lambda}{2} \int \log_e \left(1 + \frac{\epsilon m}{M - m} \right) dm$$

$$= \frac{\lambda}{2} \int \log_e \{M + (\epsilon - 1)m\} \, dm - \frac{\lambda}{2} \int \log_e (M - m) \, dm.$$

Now

$$\int \log_e (a + bx) \, dx = x \log_e (a + bx) - b \int \frac{x}{a + bx} \, dx$$

$$= x \log_e (a + bx) - \int \left(1 - \frac{a}{a + bx} \right) dx$$

$$= x \log_e (a + bx) - x + \frac{a}{b} \log_e (a + bx) + \text{constant}$$

$$= \left(x + \frac{a}{b} \right) \log_e (a + bx) - x + \text{constant}.$$

Hence,

$$\frac{2t}{\lambda} = \left(m + \frac{M}{\epsilon - 1} \right) \log_e \{M + (\epsilon - 1)m\}$$

$$- m - (m - M)\log_e (M - m) + m + C$$

where C is a constant.

Thus, since $m = 0$ when $t = 0$,

$$C = - \frac{M}{\epsilon - 1} \log_e M - M \log_e M$$

i.e.

$$C = - \frac{\epsilon M}{\epsilon - 1} \log_e M,$$

giving

$$\frac{2t}{\lambda} = \left(m + \frac{M}{\epsilon - 1} \right) \log_e \{M + (\epsilon - 1)m\}$$

$$+ (M - m)\log_e (M - m) - \frac{\epsilon M}{\epsilon - 1} \log_e M.$$

6. Definite integrals

We have seen that $\int f(x)\,dx = F(x) + C$, where C is an arbitrary constant and $F(x)$ is such that $dF(x)/dx = f(x)$.

If, having evaluated the integral of $f(x)$ with respect to x, we then put $x = a$, we obtain the result,

$$\int_{x=a} f(x)\,dx = F(a) + C. \tag{1}$$

Similarly, putting $x = b$ gives

$$\int_{x=b} f(x)\,dx = F(b) + C. \tag{2}$$

Eliminating C from equations (1) and (2) we obtain

$$\int_{x=b} f(x)\,dx - \int_{x=a} f(x)\,dx = F(b) - F(a).$$

For $b > a$, this is generally written as

$$\int_{a}^{b} f(x)\,dx = F(b) - F(a), \tag{3}$$

which is called the *definite integral* of $f(x)$ with respect to x, between $x = a$ and $x = b$.

In dealing with applications of integration in Chapter 5, we shall require the alternative definition of a definite integral in terms of a summation, which for convenience, we enunciate here.

Let the interval $x = a$ to $x = b$ be divided into n parts given by $x_1, x_2, \ldots, x_r, \ldots, x_{n+1}$, where $x_1 = a, x_{n+1} = b$.

Then the value of the sum

$$f(x_1)\,\delta x_1 + f(x_2)\,\delta x_2 + \cdots + f(x_r)\,\delta x_r + \cdots + f(x_n)\,\delta x_n$$

as $n \to \infty$, which may be expressed as

$$\lim_{\delta x_r \to 0} \sum_{r=1}^{n} f(x_r)\,\delta x_r$$

where $\delta x_r = x_{r+1} - x_r$, is equal to

$$\int_{a}^{b} f(x)\,dx.$$

This may be shown in the following way.

Since

$$f(x) = \frac{dF(x)}{dx} = \lim_{\delta x \to 0} \frac{F(x + \delta x) - F(x)}{\delta x}$$

$$\therefore \quad \lim_{\delta x \to 0} f(x)\,\delta x = \lim_{\delta x \to 0} \frac{F(x + \delta x) - F(x)}{\delta x} \lim_{\delta x \to 0} \delta x$$

$$= \lim_{\delta x \to 0} \{F(x + \delta x) - F(x)\}.$$

Hence,

$$\lim_{\delta x_r \to 0} \sum_{r=1}^{n} f(x_r)\,\delta x_r = \lim_{\delta x_r \to 0} \sum_{r=1}^{n} \{F(x_r + \delta x_r) - F(x_r)\},$$

$$= \lim_{\delta x_r \to 0} \{F(x_1 + \delta x_1) - F(x_1) + F(x_2 + \delta x_2) - F(x_2)$$

$$+ \cdots + F(x_n + \delta x_n) - F(x_n)\}$$

$$= \lim_{\delta x_r \to 0} \{-F(x_1) + F(x_n + \delta x_n)\}, \quad \text{since } x_r + \delta x_r = x_{r+1},$$

$$= F(x_{n+1}) - F(x_1)$$

$$= F(b) - F(a),$$

so that

$$\lim_{\delta x_r \to 0} \sum_{r=1}^{n} f(x_r)\,\delta x_r = \int_a^b f(x)\,dx. \tag{4}$$

From the definition given by equation (3), we obtain the following simple results:

(i) $\int_a^b f(x)\,dx = F(b) - F(a) = -\{F(a) - F(b)\}$

i.e.

$$\int_a^b f(x)\,dx = -\int_b^a f(x)\,dx.$$

(ii) for $a \leqslant c \leqslant b$,

$$\int_a^b f(x)\,dx = F(b) - F(a) = \{F(c) - F(a)\} + \{F(b) - F(c)\}$$

$$= \int_a^c f(x)\,dx + \int_c^b f(x)\,dx.$$

As the applications of integration in Chapter 5 will involve the evaluation of definite integrals, only a few illustrations will be given at this stage.

Example 44

$$\int_2^7 \frac{x+2}{(x-1)(x+3)}\,dx = \frac{3}{4}\int_2^7 \frac{1}{x-1}\,dx + \frac{1}{4}\int_2^7 \frac{1}{x+3}\,dx$$

$$= \tfrac{3}{4}\log_e(x-1)\Big|_2^7 + \tfrac{1}{4}\log_e(x+3)\Big|_2^7$$

$$= \tfrac{3}{4}\log_e 6 + \tfrac{1}{4}\log_e 2.$$

Example 45

$$\int_0^2 \frac{x+1}{\sqrt{(x^2+1)}}\,dx = \int_0^2 \frac{x}{\sqrt{(x^2+1)}}\,dx + \int_0^2 \frac{1}{\sqrt{(x^2+1)}}\,dx$$

$$= \sqrt{(x^2+1)}\,\Big|_0^2 + \sinh^{-1}x\,\Big|_0^2$$

$$= \sqrt{(5)} - 1 + \sinh^{-1}(2).$$

Example 46

$$\int_0^{\pi/6} x^2 \cos 3x\,dx = \tfrac{1}{3}x^2 \sin 3x\,\Big|_0^{\pi/6} - \frac{2}{3}\int_0^{\pi/6} x \sin 3x\,dx$$

$$= \frac{\pi^2}{108} - \frac{2}{3}\left(-\tfrac{1}{3}x\cos 3x\,\Big|_0^{\pi/6} + \frac{1}{3}\int_0^{\pi/6}\cos 3x\,dx\right)$$

$$= \frac{\pi^2}{108} - \frac{2}{9}\cdot\frac{\sin 3x}{3}\Big|_0^{\pi/6}$$

$$= \tfrac{1}{108}(\pi^2 - 8).$$

Example 47

For a gas contained in a vessel, the kinetic theory of gases gives the average molecular velocity $\bar{c}$ as

$$\bar{c} = 4\pi\left(\frac{m}{2\pi kT}\right)^{3/2}\int_0^\infty u^3 \exp\left(-\frac{mu^2}{2kT}\right)du,$$

where m is the mass of a molecule, T the temperature of the gas, and k, Boltzmann's constant.

Show that

$$\bar{c} = \sqrt{\left(\frac{8kT}{\pi m}\right)}.$$

Let $x^2 = mu^2/2kT$, so that $x\,dx = (mu/2kT)\,du$, and

$$\bar{c} = 4\pi\left(\frac{m}{2\pi kT}\right)^{3/2} \int_0^\infty u^2 \exp\left(-\frac{mu^2}{2kT}\right) u\,du$$

$$= 4\pi\left(\frac{m}{2\pi kT}\right)^{3/2} \int_0^\infty \frac{2kT}{m} x^2 \exp(-x^2) \frac{2kT}{m} x\,dx$$

$$= 4\left(\frac{2kT}{\pi m}\right)^{1/2} \int_0^\infty x^2\{x\exp(-x^2)\}\,dx$$

$$= 4\left(\frac{2kT}{\pi m}\right)^{1/2} I,$$

where

$$I = x^2\left(\frac{-\exp(-x^2)}{2}\right)\Bigg|_0^\infty + \int_0^\infty x\exp(-x^2)\,dx$$

$$= \int_0^\infty x\exp(-x^2)\,dx$$

$$= -\tfrac{1}{2}\exp(-x^2)\,|_0^\infty = \tfrac{1}{2}$$

$$\therefore\ \bar{c} = 2\left(\frac{2kT}{\pi m}\right)^{1/2} = \left(\frac{8kT}{\pi m}\right)^{1/2}.$$

N.B.

$$x^2\exp(-x^2) = \frac{x^2}{\exp(x^2)}$$

$$= \frac{x^2}{1 + x^2/1! + x^4/2! + x^6/3! + \cdots}$$

$$= \frac{1}{1/x^2 + 1 + x^2/2! + x^4/3! + \cdots} \to 0 \quad \text{as } x \to \infty.$$

7. Reduction formulae

Consider the integral $I = \int f(x, n)\,dx$, where n is a constant.

Since, in general, the value of I will be different for different values of n, it is convenient to stress this dependence by attaching to I the suffix n.

Hence,

$$I_n = \int f(x, n)\,dx,$$

and

$$I_{n-1} = \int f(x, n - 1)\,dx,$$

$$I_{n-2} = \int f(x, n - 2)\,dx,$$

and so on.

If I_n is expressed in terms of any function of $I_{n-1}, I_{n-2}, \ldots$, etc., then the expression is called a *reduction formula*.

Such reduction formulae are generally obtained by integration by parts, as we now illustrate.

Example 48

Obtain a reduction formula for

$$\int x^n \cos x \, dx.$$

Let

$$I_n = \int x^n \cos x \, dx$$

$$= x^n \sin x - n \int x^{n-1} \sin x \, dx$$

$$= x^n \sin x - n \left(-x^{n-1} \cos x + (n-1) \int x^{n-2} \cos x \, dx \right)$$

$$= x^n \sin x + nx^{n-1} \cos x - n(n-1)I_{n-2}.$$

$$\therefore \ I_n + n(n-1)I_{n-2} = x^n \sin x + nx^{n-1} \cos x.$$

Example 49

Obtain a reduction formula for

$$\int \frac{x^n}{\sqrt{(x^2+1)}} \, dx.$$

Let

$$I_n = \int \frac{x^n}{\sqrt{(x^2+1)}} \, dx$$

$$= \int x^{n-1} \frac{x}{\sqrt{(x^2+1)}} \, dx, \quad n \geqslant 1,$$

$$= x^{n-1}\sqrt{(x^2+1)} - (n-1) \int x^{n-2}\sqrt{(x^2+1)} \, dx$$

$$= x^{n-1}\sqrt{(x^2+1)} - (n-1) \int \frac{x^{n-2}(x^2+1)}{\sqrt{(x^2+1)}} \, dx$$

$$= x^{n-1}\sqrt{(x^2+1)} - (n-1) \int \frac{x^n}{\sqrt{(x^2+1)}} \, dx - (n-1) \int \frac{x^{n-2}}{\sqrt{(x^2+1)}} \, dx$$

$$\therefore \ I_n = x^{n-1}\sqrt{(x^2+1)} - (n-1)I_n - (n-1)I_{n-2}$$

and hence,

$$nI_n + (n-1)I_{n-2} = x^{n-1}\sqrt{(x^2+1)}.$$

Example 50

Obtain a reduction formula for

$$\int \sec^n x \, dx.$$

Let

$$I_n = \int \sec^n x \, dx$$

$$= \int \sec^{n-2} x \sec^2 x \, dx, \quad n \geqslant 2,$$

$$= \sec^{n-2} x \tan x - (n-2) \int \sec^{n-3} x \cdot \sec x \tan x \cdot \tan x \, dx$$

$$= \sec^{n-2} x \tan x - (n-2) \int \sec^{n-2} x \tan^2 x \, dx$$

$$= \sec^{n-2} x \tan x - (n-2) \int \sec^{n-2} x (\sec^2 x - 1) \, dx$$

$$= \sec^{n-2} x \tan x - (n-2) \int \sec^n x \, dx + (n-2) \int \sec^{n-2} x \, dx$$

$$\therefore \ I_n = \sec^{n-2} x \tan x - (n-2)I_n + (n-2)I_{n-2}.$$

Hence,

$$(n-1)I_n - (n-2)I_{n-2} = \sec^{n-2} x \tan x.$$

Example 51

Determine a reduction formula for,

$$\int \cot^n x \, dx,$$

and hence evaluate,

$$\int_0^{1/\sqrt{2}} \frac{x^5}{(1-x^2)^3} \, dx.$$

Let

$$I_n = \int \cot^n x \, dx$$

$$= \int \cot^{n-2} x (\operatorname{cosec}^2 x - 1) \, dx, \quad n \geqslant 2,$$

$$= \int \cot^{n-2} x \operatorname{cosec}^2 x \, dx - \int \cot^{n-2} x \, dx.$$

$$\therefore \ I_n = \frac{-\cot^{n-1} x}{n-1} - I_{n-2} \qquad (5)$$

D

For

$$I = \int_0^{1/\sqrt{2}} \frac{x^5}{(1-x^2)^3} \, dx,$$

let $x = \cos\theta$, then $dx = -\sin\theta \, d\theta$.

When $x = 1/\sqrt{2}$, $\cos\theta = 1/\sqrt{2}$ and so $\theta = \pi/4$, and when $x = 0$, $\cos\theta = 0$ giving $\theta = \pi/2$.

Also

$$\frac{x^5}{(1-x^2)^3} \, dx = \frac{\cos^5\theta(-\sin\theta \, d\theta)}{(1-\cos^2\theta)^3} = \frac{-\cos^5\theta}{\sin^5\theta} \, d\theta.$$

Hence,

$$I = -\int_{\pi/2}^{\pi/4} \cot^5\theta \, d\theta = \int_{\pi/4}^{\pi/2} \cot^5\theta \, d\theta \tag{6}$$

Between limits $\pi/4$ and $\pi/2$, equation (5) gives

$$I_n = \frac{-\cot^{n-1}x}{n-1}\bigg|_{\pi/4}^{\pi/2} - I_{n-2}$$

i.e.

$$I_n = \frac{1}{n-1} - I_{n-2}.$$

Putting

$$n = 5 \quad \therefore \quad I_5 = \tfrac{1}{4} - I_3$$
$$n = 3 \quad \therefore \quad I_3 = \tfrac{1}{2} - I_1$$

and

$$I_1 = \int_{\pi/4}^{\pi/2} \cot x \, dx = \log_e \sin x \bigg|_{\pi/4}^{\pi/2} = -\log_e\left(\frac{1}{\sqrt{2}}\right) = \log_e\sqrt{2}.$$

Equation (6) thus gives

$$I \equiv I_5 = \tfrac{1}{4} - (\tfrac{1}{2} - \log_e\sqrt{2}) = \log_e\sqrt{(2)} - \tfrac{1}{4}.$$

For use in Chapter 5, we obtain now a reduction formula for the integral

$$I_{n,m} = \int_0^{\pi/2} \sin^n x \cos^m x \, dx,$$

where n and m are both positive integers. We have,

$$I_{n,m} = \int_0^{\pi/2} (\sin^n x \cos x)\cos^{m-1}x \, dx$$

$$= \frac{\sin^{n+1}x \cos^{m-1}x}{n+1}\bigg|_0^{\pi/2} - \frac{m-1}{n+1}\int_0^{\pi/2}\sin^{n+1}x \cos^{m-2}x(-\sin x) \, dx$$

i.e.

$$I_{n,m} = \frac{m-1}{n+1} \int_0^{\pi/2} \sin^{n+2} x \cos^{m-2} x \, dx$$

$$= \frac{m-1}{n+1} \int_0^{\pi/2} \sin^n x (1 - \cos^2 x) \cos^{m-2} x \, dx$$

$$= \frac{m-1}{n+1} \int_0^{\pi/2} \sin^n x \cos^{m-2} x \, dx - \frac{m-1}{n+1} \int_0^{\pi/2} \sin^n x \cos^m x \, dx$$

$$\therefore \quad (n+1)I_{n,m} = (m-1)I_{n,m-2} - (m-1)I_{n,m}$$

whence,

$$I_{n,m} = \frac{m-1}{n+m} I_{n,m-2}, \qquad \text{for} \quad n \geqslant 0, \quad m \geqslant 2. \tag{7}$$

Similarly, by expressing the original integral as

$$I_{n,m} = \int_0^{\pi/2} \sin^{n-1} x (\sin x \cos^m x) \, dx$$

we would obtain the alternative result,

$$I_{n,m} = \frac{n-1}{n+m} I_{n-2,m}, \qquad \text{for} \quad n \geqslant 2, \quad m \geqslant 0. \tag{8}$$

A more useful result, known as Wallis's formula, is obtained as follows. In equation (7) replacing n by $n - 2$, gives

$$I_{n-2,m} = \frac{m-1}{n+m-2} I_{n-2,m-2} \tag{9}$$

so that equation (8) becomes

$$I_{n,m} = \frac{(n-1)(m-1)}{(n+m)(n+m-2)} I_{n-2,m-2}. \tag{10}$$

Replacing n by $n - 2$, and m by $m - 2$ in (10) now gives

$$I_{n-2,m-2} = \frac{(n-3)(m-3)}{(n+m-4)(n+m-6)} I_{n-4,m-4}$$

so that equation (10) becomes

$$I_{n,m} = \frac{(n-1)(n-3)(m-1)(m-3)}{(n+m)(n+m-2)(n+m-4)(n+m-6)} I_{n-4,m-4}.$$

Continuing this procedure, we eventually reach the stage at which $I_{n,m}$ is given in terms of $I_{0,0}, I_{1,1}, I_{1,0}$ or $I_{0,1}$ depending on whether n and m are even or odd integers.

Now

$$I_{0,0} = \int_0^{\pi/2} dx = \frac{\pi}{2}, \quad I_{1,1} = \int_0^{\pi/2} \sin x \cos x \, dx = \frac{\sin^2 x}{2} \bigg|_0^{\pi/2} = \frac{1}{2}$$

and,

$$I_{1,0} = \int_0^{\pi/2} \sin x \, dx = -\cos x \Big|_0^{\pi/2} = 1,$$

$$I_{0,1} = \int_0^{\pi/2} \cos x \, dx = \sin x \Big|_0^{\pi/2} = 1.$$

Hence, the most general result may be expressed in the form

$$I_{n,m} = \frac{(n-1)(n-3)(n-5)\ldots(m-1)(m-3)(m-5)\ldots}{(n+m)(n+m-2)(n+m-4)\ldots} \mu,$$

for $n \geq 2$, $m \geq 2$

where $n-1$, $m-1$ and $n+m$ are decreased in steps of 2 without passing through zero, and

$$\mu = \begin{cases} \pi/2, & \text{for } n, m \text{ both even} \\ 1, & \text{for all other cases.} \end{cases}$$

Applying equation (7) to the case when $n = 0$, and equation (8) to the case when $m = 0$, the following results are also obtainable:

$$I_m = \int_0^{\pi/2} \cos^m x \, dx = \frac{(m-1)(m-3)\ldots}{m(m-2)(m-4)\ldots} \mu$$

and,

$$I_n = \int_0^{\pi/2} \sin^n x \, dx = \frac{(n-1)(n-3)\ldots}{n(n-2)(n-4)\ldots} \mu.$$

Example 52

(i) $\displaystyle\int_0^{\pi/2} \sin^6 x \, dx = \frac{5 \cdot 3 \cdot 1}{6 \cdot 4 \cdot 2} \cdot \frac{\pi}{2}$

(ii) $\displaystyle\int_0^{\pi/2} \cos^5 x \, dx = \frac{4 \cdot 2}{5 \cdot 3 \cdot 1}$

(iii) $\displaystyle\int_0^{\pi/2} \sin^7 x \cos^4 x \, dx = \frac{6 \cdot 4 \cdot 2 \cdot 3 \cdot 1}{11 \cdot 9 \cdot 7 \cdot 5 \cdot 3 \cdot 1}$

(iv) $\displaystyle\int_0^{\pi/2} \sin^3 x \cos^5 x \, dx = \frac{2 \cdot 4 \cdot 2}{8 \cdot 6 \cdot 4 \cdot 2}$

(v) $\displaystyle\int_0^{\pi/2} \sin^4 x \cos^8 x \, dx = \frac{3 \cdot 1 \cdot 7 \cdot 5 \cdot 3 \cdot 1}{12 \cdot 10 \cdot 8 \cdot 6 \cdot 4 \cdot 2} \cdot \frac{\pi}{2}.$

PROBLEMS

1. Evaluate the following integrals:

(i) $\displaystyle\int \frac{3x + 2}{2x - 1} \, dx,$
(ii) $\displaystyle\int \frac{3x^2 - x - 7}{4 - 3x} \, dx,$
(iii) $\displaystyle\int \frac{x^2 + 2x - 4}{x\sqrt{x}} \, dx$

2. Integrate the following functions with respect to x:

(i) $\dfrac{x-7}{3+2x-x^2}$,

(ii) $\dfrac{4x^2-14x+15}{2x^2-5x+2}$,

(iii) $\dfrac{2+2x-x^2}{(2x+3)(x^2+1)}$,

(iv) $\dfrac{5x^2+13x-14}{(3x+1)(x-1)^2}$,

(v) $\dfrac{2x^3+9x^2+4x+2}{x(x+2)(2x^2+x+1)}$.

3. Evaluate

$$\int \frac{1}{4x^2-4x+10}\,dx,$$

and

$$\int \frac{1-2x}{3-2x-x^2}\,dx.$$

4. Evaluate

(i) $\displaystyle\int \frac{1-x}{\sqrt{(4-x^2)}}\,dx$,

(ii) $\displaystyle\int \frac{1}{\sqrt{(2+2x-x^2)}}\,dx$,

(iii) $\displaystyle\int \frac{1}{\sqrt{(9x^2+2x+2)}}\,dx$,

(iv) $\displaystyle\int \frac{6x+1}{\sqrt{(x^2+x-6)}}\,dx$.

5. Integrate the following functions with respect to x:

(i) $\tan^3 x$,

(ii) $\dfrac{\cos^3 x}{\sin^4 x}$,

(iii) $\sin^4 x$,

(iv) $\dfrac{1}{2+\cos x - 2\sin x}$,

(v) $\dfrac{1}{4\cos 2x - 1}$,

(vi) $\dfrac{2\sin x + \cos x}{\sin x - 3\cos x}$.

6. Evaluate the following integrals by using the substitution given:

(i) $\displaystyle\int \sqrt{(3-2x-x^2)}\,dx$,   using $x+1 = 2\sin\theta$,

(ii) $\displaystyle\int \sqrt{(x^2-9)}\,dx$,   using $x = 3\cosh u$,

(iii) $\displaystyle\int \frac{1}{\sqrt{\{(3-x)(x+1)\}}}\,dx$,   using $x = 3\cos^2\theta - \sin^2\theta$,

(iv) $\displaystyle\int \frac{1}{x\sqrt{(5x^2+4x+1)}}\,dx$,   using $x = \dfrac{1}{u}$,

(v) $\displaystyle\int \sqrt{\left(\dfrac{x-2}{x+2}\right)}\,dx$,   using $x = \cosh 2u$.

7. Integrate by parts, with respect to x:

(i) $x^2 \cos 3x$,

(ii) $(1+x)\log_e(x-1)^2$,

(iii) $x\tan^{-1} x$,

(iv) $e^x \cos 2x$,

(v) $(\log_e x)(\log_e 2x)$,

and write down the values of

(vi) $\displaystyle\int_0^{\pi/2} \sin^5 x \, dx,$ (vii) $\displaystyle\int_0^{\pi/2} \sin^4 x \cos^3 x \, dx,$

(viii) $\displaystyle\int_0^{\pi/2} \cos^4 x \sin^6 x \, dx.$

8. If

$$I_n = \int_0^{\pi/4} \sec\theta \tan^n \theta \, d\theta,$$

show that for $n \geqslant 2$,

$$nI_n = \sqrt{2} - (n-1)I_{n-2}.$$

Hence evaluate

$$\int_0^1 \frac{x^7}{\sqrt{(x^2+1)}} \, dx.$$

9. If

$$I_{m,n} = \int_0^1 x^m (1-x)^n \, dx,$$

show that $I_{m,n} = I_{n,m}$ and that $(m+n+1)I_{m,n} = mI_{m-1,n}$. Evaluate $I_{4,3}$.

10. Show that the work done when an ideal gas (p_1, v_1) expands adiabatically to the condition (p_2, v_2) is

$$\frac{p_1 v_1}{\gamma - 1} \left\{ \left(\frac{v_1}{v_2}\right)^{\gamma - 1} - 1 \right\},$$

where γ is the ratio of the specific heats of the gas at constant pressure and constant volume respectively.

11. For the flow of liquid through a tube of radius a and length l, the volume of the liquid passing through the tube per second, is

$$V = \frac{\pi P}{2\eta l} \int_0^a r(a^2 - r^2) \, dr,$$

where P is the pressure difference between the ends of the tube, and η is the coefficient of viscosity.

Determine the value of V.

12. In the calculation of the surface tension of a liquid by Quincke's method, a large drop of the liquid is placed upon the horizontal surface of a solid, and various measurements are taken.

The surface tension T, is given by the equation

$$\frac{g\rho}{T} \int_0^h x \, dx = \int_0^\alpha \sin\theta \, d\theta,$$

where h is the thickness of the drop, α the angle of contact of the drop with the horizontal surface, ρ the density of the liquid, and g the acceleration due to gravity.

Show that

$$T = g\rho \left(\frac{h}{2\sin(\alpha/2)}\right)^2.$$

13. A radio-active substance disintegrates with time. If the mass m, remaining after time t, satisfies the equation

$$\frac{dm}{dt} + mk = 0,$$

where k is a constant, show that $m = m_0 \exp(-kt)$, where m_0 is the initial mass.

A mixture of two radio-active substances consists initially of mass 1200 mg of one and 600 mg of the other. At the end of one day the total mass of mixture is 900 mg. At the end of two days the total mass is 550 mg. It is required to deduce the values of k_1 and k_2 for the substances, the unit of time being one day.

Show that the given data allow two different solutions.

To select the correct one, the mass remaining at the end of three days was measured and found to be 358.3 mg. Deduce the values of k_1 and k_2 correct to two decimal places.

14. The equation

$$\log_e\left(\frac{w}{w_0}\right) = \lambda \int_{m_0}^{m} \frac{dm}{m(1-m)},$$

gives the relation between w and m, the weight and molar fraction of a component remaining in a binary mixture after a finite partial distillation, w_0 and m_0 being the respective initial values, and λ a constant.

Show that

$$w = w_0 \left(\frac{m(1-m_0)}{m_0(1-m)}\right)^{\lambda}.$$

15. For a solid in equilibrium with its saturated vapour, the corresponding Clapeyron–Clausius equation is

$$\frac{dp}{dT} = \frac{ph}{RT^2},$$

where h is the heat of evaporation of one gram molecule of the solid at pressure p and temperature T, R being the gas constant.

If

$$h = h_0 + \int_{0}^{T} H \, dT,$$

where h_0 is the heat evaporation at $T = 0$, and H is the difference between the heat capacities of the vapour and solid per gram molecule, show that

$$\log_e p = \frac{-h}{RT} + \int_{0}^{T} \frac{H}{RT} \, dT + \text{constant}.$$

16. In the determination of the collision frequency of a molecule moving through homogeneous gas, the average relative velocity of colliding molecules is

$$V_A = \frac{1}{2} \int_{0}^{\pi} (\sin \theta)(V_1^2 + V_2^2 - 2V_1 V_2 \cos \theta)^{1/2} \, d\theta,$$

where V_1 and V_2 represent the velocities of two molecules moving in the gas.

Show that for $V_1 > V_2$,

$$V_A = \frac{1}{3V_1} (3V_1^2 + V_2^2).$$

17. In the production of low temperatures obtained by placing a substance in a magnetic field, the magnetic moment M of one gram of the substance is given by the equation

$$M \int_0^\pi \sin \theta \, \exp(\lambda \cos \theta) \, d\theta = \mu n \int_0^\pi \sin \theta \cos \theta \, \exp(\lambda \cos \theta) \, d\theta,$$

where λ, μ are constants and n is the number of molecules in one gram of the substance.

Show that,

$$M = \mu n \left(\coth \lambda - \frac{1}{\lambda} \right)$$

and that when λ is small,

$$M \simeq \frac{\mu n \lambda}{3}.$$

18. The Debye equation for the specific heat at constant volume, arising in the calculation of the heat capacity of solids is

$$C_V = \frac{9R}{\tau^3} \int_0^\tau \frac{x^4 e^x}{(e^x - 1)^2} \, dx, \qquad x = \frac{h\nu}{kT},$$

where h is Planck's constant, ν is the vibration frequency, k is Boltzmann's constant, T the absolute temperature, and $\tau = \theta/T$ where θ is the Debye characteristic temperature.

By integration 'by parts', show that

$$C_V = \frac{9R}{\tau^3} \left(4 \int_0^\tau \frac{x^3}{e^x - 1} \, dx - \frac{\tau^4}{e^\tau - 1} \right).$$

Show further, that if τ is a root of the equation

$$a^3 x^3 + 3a^2 x^2 + 6ax + 6 = 0,$$

where a is a constant, and

$$\sum_1^\infty \frac{1}{n^4} = \frac{\pi^4}{90},$$

then

$$C_V = \frac{9R}{5\tau^3} \left(\frac{4\pi^4}{3} - \frac{5\tau^4}{e^\tau - 1} \right).$$

CHAPTER 5

Applications of Integration

1. Areas

Let the function $f(x)$ of x be positive throughout the range $a \leqslant x \leqslant b$. The problem is to determine the area A enclosed by the curve $y = f(x)$, $x = a$, $x = b$ and the x-axis. (See Fig. 5.1.)

Divide this area into n sections each of width δx_r and consider the section PQRS where $P \equiv (x_r, 0)$ and $S \equiv (x_{r+1}, 0)$.

Then

$$\text{area of rectangle PQR'S} < \text{area PQRS} < \text{area of rectangle PQ'RS}$$

i.e.

$$f(x_r) \, \delta x_r < \text{area PQRS} < f(x_{r+1}) \, \delta x_r.$$

Carrying out the summation of all such rectangular sections throughout the range $x = a$ to $x = b$, thus gives

$$\sum_{r=1}^{n} f(x_r) \, \delta x_r < A < \sum_{r=1}^{n} f(x_{r+1}) \, \delta x_r.$$

(See Section 6 of Chapter 4.)

If the number of sections into which the area is divided, is now increased until there are an infinite number (i.e. $n \to \infty$), then the difference between

$$\sum_{r=1}^{n} f(x_r) \, \delta x_r \quad \text{and} \quad \sum_{r=1}^{n} f(x_{r+1}) \, \delta x_r$$

will tend to zero, and $\delta x_r \to 0$.

Thus,

$$A = \lim_{\delta x_r \to 0} \sum_{r=1}^{n} f(x_r) \, \delta x_r = \int_a^b f(x) \, dx. \tag{1}$$

(See equation (4) of Chapter 4.)

It should be noted that if $f(x)$ is negative throughout the range $a \leqslant x \leqslant b$ then

$$A = \int_a^b f(x) \, dx$$

is negative. A diagram should therefore be drawn when determining areas.

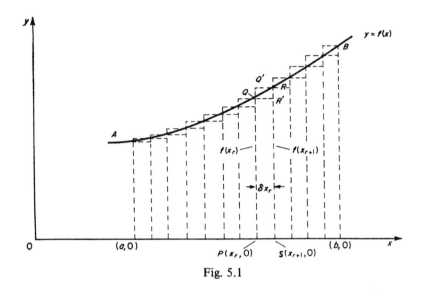

Fig. 5.1

Example 1

Determine the area cut off from the parabola $y^2 = 12x$ by the straight line $y = 2x$. (See Fig. 5.2.)

The given equations are $y^2 = 12x$ and $y = 2x$, from which it is seen that the points of intersection are $(0, 0)$ and $(3, 6)$. Thus the area required is equal to the difference between the areas beneath the parabola and the straight line, respectively, as x varies from $x = 0$ to $x = 3$.

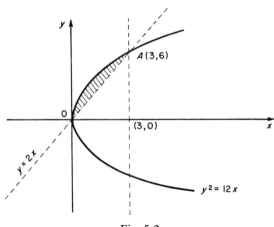

Fig. 5.2

Thus

$$\text{area} = \int_0^3 \sqrt{(12x)}\, dx - \int_0^3 (2x)\, dx$$

$$= \sqrt{(12)}\, \frac{2x^{3/2}}{3}\Big|_0^3 - x^2\Big|_0^3$$

$$= 3 \text{ square units}$$

Example 2

Find the area enclosed by the curve

$$2x^2 + y^2 - 2xy + 4x - 4y + 3 = 0.$$

Since the given equation is a quadratic in y, there will be two values of y for each value of x.

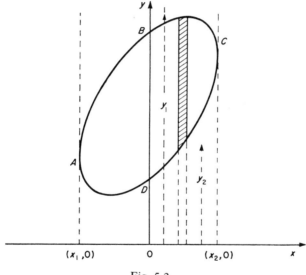

Fig. 5.3

Let the upper portion ABC of the curve, be represented by y_1, and the lower portion ADC by y_2. Then the area of the shaded segment shown in Fig. 5.3, is $(y_1 - y_2)\, \delta x$ so that the total area enclosed by the curve is

$$A = \int_{x_1}^{x_2} (y_1 - y_2)\, dx.$$

For the given curve

$$y^2 - y(2x + 4) + (2x^2 + 4x + 3) = 0$$

$$\therefore\ y = x + 2 \pm \tfrac{1}{2}\sqrt{\{(2x + 4)^2 - 4(2x^2 + 4x + 3)\}}$$

i.e.

$$y = x + 2 \pm \sqrt{(1 - x^2)}$$

i.e.

$$y_1 = x + 2 + \sqrt{(1 - x^2)} \quad \text{and} \quad y_2 = x + 2 - \sqrt{(1 - x^2)}.$$

Thus

$$y_1 - y_2 = 2\sqrt{(1 - x^2)}$$

and x_1 and x_2 are given by $1 - x^2 = 0$, i.e. $x_1 = -1, x_2 = +1$. Thus

$$A = 2 \int_{-1}^{1} \sqrt{(1 - x^2)} \, dx.$$

Let $x = \sin \theta$, so that $dx = \cos \theta \, d\theta$. Therefore

$$A = 2 \int_{-\pi/2}^{\pi/2} \cos^2 \theta \, d\theta = 4 \int_{0}^{\pi/2} \cos^2 \theta \, d\theta = \pi \text{ square units.}$$

2. Volume of revolution

We now determine the volume V generated as the area A of Section 1 is rotated through 2π radians about the x-axis (See Fig. 5.1).

Let the *volume of revolution* so formed be divided into n discs each of width δx_r, and consider the volume formed by the rotation of the area PQRS about the x-axis.

This volume, δV say, will be greater than the volume of the disc formed by the rotation of the area PQR'S about Ox and less than that formed by the rotation of the area PQ'RS about Ox.

Hence if $y_r = f(x_r)$ and $y_{r+1} = f(x_{r+1})$, then

$$\pi y_r^2 \, \delta x_r < \delta V < \pi y_{r+1}^2 \, \delta x_r.$$

Extending the process over the whole range $x = a$ to $x = b$ as in Section 1, gives

$$\sum_{r=1}^{n} \pi y_r^2 \, \delta x_r < V < \sum_{r=1}^{n} \pi y_{r+1}^2 \, \delta x_r.$$

Taking the limiting case as $n \to \infty$ and $\delta x_r \to 0$ leads to the result

$$V = \lim_{\delta x_r \to 0} \sum_{r=1}^{n} \pi y_r^2 \, \delta x_r = \int_{a}^{b} \pi y^2 \, dx. \tag{2}$$

Example 3

Determine the volume generated by the area in the first quadrant, enclosed by the curves

$$x^{2/3} + y^{2/3} = a^{2/3}, \quad x = 0, \quad y = 0,$$

as it rotates through 2π radians about the x-axis.

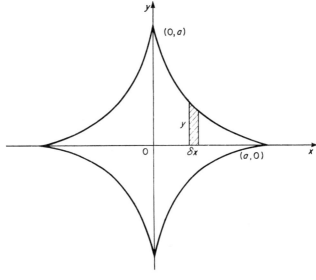

Fig. 5.4

The curve $x^{2/3} + y^{2/3} = a^{2/3}$ is called the *astroid* (Fig. 5.4).
The volume generated is

$$V = \int_0^a \pi y^2 \, dx = \pi \int_0^a (a^{2/3} - x^{2/3})^3 \, dx$$

Let $x = a \sin^3 \theta$, so that $dx = 3a \sin^2 \theta \cos \theta \, d\theta$. Then

$$V = \pi \int_0^{\pi/2} (a^2 \cos^6 \theta) \cdot 3a \sin^2 \theta \cos \theta \, d\theta$$

$$= 3\pi a^3 \int_0^{\pi/2} \cos^7 \theta \sin^2 \theta \, d\theta$$

$$= 3\pi a^3 \frac{(6 \cdot 4 \cdot 2)(1)}{9 \cdot 7 \cdot 5 \cdot 3}$$

$$= \frac{16\pi a^3}{105} \text{ cubic units.}$$

3. Length of arc

Divide the arc AB into n elements each of arc of length δs_r (see Fig. 5.5), and
consider the element QR, where $Q \equiv (x_r, y_r)$ and $R \equiv (x_r + \delta x_r, y_r + \delta y_r)$.
Then

$$(\delta s_r)^2 \simeq (\delta x_r)^2 + (\delta y_r)^2$$

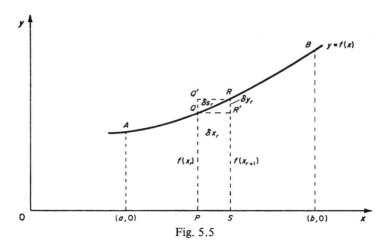

Fig. 5.5

so that

$$\delta s_r \simeq \sqrt{\left\{1 + \left(\frac{\delta y_r}{\delta x_r}\right)^2\right\}} \, \delta x_r$$

the positive value of the square root being taken.

Adding together all such elements of arc as x varies from $x = a$ to $x = b$, we obtain

$$\sum_{r=1}^{n} \delta s_r \simeq \sum_{r=1}^{n} \sqrt{\left\{1 + \left(\frac{\delta y_r}{\delta x_r}\right)^2\right\}} \, \delta x_r. \tag{3}$$

The greater the number of elements of arc δs_r chosen, the greater the accuracy of equation (3). Following the procedure of the previous two sections by letting $n \to \infty$ so that $\delta x_r \to 0$, $\delta s_r \to 0$ and $\delta y_r/\delta x_r \to dy_r/dx_r$, gives the expression for the arc length s from A to B, i.e.

$$s = \lim_{\delta s_r \to 0} \sum_{r=1}^{n} \delta s_r = \lim_{\delta x_r \to 0} \sum_{r=1}^{n} \sqrt{\left\{1 + \left(\frac{\delta y_r}{\delta x_r}\right)^2\right\}} \delta x_r$$

$$= \int_a^b \sqrt{\left\{1 + \left(\frac{dy}{dx}\right)^2\right\}} \, dx \tag{4}$$

where $dy/dx \equiv f'(x)$ is assumed to be continuous throughout the range $a \leqslant x \leqslant b$.

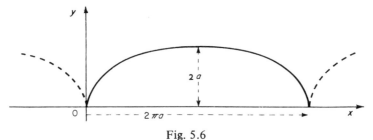

Fig. 5.6

Example 4

Determine the length of one arch of the *cycloid* (Fig. 5.6)

$$x = a(\theta - \sin \theta), \quad y = a(1 - \cos \theta).$$

$$\frac{dx}{d\theta} = a(1 - \cos \theta), \quad \frac{dy}{d\theta} = a \sin \theta$$

$$= 2a \sin^2(\theta/2) \qquad = 2a \sin(\theta/2)\cos(\theta/2).$$

$$\therefore \quad \frac{dy}{dx} = \frac{dy/d\theta}{dx/d\theta} = \cot(\theta/2).$$

Hence

$$s = \int_0^{2\pi} \sqrt{\left\{1 + \cot^2\left(\frac{\theta}{2}\right)\right\}} \cdot 2a \sin^2\left(\frac{\theta}{2}\right) d\theta$$

$$= 2a \int_0^{2\pi} \sin\left(\frac{\theta}{2}\right) d\theta = 8a.$$

4. Surface area

Referring again to Fig. 5.5, the problem now consists of determining the surface area S of the shell formed by the rotation of the arc length AB through 2π radians about the x-axis.

As the arc length δs_r is rotated about the x-axis it will trace out a frustum of a cone, the sum of all such sections representing the total surface area generated by the arc AB.

In order to obtain this sum we require an expression for the surface area of a frustum of a cone.

Figure 5.7(a) shows a frustum of a cone of slant height l, and Fig. 5.7(b) shows the surface ABB'A' of the curved section 'opened out'.

With dimensions as shown, we see that:

area of sector OAB = $\frac{1}{2}(L - l)^2\alpha$,   area of sector OA'B' = $\frac{1}{2}L^2\alpha$

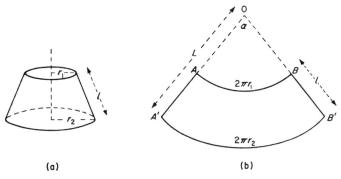

(a)   (b)

Fig. 5.7

so that

$$\text{area ABB}'\text{A}' = \frac{\alpha}{2}\{L^2 - (L - l)^2\} \tag{5}$$

Also

$$\text{arc length AB} = 2\pi r_1 = (L - l)\alpha$$

$$\text{arc length A}'\text{B}' = 2\pi r_2 = L\alpha.$$

Thus equation (5) becomes

$$\text{area ABB}'\text{A}' = \pi\{r_2 L - r_1(L - l)\}$$

$$= \pi\{r_1 l + L(r_2 - r_1)\}$$

$$= \pi l(r_1 + r_2), \tag{6}$$

since $r_1/r_2 = (L - l)/L$.

Referring to Fig. 5.5, equation (6) shows that the surface area δS_r formed by the rotation of the arc δs_r through 2π radians about the x-axis is

$$\delta S_r = \pi \delta s_r \{f(x_r) + f(x_{r+1})\}.$$

Hence the total surface area generated is given by

$$S \simeq \sum_{r=1}^{n} \delta S_r = \sum_{r=1}^{n} \pi\{f(x_r) + f(x_{r+1})\} \, \delta s_r.$$

If now $n \to \infty$ so that $\delta s_r \to 0$, then

$$S = \lim_{\delta s_r \to 0} \sum_{r=1}^{n} \pi\{f(x_r) + f(x_{r+1})\} \, \delta s_r$$

$$= \lim_{\delta s_r \to 0} \sum_{r=1}^{n} 2\pi f(x_r) \, \delta s_r, \quad \text{from equation (1)}$$

$$= \int_{x=a}^{x=b} 2\pi f(x) \, ds$$

i.e.

$$S = \int_a^b 2\pi y \sqrt{\left\{1 + \left(\frac{dy}{dx}\right)^2\right\}} \, dx, \quad \text{from equation (4)}$$

Example 5

Determine the surface area S generated by the arc of the *catenary* $y = c \cosh(x/c)$ from $x = 0$ to $x = \lambda$, when it is rotated through 2π radians about the x-axis (Fig. 5.8).

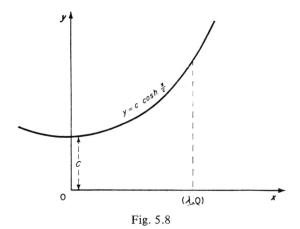

Fig. 5.8

If λ is so small that terms of the order of λ^5 and higher may be neglected, show that

$$S \simeq 2\pi\lambda c\left(1 + \frac{\lambda^2}{3c^2}\right).$$

Since

$$y = c \cosh\left(\frac{x}{c}\right)$$

$$\frac{dy}{dx} = \sinh\left(\frac{x}{c}\right).$$

$$\therefore \ S = \int_0^\lambda 2\pi y \sqrt{\left\{1 + \left(\frac{dy}{dx}\right)^2\right\}} \ dx$$

$$= \int_0^\lambda 2\pi c \cosh\left(\frac{x}{c}\right) \sqrt{\left\{1 + \sinh^2\left(\frac{x}{c}\right)\right\}} \ dx$$

$$= 2\pi c \int_0^\lambda \cosh^2\left(\frac{x}{c}\right) dx = \pi c \int_0^\lambda \left\{\cosh\left(\frac{2x}{c}\right) + 1\right\} dx$$

$$= \pi c \left[\frac{c}{2} \sinh\left(\frac{2x}{c}\right) + x\right]_0^\lambda = \pi c \left\{\frac{c}{2} \sinh\left(\frac{2\lambda}{c}\right) + \lambda\right\}.$$

But

$$\sinh\left(\frac{2\lambda}{c}\right) = \left(\frac{2\lambda}{c}\right) + \frac{1}{3!}\left(\frac{2\lambda}{c}\right)^3 + \frac{1}{5!}\left(\frac{2\lambda}{c}\right)^5 + \cdots$$

$$= \frac{2\lambda}{c} + \frac{4\lambda^3}{3c^3}, \quad \text{to } O(\lambda^4).$$

$$\therefore\ S \simeq \pi c\left(\lambda + \frac{2\lambda^3}{3c^2} + \lambda\right) = 2\pi\lambda c\left(1 + \frac{\lambda^2}{3c^2}\right).$$

5. Centroids

For a system of n particles of masses $\delta m_1, \delta m_2, \ldots, \delta m_n$ situated at points $P_1(x_1, y_1), P_2(x_2, y_2), \ldots, P_n(x_n, y_n)$ respectively (see Fig. 5.9) the *centre of mass* $G(\bar{x}, \bar{y})$ is given by the equations

$$M\bar{x} = \sum_{r=1}^{n} \delta m_r x_r, \quad M\bar{y} = \sum_{r=1}^{n} \delta m_r y_r, \quad \text{where} \quad M = \sum_{r=1}^{n} \delta m_r,$$

the equations being obtained by deriving the moment of the system about the y- and x-axes respectively.

The expressions $\Sigma\ \delta mx$, and $\Sigma\ \delta my$ are called the *first moments* of the system about Oy and Ox respectively.

For a continuous distribution in the form of an area, volume or length of arc of uniform density, the point $G(\bar{x}, \bar{y})$ is independent of the density and is called the *centroid*.

In these cases the summation sign will simply be replaced by an integral sign, and the mass M by an expression for area, volume or length of arc.

Thus

$$A\bar{x} = \int x\ \mathrm{d}A, \quad A\bar{y} = \int y\ \mathrm{d}A$$

$$V\bar{x} = \int x\ \mathrm{d}V, \quad V\bar{y} = \int y\ \mathrm{d}V \tag{7}$$

$$s\bar{x} = \int x\ \mathrm{d}s, \quad s\bar{y} = \int y\ \mathrm{d}s.$$

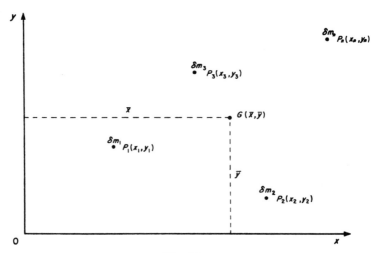

Fig. 5.9

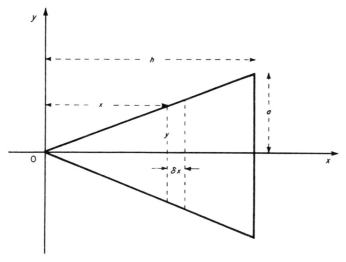

Fig. 5.10

Example 6

Determine the position of the centroid of: (i) a uniform solid cone; and (ii) a uniform hollow cone, open at the base.

Let the dimensions of the cone be as shown in Fig. 5.10.

(i) By similar triangles

$$\frac{y}{x} = \frac{a}{h}, \quad \text{i.e.} \quad y = \frac{xa}{h} \quad \text{and} \quad \frac{dy}{dx} = \frac{a}{h}$$

$$\therefore \quad V\bar{x} = \int x \, dV = \int_0^h x \cdot \pi y^2 \, dx$$

$$= \frac{\pi a^2}{h^2} \int_0^h x^3 \, dx = \frac{\pi a^2 h^2}{4}.$$

Since

$$V = \tfrac{1}{3}\pi a^2 h, \quad \therefore \quad \bar{x} = 3h/4.$$

(ii) The surface area S of the cone is given by

$$S = 2\pi \int_0^h y \sqrt{\left\{1 + \left(\frac{dy}{dx}\right)^2\right\}} \, dx$$

$$= \frac{2\pi a}{h^2} \sqrt{(h^2 + a^2)} \int_0^h x \, dx$$

$$= \pi a \sqrt{(h^2 + a^2)}.$$

Also

$$S\bar{x} = \frac{2\pi a}{h^2} \sqrt{(h^2 + a^2)} \int_0^h x^2 \, dx = \tfrac{2}{3}\pi a h \sqrt{(h^2 + a^2)};$$

and hence

$$\bar{x} = 2h/3.$$

6. Theorems of Pappus

An expression for the volume V obtained in Section 2, may be deduced by dividing the area A beneath the curve into small elements of area δA.

Let the centroid of δA be the point (x, y). Then, in the limiting case as $\delta A \to 0$, the volume V becomes the sum of the elements of volume formed by the rotation of the area δA about the x-axis.

Thus

$$V = \lim_{\delta A \to 0} \Sigma \, 2\pi y \, \delta A = \int 2\pi y \, dA = 2\pi \bar{y} \cdot A$$

from equation (7), so that:

(i) If a plane area is rotated about an axis which does not intersect the area, then the volume generated is equal to the product of the area and the distance travelled by its centroid.

Similarly, if δs is an element of arc length, of the arc s of Section 3, whose centroid is the point (x, y) then the surface area generated by rotating s about the x-axis is S, where

$$S = \lim_{\delta s \to 0} \Sigma \, 2\pi y \, \delta s = \int 2\pi y \, ds = 2\pi \bar{y} s.$$

Hence

(ii) If a length of arc is rotated about an axis which does not intersect the arc,

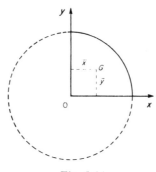

Fig. 5.11

then the surface area generated is equal to the product of the length of arc and the distance travelled by its centroid.

Example 7

Determine the position of the centroid of the area in the first quadrant enclosed by the circle $x^2 + y^2 = a^2$. Determine also the centroid of the arc length (Fig. 5.11).

If the area is rotated through 2π radians about the x-axis, it will trace out a hemisphere of volume $\frac{2}{3}\pi a^3$.

Section 6(i) thus gives

$$\frac{2}{3}\pi a^3 = 2\pi \bar{y}(\pi a^2/4)$$

i.e.

$$\bar{y} = \frac{4a}{3\pi} = \bar{x}$$

by symmetry.

If the arc length is rotated through 2π radians about the x-axis, it will trace out a hemispherical surface of area $2\pi a^2$.

Section 6(ii) thus gives

$$2\pi a^2 = 2\pi \bar{y}\left(\frac{\pi a}{2}\right)$$

i.e.

$$\bar{y} = \frac{2a}{\pi} = \bar{x}.$$

by symmetry.

7. Moments of inertia

For a system of n particles of masses $\delta m_1, \delta m_2, \ldots, \delta m_n$ situated at points distance $r_1, r_2, \ldots, r_n$ respectively, from an axis AA$'$ fixed in space, the sum

$$\delta m_1 r_1^2 + \delta m_2 r_2^2 + \cdots + \delta m_n r_n^2 \equiv \sum_{i=1}^{n} \delta m_i r_i^2$$

is called the *moment of inertia* (MI) or *second moment* of the system about this axis.

If

$$\sum_{i=1}^{n} \delta m_i r_i^2 = Mk^2$$

where M is the total mass of the system, i.e.

$$M = \sum_{i=1}^{n} \delta m_i,$$

then k is called the *radius of gyration* of the system about AA$'$.

108

For continuous distributions, the summation signs are replaced by definite integrals.

Example 8

Determine the MI of a uniform rod $(m, 2l)$ about an axis perpendicular to the axis of the rod, and passing through its centre.

Consider an element δx of the rod, distance x from OY. (See Fig. 5.12.) If ρ is the density of the rod per unit length then the mass of the element δx is $\delta m = \rho\, \delta x$.

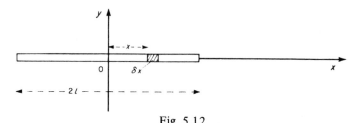

Fig. 5.12

The MI of this element about the axis OY is therefore

$$\delta I_{OY} \simeq \delta m \times x^2 = \rho x^2\, \delta x.$$

The MI of the whole rod about OY is thus

$$I_{OY} = \lim_{\delta x \to 0} \sum_{x=-l}^{l} \rho x^2\, \delta x$$

$$= \int_{-l}^{l} \rho x^2\, dx = \frac{2\rho l^3}{3}.$$

Since the mass of the rod is $M = 2l\rho$, we have

$$I_{OY} = \frac{Ml^2}{3}.$$

Example 9

Determine the MI of a uniform circular disc (M, a) about an axis perpendicular to the plane of the disc, and passing through its centre. Let δr be the thickness of a ring of radius r, centre O, lying within the disc (see Fig. 5.13).

If ρ is the density of the disc per unit area, then the MI of the ring about OY is

$$\delta I_{OY} \simeq (2\pi r\, \delta r \rho)(r^2).$$

The MI of the whole disc about OY is therefore

$$I_{OY} = \lim_{\delta r \to 0} \sum_{r=0}^{a} 2\pi \rho r^3\, \delta r$$

$$= \int_{0}^{a} 2\pi \rho r^3\, dr = \frac{\pi \rho a^4}{2}.$$

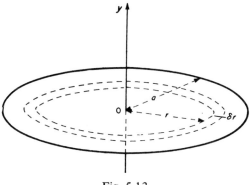

Fig. 5.13

Since the mass of the disc is $M = \pi a^2 \rho$, we have

$$I_{OY} = \frac{Ma^2}{2}.$$

Example 10

Determine the MI of a sphere (M, a) about a diameter.

Divide the sphere into discs of radius y and thickness δx (see Fig. 5.14).

If ρ is the density of the sphere per unit volume, then the mass of such a disc is $\pi y^2 \, \delta x \rho$. Thus the MI of the disc about the axis OX is, by Example 9

$$\delta I_{OX} = \tfrac{1}{2}(\text{mass})(\text{radius})^2 \simeq \tfrac{1}{2}(\pi y^2 \, \delta x \rho)(y^2).$$

The MI of the whole sphere about OX is therefore

$$I_{OX} = \lim_{\delta x \to 0} \sum_{x=-a}^{a} \tfrac{1}{2}\pi \rho y^4 \, \delta x$$

$$= \frac{1}{2} \int_{-a}^{a} \pi \rho y^4 \, dx = \frac{\pi \rho}{2} \int_{-a}^{a} (a^2 - x^2)^2 \, dx$$

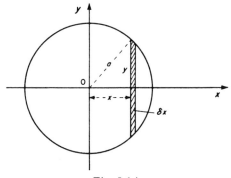

Fig. 5.14

$$= \pi \rho \int_0^a (a^4 - 2a^2 x^2 + x^4)\, dx = \tfrac{8}{15} \pi \rho a^5 .$$

Since the mass of the sphere is $M = \tfrac{4}{3} \pi a^3 \rho$, we have

$$I_{OX} = \frac{2Ma^2}{5} .$$

(i) Theorem of perpendicular axes

Let O be any point on a lamina of uniform surface density, and let OX and OY be perpendicular axes drawn in the plane of the lamina, as shown in Fig. 5.15.

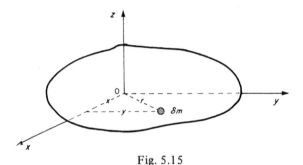

Fig. 5.15

If the lamina is divided into elements, each of mass δm, distance r from O, then the MI of this element about the axis OZ, perpendicular to the plane of the lamina is

$$\delta I_{OZ} = \delta m r^2 .$$

The total MI about OZ is therefore

$$I_{OZ} = \Sigma \, \delta m r^2$$

the summation extending over the whole lamina.

Thus

$$I_{OZ} = \Sigma (x^2 + y^2)\, \delta m = \Sigma y^2 \, \delta m + \Sigma x^2 \, \delta m$$

i.e.

$$I_{OZ} = I_{OX} + I_{OY} .$$

(ii) Theorem of parallel axes

Let G be the centre of mass of a uniform solid body of mass M, and let AA' be any axis which may or may not pass through it.

We first construct three mutually perpendicular axes GG', GX and GY such that GG' is parallel to AA' and A lies on the line GY, being distance h from G (see Fig. 5.16).

Let δm, an element of the mass M, be situated at the point $P(x, y, z)$ distance r from the axis AA'.

The MI of this element about the axis AA' is $\delta m r^2$, and combining all such

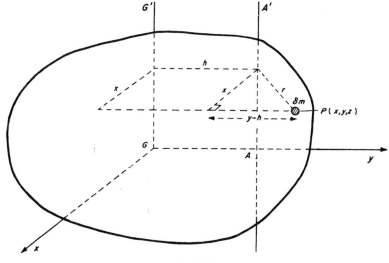

Fig. 5.16

contributions for the whole mass gives

$$I_{AA'} = \Sigma r^2 \, \delta m$$

the summation extending throughout the whole body. Hence

$$I_{AA'} = \Sigma \{x^2 + (y - h)^2\} \, \delta m$$
$$= \Sigma \{x^2 + y^2 + h^2 - 2hy\} \, \delta m$$
$$= \Sigma (x^2 + y^2) \, \delta m + h^2 \, \Sigma \, \delta m - 2h \, \Sigma y \, \delta m$$

so that

$$I_{AA'} = I_{GG'} + Mh^2$$

the third term being zero since it represents the first moment of the mass about an axis passing through its centre of mass.

Example 11

Find the MI of a uniform circular disc (M, a) about: (i) a diameter; and (ii) a tangential axis lying in the plane of the disc.

(i) Referring to Fig. 5.17(a), let OA, OB and OY be three mutually perpendicular axes as shown.

Then

$$I_{OY} = I_{OA} + I_{OB}$$
$$= 2I_{OA}$$
$$= 2I_{OB}$$

by symmetry.

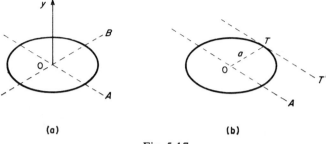

Fig. 5.17

Hence

$$I_{OY} = \frac{Ma^2}{2} = 2I_{OA}$$

$$\therefore\ I_{OA} = I_{OB} = \frac{Ma^2}{4}.$$

(ii) Let TT$'$ be a tangential axis lying in the plane of the disc (see Fig. 5.17(b)). Then

$$I_{TT'} = I_{OA} + Ma^2$$

$$= \frac{5Ma^2}{4}.$$

Example 12

The density at any point of a solid cone of mass M, height h and base radius a, varies inversely as its distance from a plane through the vertex, parallel to the base.

Show that the MI of the cone about an axis through the vertex, parallel to the base, is

$$\frac{M}{8}(a^2 + 4h^2).$$

Let the density of the cone at a point distance x from the given plane through O be $\rho = c/x$, where c is a constant.

Referring to Fig. 5.18, the mass of a disc of radius y and thickness δx, is

$$\pi y^2 \rho\, \delta x = \frac{\pi y^2 c\, \delta x}{x}.$$

The MI of this disc about the diametrical axis AA$'$ is thus

$$\delta I_{AA'} = \frac{(\text{mass})(\text{radius})^2}{4}$$

from the previous example, i.e.

$$\delta I_{AA'} \simeq \frac{1}{4}\left(\frac{\pi y^2 c\, \delta x}{x}\right)(y^2).$$

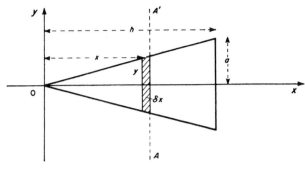

Fig. 5.18

The MI of this disc about the axis OY is, by the parallel axis theorem,

$$\delta I_{OY} = \delta I_{AA'} + (\text{mass})(x^2)$$

$$\simeq \frac{\pi y^4 c}{4x} \delta x + \pi y^2 x c\, \delta x.$$

Since

$$y = \frac{ax}{h}$$

$$\therefore\ \delta I_{OY} \simeq \frac{\pi a^2 c}{4h^4} (a^2 + 4h^2) x^3\, \delta x.$$

Hence the MI of the whole cone about OY is

$$I_{OY} = \lim_{\delta x \to 0} \sum_{x=0}^{h} \frac{\pi a^2 c}{4h^4} (a^2 + 4h^2) x^3\, \delta x$$

$$= \frac{\pi a^2 c}{4h^4} (a^2 + 4h^2) \int_0^h x^3\, dx$$

$$= \frac{\pi a^2 c}{16} (a^2 + 4h^2).$$

The mass of the cone is

$$M = \lim_{\delta x \to 0} \sum_{x=0}^{h} \frac{\pi y^2 c}{x} \delta x$$

$$= \int_0^h \left(\frac{\pi a^2 c}{h^2} \right) x\, dx = \frac{\pi a^2 c}{2}$$

so that

$$I_{OY} = \frac{M}{8} (a^2 + 4h^2).$$

8. Mean value and root mean square

If the range $a \leqslant x \leqslant b$ is divided into n parts by the abscissae $x_1, x_2, \ldots, x_r, \ldots,$ x_n, then the corresponding values of the function $f(x)$ are

$$f(x_1), f(x_2), \ldots, f(x_r), \ldots, f(x_n).$$

The *arithmetic mean* of these values is

$$\frac{1}{n}\{f(x_1) + f(x_2) + \cdots + f(x_n)\} \equiv \frac{1}{n} \sum_{r=1}^{n} f(x_r).$$

If we now make the intervals δx_r ($r = 1, 2, \ldots, n$) between two consecutive abscissae equal, so that $n \, \delta x_r = b - a$, then the arithmetic mean may be written as

$$\frac{1}{n \, \delta x_r} \sum_{r=1}^{n} f(x_r) \, \delta x_r = \frac{1}{b - a} \sum_{r=1}^{n} f(x_r) \, \delta x_r.$$

The limiting value of this expression as $n \to \infty$, i.e. as $\delta x_r \to 0$, is thus

$$\frac{1}{b - a} \lim_{\delta x_r \to 0} \sum_{r=1}^{n} f(x_r) \, \delta x_r = \frac{1}{b - a} \int_a^b f(x) \, dx$$

and is called the *mean value* of the function over the range $x = a$ to $x = b$.

The *root mean square* (RMS) of the function $f(x)$ over the range $x = a$ to $x = b$ is defined as

$$\text{RMS} = \sqrt{\left(\frac{1}{b - a} \int_a^b \{f(x)\}^2 \, dx \right)}$$

i.e. square the function, determine its mean value over the range $a \leqslant x \leqslant b$, and then take the positive square root.

Example 13

Determine the mean value and RMS of $\cos^2 \theta$ over the range $0 \leqslant \theta \leqslant \pi$.

$$\text{Mean value} = \frac{1}{\pi} \int_0^\pi \cos^2 \theta \, d\theta$$

$$= \frac{1}{2\pi} \int_0^\pi (1 + \cos 2\theta) \, d\theta = \tfrac{1}{2}$$

$$\text{RMS} = \sqrt{\left(\frac{1}{\pi} \int_0^\pi \cos^4 \theta \, d\theta \right)}$$

$$= \sqrt{\left(\frac{1}{8\pi} \int_0^\pi (\cos 4\theta + 4 \cos 2\theta + 3) \, d\theta \right)}$$

$$= \sqrt{\tfrac{3}{8}}.$$

9. Taylor's theorem

In Chapter 4 it was shown that if the functions $f(x)$ and $F(x)$ are continuous in the range $a \leqslant x \leqslant b$, and such that

$$\frac{\mathrm{d}F(x)}{\mathrm{d}x} = f(x)$$

then

$$\int_a^b f(x)\,\mathrm{d}x = F(x)\Big|_a^b = F(b) - F(a).$$

Similarly, as

$$f'(x) \equiv \frac{\mathrm{d}f(x)}{\mathrm{d}x},$$

$$\therefore \int_a^b f'(x)\,\mathrm{d}x = f(x)\Big|_a^b = f(b) - f(a).$$

Thus we can write

$$f(x + c) - f(x) = f(x + c - t)\Big|_{t=c}^{t=0}$$

$$= -f(x + c - t)\Big|_{t=0}^{t=c}$$

$$= \int_0^c f'(x + c - t)\,\mathrm{d}t \tag{8}$$

x and c being independent of t.

Integrating the right-hand side of equation (8) by parts, now gives

$$\int_0^c f'(x + c - t)\,\mathrm{d}t = tf'(x + c - t)\Big|_0^c + \int_0^c tf''(x + c - t)\,\mathrm{d}t$$

$$= cf'(x) + \int_0^c tf''(x + c - t)\,\mathrm{d}t. \tag{9}$$

Equations (8) and (9) now give

$$f(x + c) = f(x) + \frac{c}{1!}f'(x) + \int_0^c tf''(x + c - t)\,\mathrm{d}t. \tag{10}$$

Again integrating by parts

$$\int_0^c tf''(x + c - t)\,\mathrm{d}t = \frac{t^2}{2}f''(x + c - t)\Big|_0^c + \int_0^c \frac{t^2}{2}f'''(x + c - t)\,\mathrm{d}t$$

$$= \frac{c^2}{2}f''(x) + \int_0^c \frac{t^2}{2}f'''(x + c - t)\,\mathrm{d}t.$$

Equation (10) thus becomes

$$f(x + c) = f(x) + \frac{c}{1!} f'(x) + \frac{c^2}{2!} f''(x) + \int_0^c \frac{t^2}{2!} f'''(x + c - t) \, dt. \tag{11}$$

Similarly

$$\int_0^c \frac{t^2}{2} f'''(x + c - t) \, dt = \frac{t^3}{6} f'''(x + c - t) \Big|_0^c + \int_0^c \frac{t^3}{6} f^{(4)}(x + c - t) \, dt$$

$$= \frac{c^3}{6} f'''(x) + \int_0^c \frac{t^3}{6} f^{(4)}(x + c - t) \, dt$$

so that equation (11) becomes

$$f(x + c) = f(x) + \frac{c}{1!} f'(x) + \frac{c^2}{2!} f''(x) + \frac{c^3}{3!} f'''(x) + \int_0^c \frac{t^3}{3!} f^{(4)}(x + c - t) \, dt.$$

Continuing in this way gives the general result

$$f(x + c) = f(x) + \frac{c}{1!} f'(x) + \frac{c^2}{2!} f''(x) + \cdots + \frac{c^{r-1}}{(r - 1)!} f^{(r-1)}(x)$$

$$+ \int_0^c \frac{t^{r-1}}{(r - 1)!} f^{(r)}(x + c - t) \, dt \tag{12}$$

for $r = 1, 2, \ldots$, which is known as *Taylor's theorem*.

The series

$$f(x + c) = f(x) + \frac{c}{1!} f'(x) + \frac{c^2}{2!} f''(x) + \cdots + \frac{c^{r-1}}{(r - 1)!} f^{(r-1)}(x) \tag{13}$$

is called the *Taylor series*, the integral

$$\int_0^c \frac{t^{r-1}}{(r - 1)!} f^{(r)}(x + c - t) \, dt$$

being the *remainder* after r terms. It is assumed that each of the derivatives of $f(x)$ up to $f^{(r)}(x)$ is continuous throughout the range considered.

Interchanging x and c gives the alternative form

$$f(x + c) = f(c) + \frac{x}{1!} f'(c) + \frac{x^2}{2!} f''(c) + \cdots + \frac{x^{r-1}}{(r - 1)!} f^{(r-1)}(c). \tag{14}$$

The special case obtained by putting $c = 0$ in equation (14), i.e.

$$f(x) = f(0) + \frac{x}{1!} f'(0) + \frac{x^2}{2!} f''(0) + \cdots + \frac{x^{r-1}}{(r - 1)!} f^{(r-1)}(0) \tag{15}$$

is known as the *Maclaurin series*.

Example 14

If $x = \alpha$ is an approximation to a root of the equation $f(x) = 0$, show that a closer approximation is given by

$$x = \alpha - \frac{f(\alpha)}{f'(\alpha)}.$$

Let the exact value of the root considered be $x = \alpha + \epsilon$, where ϵ is small, so that $f(\alpha + \epsilon) = 0$.

Then Taylor's theorem gives

$$f(\alpha + \epsilon) = f(\alpha) + \frac{\epsilon}{1!}f'(\alpha) + \frac{\epsilon^2}{2!}f''(\alpha) + \cdots$$

$$= f(\alpha) + \frac{\epsilon}{1!}f'(\alpha)$$

to the first order.

Thus, since $f(\alpha + \epsilon) = 0$, ϵ is given by

$$\epsilon \simeq - \frac{f(\alpha)}{f'(\alpha)}$$

showing that a better approximation is

$$x = \alpha - \frac{f(\alpha)}{f'(\alpha)}.$$

Example 15

Determine the expansion of $(1 + x)^n$.

$$f(x) = (1 + x)^n \qquad \qquad \therefore \quad f(0) = 1.$$

$$f'(x) = n(1 + x)^{n-1} \qquad \qquad \therefore \quad f'(0) = n.$$

$$f''(x) = n(n - 1)(1 + x)^{n-2} \qquad \therefore \quad f''(0) = n(n - 1).$$

$$\vdots \qquad \qquad \vdots \qquad \qquad \qquad \vdots$$

$$f^{(r)}(x) = n(n - 1)\ldots(n - r + 1)x^{n-r} \quad \therefore \quad f^{(r)}(0) = n(n - 1)\ldots(n - r + 1).$$

Hence the Maclaurin series, equation (15), is

$$(1 + x)^n = 1 + n\frac{x}{1!} + \frac{n(n - 1)}{2!}x^2 + \cdots + \frac{n(n - 1)\ldots(n - r + 1)}{r!}x^r$$

as far as the term in x^r, for all values of n.

Example 16

Determine the expansion of sin x.

$$f(x) = \sin x \qquad \therefore \qquad f(0) = 0$$

$$f'(x) = \cos x \qquad \therefore \qquad f'(0) = 1$$

$$f''(x) = -\sin x \qquad \therefore \qquad f''(0) = 0$$

$$f'''(x) = -\cos x \qquad \therefore \qquad f'''(0) = -1$$

$$f^{(4)}(x) = \sin x \qquad \therefore \qquad f^{(4)}(0) = 0$$

and so on, repeating the cycle.

Thus equation (15) gives

$$\sin x = x - \frac{x^3}{3!} + \frac{x^5}{5!} - \frac{x^7}{7!} + \cdots + (-1)^{r+1} \cdot \frac{x^{2r-1}}{(2r-1)!}$$

to *r* terms.

Similarly

$$\cos x = 1 - \frac{x^2}{2!} + \frac{x^4}{4!} - \frac{x^6}{6!} + \cdots + (-1)^{r+1} \frac{x^{2r-2}}{(2r-2)!}$$

to *r* terms.

Example 17

Determine the expansion of $\tanh^{-1} x$.

$$f(x) = \tanh^{-1} x \qquad \qquad \therefore \quad f(0) = 0.$$

$$f'(x) = \frac{1}{1-x^2} = \frac{\frac{1}{2}}{1-x} + \frac{\frac{1}{2}}{1+x} \qquad \therefore \quad f'(0) = 1,$$

the expression for $f'(x)$ being expressed in partial fractions in order to facilitate further differentiation.

$$f''(x) = \frac{\frac{1}{2}}{(1-x)^2} - \frac{\frac{1}{2}}{(1+x)^2} \qquad \qquad \therefore \quad f''(0) = 0.$$

$$f'''(x) = \frac{\frac{1}{2} \cdot 2}{(1-x)^3} + \frac{\frac{1}{2} \cdot 2}{(1+x)^3} \qquad \qquad \therefore \quad f'''(0) = 2.$$

$$f^{(4)}(x) = \frac{\frac{1}{2} \cdot 2 \cdot 3}{(1-x)^4} - \frac{\frac{1}{2} \cdot 2 \cdot 3}{(1+x)^4} \qquad \qquad \therefore \quad f^{(4)}(0) = 0.$$

$$f^{(5)}(x) = \frac{\frac{1}{2} \cdot 2 \cdot 3 \cdot 4}{(1-x)^5} + \frac{\frac{1}{2} \cdot 2 \cdot 3 \cdot 4}{(1+x)^5} \qquad \therefore \quad f^{(5)}(0) = 24,$$

and so on.

$$\therefore \quad \tanh^{-1} x = x + \frac{x^3}{3} + \frac{x^5}{5} + \cdots + \frac{x^{2r-1}}{2r-1}$$

to *r* terms.

Example 18

In a gaseous mixture of H_2, D_2 and HD, several homogeneous reactions can occur.
If the ratio

$$\frac{[H]}{[D]} = \frac{a(1 - x + e) + bx}{c(1 - x) + dx}$$

where a, b, c, d and e are constants, show that

$$\frac{[H]}{[D]} = \left(\frac{b - a}{d - c}\right) + \frac{\lambda}{c} - \frac{\lambda}{c}\left(\frac{d}{c} - 1\right)x + \frac{\lambda}{c}\left(\frac{d}{c} - 1\right)^2 x^2 - \frac{\lambda}{c}\left(\frac{d}{c} - 1\right)^3 x^2 + \cdots$$

where λ is a constant.

Let $[H]/[D] = y$, then

$$y = \frac{a(1 + e) + x(b - a)}{c + x(d - c)}$$

$$= \frac{b - a}{d - c} + \frac{a(1 + e) - c\left(\dfrac{b - a}{d - c}\right)}{c + x(d - c)}$$

$$= \left(\frac{b - a}{d - c}\right) + \frac{\lambda}{c + x(d - c)}, \quad \text{say.}$$

$$\therefore \frac{dy}{dx} = \frac{-\lambda(d - c)}{\{c + x(d - c)\}^2} \quad \text{and} \quad \left(\frac{dy}{dx}\right)_{x=0} = -\frac{\lambda}{c^2}(d - c).$$

Similarly

$$\frac{d^2y}{dx^2} = \frac{2\lambda(d - c)^2}{\{c + x(d - c)\}^3}, \qquad \therefore \left(\frac{d^2y}{dx^2}\right)_{x=0} = \frac{2\lambda}{c^3}(d - c)^2$$

$$\frac{d^3y}{dx^3} = \frac{-2 \cdot 3 \cdot \lambda(d - c)^3}{\{c + x(d - c)\}^4}, \qquad \therefore \left(\frac{d^3y}{dx^3}\right)_{x=0} = -\frac{6\lambda}{c^4}(d - c)^3$$

and so on.

The Maclaurin expansion is thus

$$y = \left(\frac{b - a}{d - c}\right) + \frac{\lambda}{c} - \frac{\lambda}{c^2}(d - c) \cdot \frac{x}{1!} + \frac{2\lambda}{c^3}(d - c)^2 \cdot \frac{x^2}{2!} - \frac{6\lambda}{c^4}(d - c)^3 \cdot \frac{x^3}{3!} + \cdots$$

and the ratio

$$\frac{[H]}{[D]} = \left(\frac{b - a}{d - c}\right) + \frac{\lambda}{c} - \frac{\lambda}{c}\left(\frac{d}{c} - 1\right)x + \frac{\lambda}{c}\left(\frac{d}{c} - 1\right)^2 x^2 - \frac{\lambda}{c}\left(\frac{d}{c} - 1\right)^3 x^3 + \cdots$$

where

$$\lambda = a(1 + e) - c\left(\frac{b - a}{d - c}\right).$$

E

The same expression could have been obtained directly from the result in Example 15.

10. Simpson's rule

We now derive a numerical method which may be used for the determination of the area A considered in Section 1.

Let the range $x = a$ to $x = b$ be divided into two equal parts of width h by the line $x = c$, and let y_1, y_2, and y_3 be the values of the corresponding ordinates (see Fig. 5.19).

Then $a = c - h$ and $b = c + h$ so that the area A is given by

$$A = \int_a^b f(x)\, dx = \int_{c-h}^{c+h} f(x)\, dx. \tag{16}$$

If we now make the substitution $x = c + u$, equation (16) becomes

$$A = \int_{-h}^{h} f(c + u)\, du$$

$$= \int_{-h}^{h} \left(f(c) + \frac{u}{1!} f'(c) + \frac{u^2}{2!} f''(c) + \cdots \right) du$$

from equation (14).

Hence, if u is small, then

$$A \simeq \int_{-h}^{h} \left(f(c) + u f'(c) + \frac{u^2}{2} f''(c) \right) du$$

$$= \left[u f(c) + \frac{u^2}{2} f'(c) + \frac{u^3}{6} f''(c) \right]_{u=-h}^{u=h}$$

i.e.

$$A \simeq 2h f(c) + \frac{h^3}{3} f''(c). \tag{17}$$

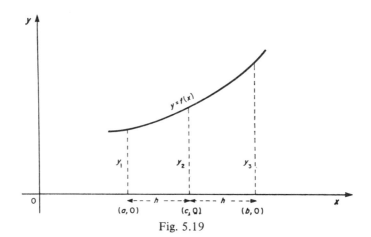

Fig. 5.19

But

$$y_1 = f(a) = f(c - h) \simeq f(c) - \frac{h}{1!} f'(c) + \frac{h^2}{2!} f''(c)$$

and

$$y_3 = f(b) = f(c + h) \simeq f(c) + \frac{h}{1!} f'(c) + \frac{h^2}{2!} f''(c) \quad \text{to } O(h^2)$$

from equation (14), so that

$$y_1 + y_3 = 2f(c) + h^2 f''(c) = 2y_2 + h^2 f''(c)$$

since $y_2 = f(c)$.

Equation (17) now becomes

$$A \simeq 2hy_2 + \frac{h}{3}(y_1 + y_3 - 2y_2)$$

i.e.

$$A \simeq \frac{h}{3}(y_1 + 4y_2 + y_3). \tag{18}$$

Hence, if the range $a \leqslant x \leqslant b$ be divided into any *even* number of equal intervals $2n$, say, by ordinates $y_1, y_2, \ldots, y_{2n+1}$ (see Fig. 5.20) then the result given in equation (18) leads to the more general result

$$A \simeq \frac{h}{3} \{ y_1 + 4(y_2 + y_4 + \cdots + y_{2n}) + 2(y_3 + y_5 + \cdots + y_{2n-1}) + y_{2n+1} \}$$

which is *Simpson's rule*.

Example 19

Obtain approximate values, to two places of decimals, of the integral

$$\int_0^1 \frac{1}{\sqrt{(9 + x^3)}} \, dx$$

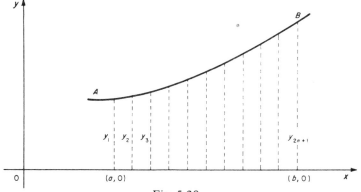

Fig. 5.20

(i) by Simpson's rule, using four strips;

(ii) by using three terms of an expansion in powers of x.

(i) Let

$$y = \frac{1}{\sqrt{(9 + x^3)}},$$

then:

x	0	$\frac{1}{4}$	$\frac{1}{2}$	$\frac{3}{4}$	1
x^3	0	0.016	0.125	0.421	1
$9 + x^3$	9.000	9.016	9.125	9.421	10.000
$\sqrt{(9 + x^3)}$	3.000	3.003	3.021	3.069	3.162
y	0.333	0.333	0.331	0.326	0.316
	y_1	y_2	y_3	y_4	y_5

Hence

$$I \approx \frac{(\frac{1}{4})}{3} \{y_1 + 4(y_2 + y_4) + 2y_3 + y_5\}$$

$$= \frac{1}{12}\{0.333 + 4(0.659) + 2(0.331) + 0.316\}$$

$$= 0.329.$$

$$\therefore \ I \approx 0.33, \quad \text{to two decimal places.}$$

(ii) $\qquad y = \frac{1}{3}\left(1 + \frac{x^3}{9}\right)^{-1/2}$

$$= \frac{1}{3}\left\{1 + \frac{(-\frac{1}{2})}{1}\left(\frac{x^3}{9}\right) + \frac{(-\frac{1}{2})(-\frac{3}{2})}{1 \cdot 2}\left(\frac{x^3}{9}\right)^2\right\} \qquad \text{to three terms}$$

$$= \frac{1}{3}\left(1 - \frac{x^3}{18} + \frac{x^6}{216}\right)$$

$$\therefore \ I = \frac{1}{3}\int_0^1 \left(1 - \frac{x^3}{18} + \frac{x^6}{216}\right)dx$$

$$= \frac{1}{3}\left[x - \frac{x^4}{72} + \frac{x^7}{1512}\right]_0^1 = 0.32(9).$$

Example 20

The specific heat s of magnesium at temperature T is given by the following table:

$T(°C)$	20	25	30	35	40	45	50
s	0.241	0.243	0.246	0.247	0.249	0.250	0.252

Calculate the total heat

$$H = \int_{20}^{50} s \, dT$$

required to raise one gram of the substance from 20°C to 50°C.

The total heat required is

$$H \approx \tfrac{5}{3}\{s_1 + 4(s_2 + s_4 + s_6) + 2(s_3 + s_5) + s_7\}$$
$$= \tfrac{5}{3}\{0.241 + 4(0.740) + 2(0.495) + 0.252\}$$
$$= \tfrac{5}{3}(4.443)$$
$$= 7.405 \text{ calories.}$$

PROBLEMS

1. Determine the area enclosed by the curves

(i) $y^2 = 4x$ and $x = 4$,

(ii) $x^2 = y - 1$ and $y = x + 3$.

2. Determine the area enclosed by the *cycloid*

$$x = a(\theta - \sin \theta), \quad y = a(1 - \cos \theta),$$

and the x-axis, between $x = 0$ and $x = 2\pi a$.

3. Show that the area enclosed by the curve $r = f(\theta)$ and the radial lines $\theta = \alpha$ and $\theta = \beta$ is

$$\frac{1}{2} \int_{\alpha}^{\beta} r^2 \, d\theta.$$

Hence determine the area enclosed by the *cardioid* $r = a(1 + \cos \theta)$.

4. Determine the area enclosed between the curve $y = \sqrt{\{(3 - x)/x\}}$ and the y-axis; and also the volume generated as this area is rotated about the y-axis through 2π radians.

5. Find the area enclosed by the *astroid*

$$x^{2/3} + y^{2/3} = a^{2/3}.$$

Determine also its total arc length.

6. Show that the area common to the circle $x^2 + y^2 = 1$ and the ellipse $x^2 + 5y^2 = 4$, is equal to

$$\frac{2}{3\sqrt{5}} \left\{ \sqrt{(5)}\pi + 12 \sin^{-1} \left(\frac{1}{4} \right) \right\} \quad \text{square units.}$$

7. Determine the surface area of a cone of height h and base radius a.

8. Find the position of the centroid of the solid generated when the area enclosed by the parabola $y^2 = 4(x - 1)$ and the line $x = 2$, is rotated about the x-axis through 2π radians.

9. Determine the radius of gyration of a uniform solid cylinder of mass M, length $2l$ and radius a, about an axis through its centroid and perpendicular to its length.

10. Find the moment of inertia of a solid cone of mass M, height h and base radius a, about:

(i) an axis through the vertex and parallel to the base; and
(ii) about a diameter of the base.

11. Show that the moment of inertia of a spherical shell of mass M and radius a, about a diameter is $2Ma^2/3$.

Hence determine the moment of inertia about a diameter, of a hollow sphere of mass M and internal and external radii a and b respectively.

12. Find the Maclaurin expansions for the following functions:

(i) $\tan x$,   (ii) $\sinh^{-1} x$,   (iii) $\log_e \cos x$.

13. Given that $x = f(y) = y - 2y^2 + 5y^3$, obtain the first three derivatives of y with respect to x at $x = 0$. Hence show that the Maclaurin expansion is

$$y = x + 2x^2 + 3x^3 + \cdots.$$

Denoting $(x + 2x^2 + 3x^3)$ by $g(x)$, deduce that

$$\lim_{x \to 0} \frac{g(x)f(x) - x^2}{x^4} = 4.$$

14. Determine the value of

$$\int_0^{0.6} e^{x^2} \, dx$$

correct to three places of decimals:
(i)   by Simpson's rule using six strips; and
(ii) by using the first four terms of an expansion in powers of x.

15. Determine the value of

$$\int_0^1 \frac{x}{16 - x^4} \, dx$$

(i)   by Simpson's rule using eight strips;
(ii) by using the first four terms of an expansion in powers of x; and
(iii) by direct integration.

16. An inverted spherical flask (radius 10 cm) of acid is used to give a continuous flow during an experiment. If, when the flask is half full, the depth of the acid is decreasing at the rate of $\frac{1}{3}$ cm per minute, determine the rate of flow from the flask (assumed constant).

17. The specific heat s, of a substance at temperature T, is given in the following table:

$T(°C)$	0	3	6	9	12	15	18	21	24
s	1.082	1.079	1.075	1.071	1.069	1.066	1.062	1.058	1.053

Determine, using Simpson's rule, the total heat H required to raise one gram of the substance from $0 °C$ to $24 °C$, given that

$$H = \int_0^{24} s \, dT.$$

CHAPTER 6

Determinants and Matrices

1. Determinants

Consider the two equations

$$a_{11}x_1 + a_{12}x_2 = 0$$
$$a_{21}x_1 + a_{22}x_2 = 0$$

in the unknowns x_1 and x_2. Elimination of x_2 leads to

$$x_1(a_{11}a_{22} - a_{12}a_{21}) = 0,$$

the coefficient of x_1 being equal to a *second order determinant* defined by

$$\begin{vmatrix} a_{11} & a_{12} \\ a_{21} & a_{22} \end{vmatrix} = a_{11}a_{22} - a_{12}a_{21}.$$

Suppose now that we wish to solve the equations

$$a_{11}x_1 + a_{12}x_2 + k_1 = 0,$$
$$a_{21}x_1 + a_{22}x_2 + k_2 = 0.$$

The usual elimination technique yields

$$\frac{x_1}{a_{12}k_2 - a_{22}k_1} = \frac{-x_2}{a_{11}k_2 - a_{21}k_1} = \frac{1}{a_{11}a_{22} - a_{12}a_{21}},$$

which may be expressed in determinantal form as

$$\frac{x_1}{\begin{vmatrix} a_{12} & k_1 \\ a_{22} & k_2 \end{vmatrix}} = \frac{-x_2}{\begin{vmatrix} a_{11} & k_1 \\ a_{21} & k_2 \end{vmatrix}} = \frac{1}{\begin{vmatrix} a_{11} & a_{12} \\ a_{21} & a_{22} \end{vmatrix}}.$$

Example 1

Solve the equations

$$2x_1 + 3x_2 - 4 = 0,$$
$$-x_1 + 4x_1 + 13 = 0.$$

In this case

$$\frac{x_1}{\begin{vmatrix} 3 & -4 \\ 4 & 13 \end{vmatrix}} = \frac{-x_2}{\begin{vmatrix} 2 & -4 \\ -1 & 13 \end{vmatrix}} = \frac{1}{\begin{vmatrix} 2 & 3 \\ -1 & 4 \end{vmatrix}}$$

$$\therefore \frac{x_1}{55} = \frac{-x_2}{22} = \frac{1}{11}$$

giving $x_1 = 5$ and $x_2 = -2$.

Consider next the following homogeneous equations:

$$a_{21}x_1 + a_{22}x_2 + a_{23}x_3 = 0$$
$$a_{31}x_1 + a_{32}x_2 + a_{33}x_3 = 0$$

or, assuming $x_3 \neq 0$,

$$a_{21}(x_1/x_3) + a_{22}(x_2/x_3) + a_{23} = 0,$$
$$a_{31}(x_1/x_3) + a_{32}(x_2/x_3) + a_{33} = 0.$$

Solving for x_1/x_3, x_2/x_3, we obtain

$$\frac{x_1/x_3}{\begin{vmatrix} a_{22} & a_{23} \\ a_{32} & a_{33} \end{vmatrix}} = \frac{-x_2/x_3}{\begin{vmatrix} a_{21} & a_{23} \\ a_{31} & a_{33} \end{vmatrix}} = \frac{1}{\begin{vmatrix} a_{21} & a_{22} \\ a_{31} & a_{32} \end{vmatrix}},$$

so that

$$\frac{x_1}{\begin{vmatrix} a_{22} & a_{23} \\ a_{32} & a_{33} \end{vmatrix}} = \frac{-x_2}{\begin{vmatrix} a_{21} & a_{23} \\ a_{31} & a_{33} \end{vmatrix}} = \frac{x_3}{\begin{vmatrix} a_{21} & a_{22} \\ a_{31} & a_{32} \end{vmatrix}}.$$

If these solutions have to satisfy a further equation

$$a_{11}x_1 + a_{12}x_2 + a_{13}x_3 = 0,$$

then

$$a_{11}\begin{vmatrix} a_{22} & a_{23} \\ a_{32} & a_{33} \end{vmatrix} - a_{12}\begin{vmatrix} a_{21} & a_{23} \\ a_{31} & a_{33} \end{vmatrix} + a_{13}\begin{vmatrix} a_{21} & a_{22} \\ a_{31} & a_{32} \end{vmatrix} = 0.$$

The left-hand side of the equation is the expanded form of a *third order determinant*, defined by

$$\begin{vmatrix} a_{11} & a_{12} & a_{13} \\ a_{21} & a_{22} & a_{23} \\ a_{31} & a_{32} & a_{33} \end{vmatrix} = a_{11}\begin{vmatrix} a_{22} & a_{23} \\ a_{32} & a_{33} \end{vmatrix} - a_{12}\begin{vmatrix} a_{21} & a_{23} \\ a_{31} & a_{33} \end{vmatrix} + a_{13}\begin{vmatrix} a_{21} & a_{22} \\ a_{31} & a_{32} \end{vmatrix}$$

$$= a_{11}(a_{22}a_{33} - a_{23}a_{32}) - a_{12}(a_{21}a_{33} - a_{23}a_{31})$$
$$+ a_{31}(a_{21}a_{32} - a_{22}a_{31}).$$

If we associate a sign, $(-1)^{i+j}$, with each element a_{ij} of the determinant, alternative, but equivalent expressions, may be derived by appropriate manipulation of the right-hand side of the last equation.

For example,

$$\begin{vmatrix} a_{11} & a_{12} & a_{13} \\ a_{21} & a_{22} & a_{23} \\ a_{31} & a_{32} & a_{33} \end{vmatrix} = -a_{12}\begin{vmatrix} a_{21} & a_{23} \\ a_{31} & a_{33} \end{vmatrix} + a_{22}\begin{vmatrix} a_{11} & a_{13} \\ a_{31} & a_{33} \end{vmatrix} - a_{32}\begin{vmatrix} a_{11} & a_{13} \\ a_{21} & a_{23} \end{vmatrix},$$

and so on, expansion being valid along any row, or down any column.

Thus, for example, expanding along the first row,

$$\begin{vmatrix} 1 & 2 & -2 \\ 3 & -1 & 5 \\ 4 & -2 & 0 \end{vmatrix} = (1)\begin{vmatrix} -1 & 5 \\ -2 & 0 \end{vmatrix} - (2)\begin{vmatrix} 3 & 5 \\ 4 & 0 \end{vmatrix} + (-2)\begin{vmatrix} 3 & -1 \\ 4 & -2 \end{vmatrix}$$

$$= (1)(10) - (2)(-20) + (-2)(-2)$$

$$= 10 + 40 + 4$$

$$= 54.$$

Alternatively, expanding down the second column, gives

$$\begin{vmatrix} 1 & 2 & -2 \\ 3 & -1 & 5 \\ 4 & -2 & 0 \end{vmatrix} = -(2)\begin{vmatrix} 3 & 5 \\ 4 & 0 \end{vmatrix} + (-1)\begin{vmatrix} 1 & -2 \\ 4 & 0 \end{vmatrix} - (-2)\begin{vmatrix} 1 & -2 \\ 3 & 5 \end{vmatrix}$$

$$= -(2)(-20) + (-1)(8) - (-2)(11)$$

$$= 40 - 8 + 22$$

$$= 54,$$

and so on.

2. Elimination

When dealing with the sets of equations

$$a_{11}x_1 + a_{12}x_2 = 0, \quad a_{11}x_1 + a_{12}x_2 + a_{13}x_3 = 0,$$
$$a_{21}x_1 + a_{22}x_2 = 0; \quad a_{21}x_1 + a_{22}x_2 + a_{23}x_3 = 0,$$
$$a_{31}x_1 + a_{32}x_2 + a_{33}x_3 = 0;$$

it was found that in order to obtain unique values of x_1, x_2 and x_3 to satisfy each set, a certain relationship between the coefficients was necessary, viz.,

$$\begin{vmatrix} a_{11} & a_{12} \\ a_{21} & a_{22} \end{vmatrix} = 0, \quad \begin{vmatrix} a_{11} & a_{12} & a_{13} \\ a_{21} & a_{22} & a_{23} \\ a_{31} & a_{32} & a_{33} \end{vmatrix} = 0,$$

respectively.

These determinants are known as the *eliminants* of the respective sets of equations, and we note that there is effectively only one unknown, i.e. x_1/x_2 in the first set, and two, i.e. $x_1/x_3, x_2/x_3$ in the second set.

A set of linear equations in which the number of equations is greater than the number of unknowns is said to be *consistent* if the corresponding eliminant is zero.

Example 2

Determine the real value of λ for which the equations

$$\lambda x_1 + 7x_2 - x_3 = 0$$
$$4x_1 - x_2 + 3\lambda x_3 = 0$$
$$x_1 + \lambda x_2 - x_3 = 0$$

are consistent.

For consistency, we require $\Delta = 0$, where

$$\Delta = \begin{vmatrix} \lambda & 7 & -1 \\ 4 & -1 & 3\lambda \\ 1 & \lambda & -1 \end{vmatrix} = \lambda \begin{vmatrix} -1 & 3\lambda \\ \lambda & -1 \end{vmatrix} - 7 \begin{vmatrix} 4 & 3\lambda \\ 1 & -1 \end{vmatrix} - \begin{vmatrix} 4 & -1 \\ 1 & \lambda \end{vmatrix}$$

$$= \lambda(1 - 3\lambda^2) - 7(-4 - 3\lambda) - (4\lambda + 1)$$
$$= \lambda^3 - 6\lambda - 9$$
$$= (\lambda - 3)(\lambda^2 + 3\lambda + 3),$$

and hence $\lambda = 3$.

3. Solution of equations by determinants

Extending the above ideas to the set of linear equations

$$a_{11}x_1 + a_{12}x_2 + a_{13}x_3 + k_1 = 0$$
$$a_{21}x_1 + a_{22}x_2 + a_{23}x_3 + k_2 = 0$$
$$a_{31}x_1 + a_{32}x_2 + a_{33}x_3 + k_3 = 0$$

we find

$$\frac{x_1}{\begin{vmatrix} a_{12} & a_{13} & k_1 \\ a_{22} & a_{23} & k_2 \\ a_{32} & a_{33} & k_3 \end{vmatrix}} = \frac{-x_2}{\begin{vmatrix} a_{11} & a_{13} & k_1 \\ a_{21} & a_{23} & k_2 \\ a_{31} & a_{33} & k_3 \end{vmatrix}} = \frac{x_3}{\begin{vmatrix} a_{11} & a_{12} & k_1 \\ a_{21} & a_{22} & k_2 \\ a_{31} & a_{32} & k_3 \end{vmatrix}} = \frac{-1}{\begin{vmatrix} a_{11} & a_{12} & a_{13} \\ a_{21} & a_{22} & a_{23} \\ a_{31} & a_{32} & a_{33} \end{vmatrix}}$$

and similarly for higher orders. In general, we find that

$$a_{11}x_1 + a_{12}x_2 + a_{13}x_3 + \cdots + a_{1n}x_n + k_1 = 0$$
$$a_{21}x_1 + a_{22}x_2 + a_{23}x_3 + \cdots + a_{2n}x_n + k_2 = 0$$
$$\vdots \qquad \vdots \qquad \vdots \qquad \qquad \vdots \qquad \vdots$$
$$a_{n1}x_1 + a_{n2}x_2 + a_{n3}x_3 + \cdots + a_{nn}x_n + k_n = 0$$

i.e. n equations in n unknowns $x_1, x_2, x_3, \ldots, x_n$, leads to

$$\frac{x_1}{\Delta_1} = \frac{-x_2}{\Delta_2} = \frac{x_3}{\Delta_3} = \cdots = \frac{(-1)^n x_n}{\Delta_n} = \frac{(-1)^{n-1}}{\Delta},$$

$\Delta_1, \Delta_2, \Delta_3, \ldots, \Delta_n$ and Δ being determinants of the *nth order*.

For example,

$$\Delta_2 = \begin{vmatrix} a_{11} & a_{13} & a_{14} & \cdots & a_{1n} \\ a_{21} & a_{23} & a_{24} & \cdots & a_{2n} \\ \vdots & \vdots & \vdots & & \vdots \\ a_{n1} & a_{n3} & a_{n4} & \cdots & a_{nn} \end{vmatrix}, \quad \Delta_3 = \begin{vmatrix} a_{11} & a_{12} & a_{14} & \cdots & a_{1n} \\ a_{21} & a_{22} & a_{24} & \cdots & a_{2n} \\ \vdots & \vdots & \vdots & & \vdots \\ a_{n1} & a_{n2} & a_{n4} & \cdots & a_{nn} \end{vmatrix},$$

and so on.

Example 3

Solve the equations

$$x_1 - 2x_2 + x_3 - 6 = 0$$
$$2x_1 + x_2 - x_3 + 2 = 0$$
$$-x_1 + 3x_2 + 2x_3 - 2 = 0.$$

The solutions are given by

$$\frac{x_1}{\begin{vmatrix} -2 & 1 & -6 \\ 1 & -1 & 2 \\ 3 & 2 & -2 \end{vmatrix}} = \frac{-x_2}{\begin{vmatrix} 1 & 1 & -6 \\ 2 & -1 & 2 \\ -1 & 2 & -2 \end{vmatrix}} = \frac{x_3}{\begin{vmatrix} 1 & -2 & -6 \\ 2 & 1 & 2 \\ -1 & 3 & -2 \end{vmatrix}} = \frac{-1}{\begin{vmatrix} 1 & -2 & 1 \\ 2 & 1 & -1 \\ -1 & 3 & 2 \end{vmatrix}}$$

$$\therefore \frac{x_1}{-18} = \frac{-x_2}{-18} = \frac{x_3}{-54} = \frac{-1}{18}.$$

$$\therefore x_1 = 1, \quad x_2 = -1, \quad x_3 = 3.$$

4. Minors and cofactors

The quantities $a_{11}, a_{12}, a_{13}, a_{21}, \ldots$, etc. are known as the *elements* of the determinant. The determinant obtained by omitting the row and column containing a particular element is known as the *minor* of that element.

Thus, if

$$\Delta = \begin{vmatrix} a_{11} & a_{12} & a_{13} \\ a_{21} & a_{22} & a_{23} \\ a_{31} & a_{32} & a_{33} \end{vmatrix}$$

then

$$\begin{vmatrix} a_{22} & a_{23} \\ a_{32} & a_{33} \end{vmatrix} \quad \text{is the minor of } a_{11},$$

$$\begin{vmatrix} a_{21} & a_{23} \\ a_{31} & a_{33} \end{vmatrix} \quad \text{is the minor of } a_{12},$$

$$\begin{vmatrix} a_{21} & a_{22} \\ a_{31} & a_{32} \end{vmatrix} \quad \text{is the minor of } a_{13},$$

and so on.

Since each minor is associated with elements which in turn are associated with a ± sign, we define the *cofactors* of the corresponding elements to be the signed minors.

Thus, in the example cited, the cofactors of a_{11}, a_{12} and a_{13}, are defined to be

$$A_{11} = + \begin{vmatrix} a_{22} & a_{23} \\ a_{32} & a_{33} \end{vmatrix}, \qquad A_{12} = - \begin{vmatrix} a_{21} & a_{23} \\ a_{31} & a_{33} \end{vmatrix}, \qquad A_{13} = + \begin{vmatrix} a_{21} & a_{22} \\ a_{31} & a_{32} \end{vmatrix},$$

respectively, and we may write

$$\Delta = a_{11}A_{11} + a_{12}A_{12} + a_{13}A_{13}.$$

Similar definitions exist for determinants of order n. For example, if

$$\Delta = \begin{vmatrix} a_{11} & a_{12} & a_{13} & \cdots & a_{1n} \\ a_{21} & a_{22} & a_{23} & \cdots & a_{2n} \\ \vdots & \vdots & \vdots & & \vdots \\ a_{n1} & a_{n2} & a_{n3} & \cdots & a_{nn} \end{vmatrix},$$

then the minor of the general element a_{ij} is defined to be the determinant obtained by omitting the ith row and jth column, the sign of the corresponding cofactor A_{ij} being $(-1)^{i+j}$.

Expansion along the ith row gives

$$\Delta = a_{i1}A_{i1} + a_{i2}A_{i2} + \cdots + a_{in}A_{in} = \sum_{j=1}^{n} a_{ij}A_{ij},$$

and down the jth column

$$\Delta = a_{1j}A_{1j} + a_{2j}A_{2j} + \cdots + a_{nj}A_{nj} = \sum_{i=1}^{n} a_{ij}A_{ij}.$$

Example 4

Evaluate

$$\Delta = \begin{vmatrix} 2 & 1 & -1 & 3 \\ 5 & 0 & 1 & 0 \\ -1 & 2 & 4 & 1 \\ 0 & 3 & 1 & 2 \end{vmatrix}.$$

We expand along the second row since the zero elements reduce the calculations to a minimum.

Since the element 5 is in the second row and first column, its cofactor is given by

$$(-1)^{2+1} \begin{vmatrix} 2 & 1 & -1 & 3 \\ 5 & 0 & 1 & 0 \\ -1 & 2 & 4 & 1 \\ 0 & 3 & 1 & 2 \end{vmatrix} = (-1)^3 \begin{vmatrix} 1 & -1 & 3 \\ 2 & 4 & 1 \\ 3 & 1 & 2 \end{vmatrix}.$$

With similar calculations for the cofactors of the remaining elements of row two, it follows that

$$\Delta = -(5) \begin{vmatrix} 1 & -1 & 3 \\ 2 & 4 & 1 \\ 3 & 1 & 2 \end{vmatrix} + (0) \begin{vmatrix} 2 & -1 & 3 \\ -1 & 4 & 1 \\ 0 & 1 & 2 \end{vmatrix} - (1) \begin{vmatrix} 2 & 1 & 3 \\ -1 & 2 & 1 \\ 0 & 3 & 2 \end{vmatrix}$$

$$+ (0) \begin{vmatrix} 2 & 1 & -1 \\ -1 & 2 & 4 \\ 0 & 3 & 1 \end{vmatrix}$$

and so

$$\Delta = -(5)(-22) + (0)(9) - (1)(-5) + (0)(-16) = 115.$$

The following theorems are proved only for third order determinants but can be extended to cover all orders of determinant.

THEOREM I: A determinant is unaltered in value if the rows are changed to columns (and columns to rows).

Let

$$\Delta = \begin{vmatrix} a_{11} & a_{12} & a_{13} \\ a_{21} & a_{22} & a_{23} \\ a_{31} & a_{32} & a_{33} \end{vmatrix}$$

$$= a_{11}(a_{22}a_{33} - a_{23}a_{32}) - a_{12}(a_{21}a_{33} - a_{23}a_{31}) + a_{13}(a_{21}a_{32} - a_{22}a_{31}).$$

Interchanging rows and columns,

$$\Delta' = \begin{vmatrix} a_{11} & a_{21} & a_{31} \\ a_{12} & a_{22} & a_{32} \\ a_{13} & a_{23} & a_{33} \end{vmatrix}$$

$$= a_{11}(a_{22}a_{33} - a_{23}a_{32}) - a_{21}(a_{12}a_{33} - a_{32}a_{13}) + a_{31}(a_{12}a_{23} - a_{22}a_{13})$$

$$= a_{11}(a_{22}a_{33} - a_{23}a_{32}) - a_{12}(a_{21}a_{33} - a_{23}a_{31}) + a_{13}(a_{21}a_{32} - a_{22}a_{31})$$

$$= \Delta.$$

THEOREM II: If two rows (or columns) are interchanged the determinant changes sign.

For,

$$\begin{vmatrix} a_{21} & a_{22} & a_{23} \\ a_{11} & a_{12} & a_{13} \\ a_{31} & a_{32} & a_{33} \end{vmatrix} = a_{21}(a_{12}a_{33} - a_{13}a_{32}) - a_{22}(a_{11}a_{33} - a_{13}a_{31})$$
$$+ a_{23}(a_{11}a_{32} - a_{12}a_{31})$$
$$= -a_{11}(a_{22}a_{33} - a_{23}a_{32}) + a_{12}(a_{21}a_{33} - a_{23}a_{31})$$
$$- a_{13}(a_{21}a_{32} - a_{22}a_{31})$$
$$= -\Delta.$$

THEOREM III: If two rows (or columns) are identical or proportional the value of the determinant is zero.

Suppose, for example, that the second and third rows are proportional, i.e.

$$a_{21} = ka_{31}, \quad a_{22} = ka_{32}, \quad a_{23} = ka_{33}.$$

Then,

$$\Delta = a_{11}(a_{22}a_{33} - a_{23}a_{32}) - a_{12}(a_{21}a_{33} - a_{23}a_{31}) + a_{13}(a_{21}a_{32} - a_{22}a_{31})$$
$$= a_{11}(ka_{32}a_{33} - ka_{33}a_{32}) - a_{12}(ka_{31}a_{33} - ka_{33}a_{31})$$
$$+ a_{13}(ka_{31}a_{32} - ka_{32}a_{31})$$
$$= 0.$$

THEOREM IV: If the elements of any row (or column) are each multiplied by the same factor k, the value of the new determinant is k times that of the original.

In this case,

$$\begin{vmatrix} ka_{11} & ka_{12} & ka_{13} \\ a_{21} & a_{22} & a_{23} \\ a_{31} & a_{32} & a_{33} \end{vmatrix} = ka_{11}A_{11} + ka_{12}A_{12} + ka_{13}A_{13}$$
$$= k(a_{11}A_{11} + a_{12}A_{12} + a_{13}A_{13})$$
$$= k\begin{vmatrix} a_{11} & a_{12} & a_{13} \\ a_{21} & a_{22} & a_{23} \\ a_{31} & a_{32} & a_{33} \end{vmatrix}.$$

THEOREM V: If each element of any row (or column) consists of the algebraic sum of r terms, the determinant is equivalent to the sum of r other determinants, the elements of which consist of single terms.

For,

$$
\begin{vmatrix}
a_{11}+b_{11}+c_{11} & a_{12} & a_{13} \\
a_{21}+b_{21}+c_{21} & a_{22} & a_{23} \\
a_{31}+b_{31}+c_{31} & a_{32} & a_{33}
\end{vmatrix}
$$

$$= (a_{11}+b_{11}+c_{11})A_{11}+(a_{21}+b_{21}+c_{21})A_{21}+(a_{31}+b_{31}+c_{31})A_{31}$$

$$= (a_{11}A_{11}+a_{21}A_{21}+a_{31}A_{31})+(b_{11}A_{11}+b_{21}A_{21}+b_{31}A_{31})$$

$$+ (c_{11}A_{11}+c_{21}A_{21}+c_{31}A_{31})$$

$$
=
\begin{vmatrix}
a_{11} & a_{12} & a_{13} \\
a_{21} & a_{22} & a_{23} \\
a_{31} & a_{32} & a_{33}
\end{vmatrix}
+
\begin{vmatrix}
b_{11} & a_{12} & a_{13} \\
b_{21} & a_{22} & a_{23} \\
b_{31} & a_{32} & a_{33}
\end{vmatrix}
+
\begin{vmatrix}
c_{11} & a_{12} & a_{13} \\
c_{21} & a_{22} & a_{23} \\
c_{31} & a_{32} & a_{33}
\end{vmatrix}.
$$

Similarly, if the elements of two different rows (or columns) consist of two or more terms,

$$
\begin{vmatrix}
a_{11}+b_{11} & a_{12}+c_{12} & a_{13} \\
a_{21}+b_{21} & a_{22}+c_{22} & a_{23} \\
a_{31}+b_{31} & a_{32}+c_{32} & a_{33}
\end{vmatrix}
$$

$$
=
\begin{vmatrix}
a_{11} & a_{12}+c_{12} & a_{13} \\
a_{21} & a_{22}+c_{22} & a_{23} \\
a_{31} & a_{32}+c_{32} & a_{33}
\end{vmatrix}
+
\begin{vmatrix}
b_{11} & a_{12}+c_{12} & a_{13} \\
b_{21} & a_{22}+c_{22} & a_{23} \\
b_{31} & a_{32}+c_{32} & a_{33}
\end{vmatrix}
$$

$$
= -
\begin{vmatrix}
a_{12}+c_{12} & a_{11} & a_{13} \\
a_{22}+c_{22} & a_{21} & a_{23} \\
a_{32}+c_{32} & a_{31} & a_{33}
\end{vmatrix}
-
\begin{vmatrix}
a_{12}+c_{12} & b_{11} & a_{13} \\
a_{22}+c_{22} & b_{21} & a_{23} \\
a_{32}+c_{32} & b_{31} & a_{33}
\end{vmatrix}
$$

$$
= -
\begin{vmatrix}
a_{12} & a_{11} & a_{13} \\
a_{22} & a_{21} & a_{23} \\
a_{32} & a_{31} & a_{33}
\end{vmatrix}
-
\begin{vmatrix}
c_{12} & a_{11} & a_{13} \\
c_{22} & a_{21} & a_{23} \\
c_{32} & a_{31} & a_{33}
\end{vmatrix}
-
\begin{vmatrix}
a_{12} & b_{11} & a_{13} \\
a_{22} & b_{21} & a_{23} \\
a_{32} & b_{31} & a_{33}
\end{vmatrix}
$$

$$
-
\begin{vmatrix}
c_{12} & b_{11} & a_{13} \\
c_{22} & b_{21} & a_{23} \\
c_{32} & b_{31} & a_{33}
\end{vmatrix}
$$

$$
=
\begin{vmatrix}
a_{11} & a_{12} & a_{13} \\
a_{21} & a_{22} & a_{23} \\
a_{31} & a_{32} & a_{33}
\end{vmatrix}
+
\begin{vmatrix}
a_{11} & c_{12} & a_{13} \\
a_{21} & c_{22} & a_{23} \\
a_{31} & c_{32} & a_{33}
\end{vmatrix}
+
\begin{vmatrix}
b_{11} & a_{12} & a_{13} \\
b_{21} & a_{22} & a_{23} \\
b_{31} & a_{32} & a_{33}
\end{vmatrix}
$$

$$
+
\begin{vmatrix}
b_{11} & c_{12} & a_{13} \\
b_{21} & c_{22} & a_{23} \\
b_{31} & c_{32} & a_{33}
\end{vmatrix}.
$$

THEOREM VI: If, to the elements of any row (or column) are added k times the corresponding elements of any other row (or column), the value of the determinant is unaltered.

For,

$$
\begin{vmatrix} a_{11} + ka_{12} & a_{12} & a_{13} \\ a_{21} + ka_{22} & a_{22} & a_{23} \\ a_{31} + ka_{32} & a_{32} & a_{33} \end{vmatrix} = \begin{vmatrix} a_{11} & a_{12} & a_{13} \\ a_{21} & a_{22} & a_{23} \\ a_{31} & a_{32} & a_{33} \end{vmatrix} + k \begin{vmatrix} a_{12} & a_{12} & a_{13} \\ a_{22} & a_{22} & a_{23} \\ a_{32} & a_{32} & a_{33} \end{vmatrix}
$$

$$
= \begin{vmatrix} a_{11} & a_{12} & a_{13} \\ a_{21} & a_{22} & a_{23} \\ a_{31} & a_{32} & a_{33} \end{vmatrix},
$$

a result which is particularly useful in the evaluation of determinants.

Example 5

Evaluate

$$
\Delta = \begin{vmatrix} 6 & 4 & 2 & 8 \\ 17 & 12 & 5 & 20 \\ 17 & 15 & 8 & -3 \\ 8 & 7 & 4 & 6 \end{vmatrix}.
$$

Taking out the common factor 2 from row one,

$$
\Delta = 2 \begin{vmatrix} 3 & 2 & 1 & 4 \\ 17 & 12 & 5 & 20 \\ 17 & 15 & 8 & -3 \\ 8 & 7 & 4 & 6 \end{vmatrix}.
$$

Subtracting five times the elements of row one from corresponding elements of row two,

$$
\Delta = 2 \begin{vmatrix} 3 & 2 & 1 & 4 \\ 2 & 2 & 0 & 0 \\ 17 & 15 & 8 & -3 \\ 8 & 7 & 4 & 6 \end{vmatrix}.
$$

Subtracting the elements of column two from corresponding elements of column one,

$$
\Delta = 2 \begin{vmatrix} 1 & 2 & 1 & 4 \\ 0 & 2 & 0 & 0 \\ 2 & 15 & 8 & -3 \\ 1 & 7 & 4 & 6 \end{vmatrix}.
$$

Expanding along row two,

$$
\Delta = (2)(2) \begin{vmatrix} 1 & 1 & 4 \\ 2 & 8 & -3 \\ 1 & 4 & 6 \end{vmatrix}.
$$

Subtracting four times the elements of column one from corresponding elements of column two,

$$\Delta = 4 \begin{vmatrix} 1 & -3 & 4 \\ 2 & 0 & -3 \\ 1 & 0 & 6 \end{vmatrix}.$$

Finally, expansion down column two gives,

$$\Delta = 4(-1)(-3) \begin{vmatrix} 2 & -3 \\ 1 & 6 \end{vmatrix} = 4(-1)(-3)(15) = 180.$$

Example 6

Solve the equation $\Delta = 0$, where

$$\Delta = \begin{vmatrix} 1-x & 2 & 1 \\ 3 & 3-x & -4 \\ -2 & -3 & 5-x \end{vmatrix}.$$

Adding corresponding elements of the second and third rows to row one,

$$\Delta = \begin{vmatrix} 2-x & 2-x & 2-x \\ 3 & 3-x & -4 \\ -2 & -3 & 5-x \end{vmatrix} = (2-x) \begin{vmatrix} 1 & 1 & 1 \\ 3 & 3-x & -4 \\ -2 & -3 & 5-x \end{vmatrix}.$$

Subtracting the elements of column one from corresponding elements of both column two and column three,

$$\Delta = (2-x) \begin{vmatrix} 1 & 0 & 0 \\ 3 & -x & -7 \\ -2 & -1 & 7-x \end{vmatrix} = (2-x)(1) \begin{vmatrix} -x & -7 \\ -1 & 7-x \end{vmatrix}.$$

Hence,

$$(2-x)(-7x + x^2 - 7) = 0,$$

giving

$$x = 2, \quad \frac{7 \pm \sqrt{(77)}}{2}.$$

Example 7

In determining the rotational energy levels of a non-linear molecule, the following equation arises:

$$\begin{vmatrix} \tfrac{1}{2}a + b + \tfrac{1}{2}c - E & \tfrac{1}{2}a - \tfrac{1}{2}c \\ \tfrac{1}{2}a - \tfrac{1}{2}c & \tfrac{1}{2}a + b + \tfrac{1}{2}c - E \end{vmatrix} = 0,$$

where a, b and c are constants related to certain moments of inertia of the molecule. Show that $E = a + b$ or $E = b + c$.

Adding the elements of row two to corresponding elements of row one,

$$\begin{vmatrix} a + b - E & a + b - E \\ \frac{1}{2}a - \frac{1}{2}c & \frac{1}{2}a + b + \frac{1}{2}c - E \end{vmatrix} = 0$$

i.e.

$$(a + b - E) \begin{vmatrix} 1 & 1 \\ \frac{1}{2}a - \frac{1}{2}c & \frac{1}{2}a + b + \frac{1}{2}c - E \end{vmatrix} = 0.$$

Subtracting the elements of column one from corresponding elements of column two,

$$(a + b - E) \begin{vmatrix} 1 & 0 \\ \frac{1}{2}a - \frac{1}{2}c & b + c - E \end{vmatrix} = 0,$$

giving

$$(a + b - E)(b + c - E) = 0,$$

and hence

$$E = a + b \quad \text{or} \quad E = b + c.$$

Example 8

Solve the equations

$$x + 2y + z + 2s - 1 = 0$$

$$-2x + y - z + s + 5 = 0$$

$$4x - y + 2z + 2s + 1 = 0$$

$$- x + 3y + z + 3s - 3 = 0.$$

Using the notation of Section 3,

$$\frac{x}{\Delta_1} = \frac{-y}{\Delta_2} = \frac{z}{\Delta_3} = \frac{-s}{\Delta_4} = \frac{1}{\Delta},$$

where, for example,

$$\Delta = \begin{vmatrix} 1 & 2 & 1 & 2 \\ -2 & 1 & -1 & 1 \\ 4 & -1 & 2 & 2 \\ -1 & 3 & 1 & 3 \end{vmatrix}.$$

Carrying out the operations: column 1 + twice column 4, column 2 − column 4, column 3 + column 4, gives

$$\Delta = \begin{vmatrix} 5 & 0 & 3 & 2 \\ 0 & 0 & 0 & 1 \\ 8 & -3 & 4 & 2 \\ 5 & 0 & 4 & 3 \end{vmatrix} = -(-3) \begin{vmatrix} 5 & 3 & 2 \\ 0 & 0 & 1 \\ 5 & 4 & 3 \end{vmatrix} = 3(-1) \begin{vmatrix} 5 & 3 \\ 5 & 4 \end{vmatrix} = -15.$$

Similarly, we find

$$\Delta_1 = \begin{vmatrix} 2 & 1 & 2 & -1 \\ 1 & -1 & 1 & 5 \\ -1 & 2 & 2 & 1 \\ 3 & 1 & 3 & -3 \end{vmatrix} = 30, \quad \Delta_2 = \begin{vmatrix} 1 & 1 & 2 & -1 \\ -2 & -1 & 1 & 5 \\ 4 & 2 & 2 & 1 \\ -1 & 1 & 3 & -3 \end{vmatrix} = 15,$$

$$\Delta_3 = \begin{vmatrix} 1 & 2 & 2 & -1 \\ -2 & 1 & 1 & 5 \\ 4 & -1 & 2 & 1 \\ -1 & 3 & 3 & -3 \end{vmatrix} = -105, \quad \Delta_4 = \begin{vmatrix} 1 & 2 & 1 & -1 \\ -2 & 1 & -1 & 5 \\ 4 & -1 & 2 & 1 \\ -1 & 3 & 1 & -3 \end{vmatrix} = -45,$$

giving $x = -2$, $y = 1$, $z = 7$ and $s = -3$.

5. Matrices

A *matrix* is a *rectangular* array of elements which is subject to certain specified laws. It is usually represented by a single symbol.

Thus, a matrix of *order* $m \times n$, having m rows and n columns may be written in the form

$$\mathbf{A} = [a_{ij}] = \begin{bmatrix} a_{11} & a_{12} & a_{13} & \cdots & a_{1n} \\ a_{21} & a_{22} & a_{23} & \cdots & a_{2n} \\ \vdots & \vdots & \vdots & & \vdots \\ a_{m1} & a_{m2} & a_{m3} & \cdots & a_{mn} \end{bmatrix},$$

a_{ij} being the element in the ith row and jth column.

The particular case in which the number of rows and columns are equal is referred to as a *square matrix*, whilst those matrices consisting of a single row or single column, are called *row vector matrices* (or simply *row matrices*) and *column vector matrices* (or simply *column matrices*) respectively.

A square matrix with n rows and n columns is said to be of *order n*. A *unit matrix* is a square matrix in which

$$a_{ii} = 1, \quad \text{for all } i$$

and

$$a_{ij} = 0, \quad \text{for } i \neq j.$$

It is usually denoted by $\mathbf{I}$. For example,

$$\mathbf{I} = \begin{bmatrix} 1 & 0 & 0 & 0 \\ 0 & 1 & 0 & 0 \\ 0 & 0 & 1 & 0 \\ 0 & 0 & 0 & 1 \end{bmatrix}$$

is a unit matrix of order 4.

A matrix with all elements equal to zero is called a *null matrix* and is denoted by **0**.

Two matrices $A = [a_{ij}]$ and $B = [b_{ij}]$ are said to be equal if corresponding elements are equal, i.e. $a_{ij} = b_{ij}$.

It is important to realize that the elements of a matrix cannot be manipulated as in determinants.

6. Rules of matrix algebra

(i) Matrix addition

The sum of two matrices is obtained by adding together corresponding elements, so that if $A = [a_{ij}]$ and $B = [b_{ij}]$, then $C = A + B$ is the matrix having a general element of the form $c_{ij} = a_{ij} + b_{ij}$.

Thus, if

$$A = \begin{bmatrix} 0 & 1 \\ 2 & -2 \\ 3 & 1 \end{bmatrix}, \quad B = \begin{bmatrix} 1 & 0 \\ -1 & -2 \\ 1 & 4 \end{bmatrix},$$

then

$$C = A + B = \begin{bmatrix} 1 & 1 \\ 1 & -4 \\ 4 & 5 \end{bmatrix}.$$

We observe that addition may be applied only to matrices of equal order.

(ii) Multiplication by a constant

Multiplication of a matrix by a constant is obtained by multiplying each element of the matrix by that constant, i.e. $kA = k[a_{ij}] = [ka_{ij}]$. For example

$$3 \begin{bmatrix} 2 & -1 \\ 2 & 4 \end{bmatrix} = \begin{bmatrix} 6 & -3 \\ 6 & 12 \end{bmatrix}.$$

It follows from (i) and (ii), that the addition of k equal matrices is equal to k times that matrix, i.e. $A + A + \cdots + A = kA$.

(iii) Matrix subtraction

The difference of two matrices $A = [a_{ij}]$ and $B = [b_{ij}]$, is obtained by forming the sum of A with $(-1)B$, so that if $C = A - B$, then $c_{ij} = a_{ij} - b_{ij}$.

Thus, if

$$A = \begin{bmatrix} 1 & 2 & -1 \\ 0 & 3 & 2 \end{bmatrix}, \quad B = \begin{bmatrix} 0 & 1 & -1 \\ 2 & 1 & 1 \end{bmatrix},$$

then

$$C = A - B = \begin{bmatrix} 1 & 1 & 0 \\ -2 & 2 & 1 \end{bmatrix}.$$

Clearly the two matrices to be subtracted must have an equal number of rows and an equal number of columns.

(iv) *Matrix multiplication*

Two matrices **A** and **B** can be multiplied together only if the number of columns of the first matrix is equal to the number of rows of the second matrix.

If $A = [a_{ik}]$ and $B = [b_{kj}]$, then the general element of the product $C = AB$ is given by

$$c_{ij} = \sum_{k=1}^{n} a_{ik} b_{kj},$$

n being the number of columns of **A** (and the number of rows of **B**). Thus,

$$\begin{bmatrix} a_{11} & a_{12} & a_{13} \\ a_{21} & a_{22} & a_{23} \end{bmatrix} \begin{bmatrix} b_{11} & b_{12} \\ b_{21} & b_{22} \\ b_{31} & b_{32} \end{bmatrix}$$

$$= \begin{bmatrix} (a_{11}b_{11} + a_{12}b_{21} + a_{13}b_{31}) & (a_{11}b_{12} + a_{12}b_{22} + a_{13}b_{32}) \\ (a_{21}b_{11} + a_{22}b_{21} + a_{23}b_{31}) & (a_{21}b_{12} + a_{22}b_{22} + a_{23}b_{32}) \end{bmatrix}$$

and so on.

Example 9

Determine **AB**, if

$$A = \begin{bmatrix} 2 & 3 \\ -1 & 2 \end{bmatrix}, \quad B = \begin{bmatrix} 0 & 2 & 1 \\ -1 & 0 & 3 \end{bmatrix}.$$

$$AB = \begin{bmatrix} 2 & 3 \\ -1 & 2 \end{bmatrix} \begin{bmatrix} 0 & 2 & 1 \\ -1 & 0 & 3 \end{bmatrix} = \begin{bmatrix} (0-3) & (4+0) & (2+9) \\ (0-2) & (-2+0) & (-1+6) \end{bmatrix}$$

$$= \begin{bmatrix} -3 & 4 & 11 \\ -2 & -2 & 5 \end{bmatrix}.$$

Note that in this example, the product **BA** does not exist since the number of columns of **B** is not equal to the number of rows of **A**. Even if both **AB** and **BA** exist, in general $AB \neq BA$.

For example, if

$$A = \begin{bmatrix} 2 & 1 \\ 1 & 0 \end{bmatrix}, \quad B = \begin{bmatrix} 1 & 2 \\ 0 & 1 \end{bmatrix},$$

then

$$AB = \begin{bmatrix} 2 & 1 \\ 1 & 0 \end{bmatrix} \begin{bmatrix} 1 & 2 \\ 0 & 1 \end{bmatrix} = \begin{bmatrix} (2+0) & (4+1) \\ (1+0) & (2+0) \end{bmatrix} = \begin{bmatrix} 2 & 5 \\ 1 & 2 \end{bmatrix}$$

whereas,

$$\mathbf{BA} = \begin{bmatrix} 1 & 2 \\ 0 & 1 \end{bmatrix} \begin{bmatrix} 2 & 1 \\ 1 & 0 \end{bmatrix} = \begin{bmatrix} (2+2) & (1+0) \\ (0+1) & (0+0) \end{bmatrix} = \begin{bmatrix} 4 & 1 \\ 1 & 0 \end{bmatrix}.$$

It is easily verified that if $\mathbf{I}$ is the unit matrix of order n, then $\mathbf{I} \times \mathbf{I} = \mathbf{I}$, i.e. $\mathbf{I}^2 = \mathbf{I}$, and similarly $\mathbf{I} = \mathbf{I}^2 = \mathbf{I}^3 = \cdots = \mathbf{I}^m$. Also, if the matrix $\mathbf{A}$ is multiplied by a unit matrix $\mathbf{I}$ of the correct order (so that multiplication is possible) then

$$\mathbf{IA} = \mathbf{AI} = \mathbf{A}.$$

Multiplication by a null matrix gives a null matrix.

Example 10

The following matrices, known as Pauli spin matrices, are used in elementary particle physics:

$$\sigma_1 = \begin{bmatrix} 0 & 1 \\ 1 & 0 \end{bmatrix}, \quad \sigma_2 = \begin{bmatrix} 0 & -i \\ i & 0 \end{bmatrix}, \quad \sigma_3 = \begin{bmatrix} 1 & 0 \\ 0 & -1 \end{bmatrix}$$

where $i^2 = -1$. Show that:

(i) $\quad \sigma_1^2 = \sigma_2^2 = \sigma_3^2 = \mathbf{I}$,

(ii) $\quad \sigma_1\sigma_2 = -\sigma_2\sigma_1 = i\sigma_3$.

(i) $\sigma_1^2 = \begin{bmatrix} 0 & 1 \\ 1 & 0 \end{bmatrix} \begin{bmatrix} 0 & 1 \\ 1 & 0 \end{bmatrix} = \begin{bmatrix} 1 & 0 \\ 0 & 1 \end{bmatrix} = \mathbf{I}$,

$\sigma_2^2 = \begin{bmatrix} 0 & -i \\ i & 0 \end{bmatrix} \begin{bmatrix} 0 & -i \\ i & 0 \end{bmatrix} = \begin{bmatrix} -i^2 & 0 \\ 0 & -i^2 \end{bmatrix} = \begin{bmatrix} 1 & 0 \\ 0 & 1 \end{bmatrix} = \mathbf{I}$,

$\sigma_3^2 = \begin{bmatrix} 1 & 0 \\ 0 & -1 \end{bmatrix} \begin{bmatrix} 1 & 0 \\ 0 & -1 \end{bmatrix} = \begin{bmatrix} 1 & 0 \\ 0 & 1 \end{bmatrix} = \mathbf{I}$.

(ii) $\sigma_1\sigma_2 = \begin{bmatrix} 0 & 1 \\ 1 & 0 \end{bmatrix} \begin{bmatrix} 0 & -i \\ i & 0 \end{bmatrix} = \begin{bmatrix} i & 0 \\ 0 & -i \end{bmatrix} = i\sigma_3$,

$\sigma_2\sigma_1 = \begin{bmatrix} 0 & -i \\ i & 0 \end{bmatrix} \begin{bmatrix} 0 & 1 \\ 1 & 0 \end{bmatrix} = \begin{bmatrix} -i & 0 \\ 0 & i \end{bmatrix} = -i\sigma_3$.

7. Transpose of a matrix

If a given matrix $\mathbf{A} = [a_{ij}]$ is transposed by interchanging rows and columns, then the new matrix so formed, is called the *transpose* of $\mathbf{A}$ and is denoted by $\mathbf{A}' = [a_{ji}]$. For example, if

$$\mathbf{A} = \begin{bmatrix} 1 & 2 & -1 \\ 0 & -1 & 2 \end{bmatrix}, \quad \text{then} \quad \mathbf{A}' = \begin{bmatrix} 1 & 0 \\ 2 & -1 \\ -1 & 2 \end{bmatrix}$$

Note that $\mathbf{AA}'$ always exists and that if $\mathbf{AA}' = \mathbf{I}$, then the matrix $\mathbf{A}$ is said to be *orthogonal*. For example,

$$\mathbf{A} = \begin{bmatrix} \cos\theta & \sin\theta \\ -\sin\theta & \cos\theta \end{bmatrix}$$

is orthogonal, since

$$\mathbf{AA}' = \begin{bmatrix} \cos\theta & \sin\theta \\ -\sin\theta & \cos\theta \end{bmatrix}\begin{bmatrix} \cos\theta & -\sin\theta \\ \sin\theta & \cos\theta \end{bmatrix}$$

$$= \begin{bmatrix} \cos^2\theta + \sin^2\theta & -\cos\theta\sin\theta + \sin\theta\cos\theta \\ -\sin\theta\cos\theta + \cos\theta\sin\theta & \sin^2\theta + \cos^2\theta \end{bmatrix}$$

$$= \begin{bmatrix} 1 & 0 \\ 0 & 1 \end{bmatrix},$$

i.e. the unit matrix of order two.

8. The inverse matrix

Consider a square matrix of order 3, i.e.

$$\mathbf{A} = \begin{bmatrix} a_{11} & a_{12} & a_{13} \\ a_{21} & a_{22} & a_{23} \\ a_{31} & a_{32} & a_{33} \end{bmatrix},$$

for which there is a corresponding determinant $|\mathbf{A}|$ having the same elements as in the matrix, viz.,

$$|\mathbf{A}| = \begin{vmatrix} a_{11} & a_{12} & a_{13} \\ a_{21} & a_{22} & a_{23} \\ a_{31} & a_{32} & a_{33} \end{vmatrix}.$$

Suppose that we construct a matrix consisting of the cofactors A_{ij} of the elements a_{ij} of $\mathbf{A}$, as follows:

$$\begin{bmatrix} A_{11} & A_{12} & A_{13} \\ A_{21} & A_{22} & A_{23} \\ A_{31} & A_{32} & A_{33} \end{bmatrix}.$$

If we form the transpose of this matrix and at the same time divide all its elements by $|\mathbf{A}|$, then the new matrix so formed, is called the *inverse matrix* of $\mathbf{A}$, or the *reciprocal* matrix of $\mathbf{A}$, and is denoted by $\mathbf{A}^{-1}$. Thus

$$\mathbf{A}^{-1} = \begin{bmatrix} \dfrac{A_{11}}{|\mathbf{A}|} & \dfrac{A_{21}}{|\mathbf{A}|} & \dfrac{A_{31}}{|\mathbf{A}|} \\ \dfrac{A_{12}}{|\mathbf{A}|} & \dfrac{A_{22}}{|\mathbf{A}|} & \dfrac{A_{32}}{|\mathbf{A}|} \\ \dfrac{A_{13}}{|\mathbf{A}|} & \dfrac{A_{23}}{|\mathbf{A}|} & \dfrac{A_{33}}{|\mathbf{A}|} \end{bmatrix} = \frac{1}{|\mathbf{A}|}\begin{bmatrix} A_{11} & A_{21} & A_{31} \\ A_{12} & A_{22} & A_{32} \\ A_{13} & A_{23} & A_{33} \end{bmatrix}.$$

The inverse of a square matrix of any order is similarly defined, but if $|A| = 0$, no inverse exists.

Suppose, for example, we wish to determine the inverse matrix of

$$A = \begin{bmatrix} 2 & 1 & 0 \\ 0 & 2 & 1 \\ 0 & 3 & 1 \end{bmatrix}.$$

First, we determine

$$|A| = \begin{vmatrix} 2 & 1 & 0 \\ 0 & 2 & 1 \\ 0 & 3 & 1 \end{vmatrix} = 2 \begin{vmatrix} 2 & 1 \\ 3 & 1 \end{vmatrix} = -2,$$

and

$$\begin{bmatrix} A_{11} & A_{12} & A_{13} \\ A_{21} & A_{22} & A_{23} \\ A_{31} & A_{32} & A_{33} \end{bmatrix} = \begin{bmatrix} -1 & 0 & 0 \\ -1 & 2 & -6 \\ 1 & -2 & 4 \end{bmatrix}.$$

Then,

$$A^{-1} = \frac{1}{(-2)} \begin{bmatrix} -1 & -1 & 1 \\ 0 & 2 & -2 \\ 0 & -6 & 4 \end{bmatrix} = \begin{bmatrix} \frac{1}{2} & \frac{1}{2} & -\frac{1}{2} \\ 0 & -1 & 1 \\ 0 & 3 & -2 \end{bmatrix}.$$

It is easily verified that $AA^{-1} = A^{-1}A = I$, which can be used to check the calculations. For example, in this case,

$$\begin{bmatrix} 2 & 1 & 0 \\ 0 & 2 & 1 \\ 0 & 3 & 1 \end{bmatrix} \begin{bmatrix} \frac{1}{2} & \frac{1}{2} & -\frac{1}{2} \\ 0 & -1 & 1 \\ 0 & 3 & -2 \end{bmatrix} = \begin{bmatrix} 1 & 0 & 0 \\ 0 & 1 & 0 \\ 0 & 0 & 1 \end{bmatrix}.$$

In the examples that follow, we assume (without proof) the associative, commutative and distributive laws of matrix algebra, viz.,

$$A + B = B + A,$$
$$A + (B + C) = (A + B) + C,$$
$$\lambda(A + B) = \lambda A + \lambda B,$$
$$A(B + C) = AB + AC,$$
$$(A + B)C = AC + BC,$$
$$A(BC) = (AB)C,$$

distinguishing only between *pre-* and *post-multiplication*.

Example 11

Given two square matrices A, B of the same order and with $|A|$, $|B| \neq 0$, show that $(AB)^{-1} = B^{-1}A^{-1}$.

Now,

$$(AB)(B^{-1}A^{-1}) = A(BB^{-1})A^{-1} = AI\ A^{-1} = AA^{-1} = I,$$

and pre-multiplying each side of this equation by $(AB)^{-1}$, gives

$$(AB)^{-1}(AB)(B^{-1}A^{-1}) = (AB)^{-1}I.$$

But,

$$(AB)^{-1}(AB) = I,$$
$$\therefore\ B^{-1}A^{-1} = (AB)^{-1}I = (AB)^{-1}.$$

Note that ABA^{-1} cannot be simplified by combining A and A^{-1}, since the order of the matrices cannot be altered. BA^{-1} will not, in general, be equal to $A^{-1}B$.

Example 12

If A is an orthogonal matrix, show that $A^{-1} = A'$. Since A is orthogonal, $AA' = I$. Therefore

$$A^{-1}AA' = A^{-1}I.$$
$$\therefore\ (A^{-1}A)A' = A^{-1}.$$
$$\therefore\ A' = A^{-1}.$$

9. Solution of equations by matrices

Suppose that $x_i, y_i\ (i = 1, 2, 3)$ are variables satisfying the following equations:

$$a_{11}x_1 + a_{12}x_2 + a_{13}x_3 = y_1,$$
$$a_{21}x_1 + a_{22}x_2 + a_{23}x_3 = y_2,$$
$$a_{31}x_1 + a_{32}x_2 + a_{33}x_3 = y_3,$$

in which y_i is expressed in terms of x_1, x_2, x_3, and that we wish to determine x_i in terms of y_1, y_2, y_3.

Writing

$$A = \begin{bmatrix} a_{11} & a_{12} & a_{13} \\ a_{21} & a_{22} & a_{23} \\ a_{31} & a_{32} & a_{33} \end{bmatrix}, \quad X = \begin{bmatrix} x_1 \\ x_2 \\ x_3 \end{bmatrix}, \quad Y = \begin{bmatrix} y_1 \\ y_2 \\ y_3 \end{bmatrix},$$

then

$$\begin{bmatrix} a_{11} & a_{12} & a_{13} \\ a_{21} & a_{22} & a_{23} \\ a_{31} & a_{32} & a_{33} \end{bmatrix} \begin{bmatrix} x_1 \\ x_2 \\ x_3 \end{bmatrix} = \begin{bmatrix} a_{11}x_1 + a_{12}x_2 + a_{13}x_3 \\ a_{21}x_1 + a_{22}x_2 + a_{23}x_3 \\ a_{31}x_1 + a_{32}x_2 + a_{33}x_3 \end{bmatrix} = \begin{bmatrix} y_1 \\ y_2 \\ y_3 \end{bmatrix},$$

i.e.

$$AX = Y.$$

Pre-multiplying each side by A^{-1}, assuming A^{-1} exists, i.e. $|A| \neq 0$, gives $A^{-1}AX = A^{-1}Y$. Thus

$$X = A^{-1}Y$$

and by determining the inverse matrix A^{-1} it is possible to find x_i in terms of y_1, y_2 and y_3.

The matrix form of the solutions to the set of linear equations of Section 3, is obtained by replacing the variables y_1, y_2, y_3 by the constants $-k_1, -k_2, -k_3$. Thus if,

$$K = \begin{bmatrix} -k_1 \\ -k_2 \\ -k_3 \end{bmatrix},$$

then $AX = K$ and the solutions are given by $X = A^{-1}K$.

Example 13

Find y_1, y_2, y_3 in terms of x_1, x_2, x_3, given that

$$x_1 + x_2 - 2x_3 = y_1,$$

$$-2x_1 + 4x_2 + 5x_3 = y_2,$$

$$- x_1 + 3x_2 + 3x_3 = y_3.$$

We have,

$$|A| = \begin{vmatrix} 1 & 1 & -2 \\ -2 & 4 & 5 \\ -1 & 3 & 3 \end{vmatrix} = \begin{vmatrix} 1 & 0 & 0 \\ -2 & 6 & 1 \\ -1 & 4 & 1 \end{vmatrix} = \begin{vmatrix} 6 & 1 \\ 4 & 1 \end{vmatrix} = 2,$$

$$\begin{bmatrix} A_{11} & A_{12} & A_{13} \\ A_{21} & A_{22} & A_{23} \\ A_{31} & A_{32} & A_{33} \end{bmatrix} = \begin{bmatrix} -3 & 1 & -2 \\ -9 & 1 & -4 \\ 13 & -1 & 6 \end{bmatrix},$$

and

$$A^{-1} = \tfrac{1}{2} \begin{bmatrix} -3 & -9 & 13 \\ 1 & 1 & -1 \\ -2 & -4 & 6 \end{bmatrix}.$$

Since $X = A^{-1}Y$, we have

$$\begin{bmatrix} x_1 \\ x_2 \\ x_3 \end{bmatrix} = \tfrac{1}{2} \begin{bmatrix} -3 & -9 & 13 \\ 1 & 1 & -1 \\ -2 & -4 & 6 \end{bmatrix} \begin{bmatrix} y_1 \\ y_2 \\ y_3 \end{bmatrix} = \tfrac{1}{2} \begin{bmatrix} -3y_1 - 9y_2 + 13y_3 \\ y_1 + y_2 - y_3 \\ -2y_1 - 4y_2 + 6y_3 \end{bmatrix}$$

giving,

$$-3y_1 - 9y_2 + 13y_3 = 2x_1,$$
$$y_1 + y_2 - y_3 = 2x_2,$$
$$-2y_1 - 4y_2 + 6y_3 = 2x_3.$$

PROBLEMS

1. Evaluate the determinants

(i) $\begin{vmatrix} 3 & 2 \\ 4 & 5 \end{vmatrix}$,
(ii) $\begin{vmatrix} \cos\theta & -\sin\theta \\ \sin\theta & \cos\theta \end{vmatrix}$,
(iii) $\begin{vmatrix} 3 & 1 & 2 \\ 4 & 0 & 1 \\ 3 & 2 & 2 \end{vmatrix}$

2. Solve the equation

$$\begin{vmatrix} 3x - 2 & 6x - 4 \\ x + 3 & 2x + 1 \end{vmatrix} = 0$$

3. The determinant

$$\begin{vmatrix} \alpha - E & \beta \\ \beta & \alpha - E \end{vmatrix} = 0$$

arises in the determination of the energies of the two molecular orbits of the H_2 molecule. Show that the solution is

$$E = \alpha \pm \beta$$

4. To find the molecular orbital energies of trimethylenemethane, the following determinant has to be solved:

$$\begin{vmatrix} x & 0 & 0 & 1 \\ 0 & x & 0 & 1 \\ 0 & 0 & x & 1 \\ 1 & 1 & 1 & x \end{vmatrix} = 0,$$

Find the values of x.

5. In the study of the vibration of a diatomic molecule with nuclear masses m_1 and m_2, and k the restoring force constant, the following determinant arises:

$$\begin{vmatrix} \dfrac{k}{m_1} - \lambda & \dfrac{-k}{\sqrt{(m_1 m_2)}} \\ \dfrac{-k}{\sqrt{(m_1 m_2)}} & \dfrac{k}{m_2} - \lambda \end{vmatrix} = 0.$$

Find λ in terms of k, m_1 and m_2.

6. Evaluate the determinants

(i) $\begin{vmatrix} 5 & 2 & 3 & 0 \\ 15 & 7 & 2 & -2 \\ 10 & 5 & 3 & -1 \\ 5 & 4 & 6 & 2 \end{vmatrix}$,
(ii) $\begin{vmatrix} 7 & 3 & 0 & 4 & 5 \\ 2 & 6 & 1 & 2 & 4 \\ 8 & 4 & 0 & 5 & 6 \\ 4 & 12 & 2 & 5 & 10 \\ 6 & -1 & 3 & -4 & 7 \end{vmatrix}$.

7. Find λ in terms of m, M (the masses of atoms) and k (the force constant) in the following determinant which arises in the determination of the normal modes of vibration of a particular molecule:

$$\begin{vmatrix} \dfrac{k}{m} - \lambda & \dfrac{-6}{\sqrt{(mM)}} & 0 \\ \dfrac{-k`}{\sqrt{(mM)}} & \dfrac{2k}{M} - \lambda & -\dfrac{k}{\sqrt{(mM)}} \\ 0 & \dfrac{-k}{\sqrt{(mM)}} & \dfrac{k}{m} - \lambda \end{vmatrix} = 0.$$

8. Solve, using determinants, the simultaneous equations

$$3x + 2y - z = 8$$
$$2x - 3y - 2z = -3$$
$$x + y + 3z = 2.$$

9. The yield of a certain crop (z) depends upon the amounts of phosphates (x) and nitrogen (y) used, according to the equation

$$z = a + bx + cy$$

over the range considered.

Experimental data are obtained in order to find a, b and c by the method of least squares (see Chapter 15). Summing over the relevant values gives the following three equations which have to be solved:

$$27 - 3a - 10b - 4c = 0$$
$$19 - 2a - 5b - 4c = 0$$
$$25 - 2a - 10b - 6c = 0.$$

Use determinants to find a, b and c.

10. If

$$A = \begin{bmatrix} 2 & 1 & -3 \\ 3 & 1 & 2 \end{bmatrix}, \quad B = \begin{bmatrix} 6 & 2 \\ 1 & 3 \\ 2 & 4 \end{bmatrix}, \quad C = \begin{bmatrix} 1 & 5 & 2 \\ -1 & 0 & 3 \end{bmatrix}, \quad D = [1, 0, -3]$$

find

(i) $A + C$,   (ii) $A - C$,   (iii) $2A + 3C$,   (iv) AB,   (v) BA,   (vi) DB.

11. If

$$A = \begin{bmatrix} 3 & 2 \\ 7 & 5 \end{bmatrix} \quad \text{and} \quad B = \begin{bmatrix} 5 & -2 \\ -7 & 3 \end{bmatrix}$$

show that $AB = I$.

What are the values of (i) A^{-1}, (ii) BA, (iii) $B^{-1}A^{-1}B^2A$?

12. E_1, E_2 and E_3 are three of the Dirac matrices used in the description of

elementary particle physics:

$$E_1 = \begin{bmatrix} 0 & 1 & 0 & 0 \\ 1 & 0 & 0 & 0 \\ 0 & 0 & 0 & 1 \\ 0 & 0 & 1 & 0 \end{bmatrix}, \quad E_2 = \begin{bmatrix} 0 & -i & 0 & 0 \\ i & 0 & 0 & 0 \\ 0 & 0 & 0 & -i \\ 0 & 0 & i & 0 \end{bmatrix}, \quad E_3 = \begin{bmatrix} 1 & 0 & 0 & 0 \\ 0 & -1 & 0 & 0 \\ 0 & 0 & 1 & 0 \\ 0 & 0 & 0 & -1 \end{bmatrix}.$$

Show that

(i) $E_i^2 = I$ for $i = 1, 2, 3$,

(ii) $E_i = \pm E_i'$ for $i = 1, 2, 3$,

(iii) $E_i E_j = \pm i E_k$ for $i, j, k = 1, 2, 3$, $i \neq j \neq k$.

What is the value of $E_1 E_2^2 E_3 E_1^3$?

13. Write down the transposes of

$$A = [2 \quad 3 \quad -4] \quad \text{and} \quad B = \begin{bmatrix} 2 & 6 & 1 \\ -1 & 0 & 2 \end{bmatrix};$$

and find AA'.

14. If

$$A = \begin{bmatrix} 2 & 1 \\ 3 & 2 \end{bmatrix}, \quad B = \begin{bmatrix} 1 & 2 \\ 3 & 1 \end{bmatrix},$$

verify that $(AB)' = B'A'$. (*N.B.* this is always true).

15. Show that

$$A = \tfrac{1}{3} \begin{bmatrix} 2 & 1 & 2 \\ 1 & 2 & -2 \\ 2 & -2 & -1 \end{bmatrix}$$

is an orthogonal matrix.

16. Find the inverse matrix of

$$A = \begin{bmatrix} 3 & 1 & -2 \\ 1 & 2 & 3 \\ 2 & 0 & 1 \end{bmatrix}$$

and check your answer by calculating AA^{-1}.

17. The transformations used for rotating the p_x, p_y orbitals of a molecule, in one particular case, are

$$p_x' = 0p_x + 1p_y$$
$$p_y' = -1p_x + 0p_y.$$

Write these equations in matrix form.

18. The following matrices occur in the theory concerned with energy levels when considering the rotation of molecules:

$$M = \begin{bmatrix} A + \tfrac{1}{2}(B + C) & 0 & -\tfrac{1}{2}(B - C) \\ 0 & B + C & 0 \\ -\tfrac{1}{2}(B - C) & 0 & A + \tfrac{1}{2}(B + C) \end{bmatrix}$$

where A, B, C are related to the moments of inertia of the molecule about three axes.

Also

$$\mathbf{X} = \frac{1}{\sqrt{2}} \begin{bmatrix} -1 & 0 & 1 \\ 0 & \sqrt{2} & 0 \\ 1 & 0 & 1 \end{bmatrix}.$$

Show that

$$\mathbf{X'MX} = \begin{bmatrix} A + B & 0 & 0 \\ 0 & B + C & 0 \\ 0 & 0 & A + C \end{bmatrix}$$

19. Solve the simultaneous equations in Question 8 by calculating the inverse of of the matrix of the coefficients.

CHAPTER 7

Vectors

The majority of physical quantities fall into two categories. The first group may be specified completely in terms of a single number or *magnitude*, a member of the group being referred to as a *scalar* quantity or simply a scalar. Thus, length, mass, energy and temperature are all examples of scalar quantities. Quantities belonging to the second group require an associated *direction* in addition to the magnitude, in order to complete their specification, and are called *vector* quantities or simply vectors. Examples are displacement, velocity, force and angular momentum.

1. Representation of a vector

The characteristics of a vector quantity, described above, may be represented by a line segment of appropriate length and direction. Thus, if A, B and C represent any points in space, then the line segments OA, OB, OC (see Fig. 7.1) having directions say from O to A, O to B and O to C respectively, symbolize vectors and are usually denoted by

$$\overline{OA} = \mathbf{a}, \quad \overline{OB} = \mathbf{b}, \quad \overline{OC} = \mathbf{c}.$$

The length OA is the magnitude of **a** and is generally denoted by $|\mathbf{a}|$ or a. If, therefore, we introduce a *unit vector* **â** having the same direction as **a** but of unit length, then $\mathbf{a} = |\mathbf{a}|\,\hat{\mathbf{a}} = a\hat{\mathbf{a}}$. Similarly, we write

$$\mathbf{b} = |\mathbf{b}|\,\hat{\mathbf{b}} = b\hat{\mathbf{b}}, \quad \mathbf{c} = |\mathbf{c}|\,\hat{\mathbf{c}} = c\hat{\mathbf{c}}.$$

We note that two vectors **a** and **b** are equal if, and only if, they have the same magnitude and direction irrespective of their relative positions, and we write **a** = **b**. If **a** and **b** have the same magnitude but opposite directions, then **a** = −**b** or **b** = −**a** (see Fig. 7.2).

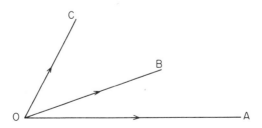

Fig. 7.1

Fig. 7.2

2. Vector addition

To determine the sum of *any* two vectors **a** and **b**, equivalent vectors are *localized* at a common point O and compounded according to the *parallelogram law of addition* (see Fig. 7.3).

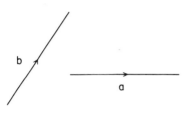 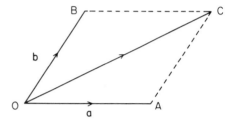

Fig. 7.3

With sides $\overline{OA} = \mathbf{a}$, $\overline{OB} = \mathbf{b}$, we construct the parallelogram OACB and define

$$\overline{OA} + \overline{OB} = \overline{OC}.$$

Moreover, since $\overline{AC} = \overline{OB}$, it follows that

$$\overline{OA} + \overline{AC} = \overline{OA} + \overline{OB} = \overline{OC}$$

which is the *triangle law of addition*. Similarly, $\overline{OB} + \overline{BC} = \overline{OC}$, verifying that

$$\mathbf{a} + \mathbf{b} = \mathbf{b} + \mathbf{a}$$

and the vector addition is *commutative*.

From these definitions, if follows that if **a**, **b** and **c** are any three vectors, then

$$\mathbf{a} + (\mathbf{b} + \mathbf{c}) = (\mathbf{a} + \mathbf{b}) + \mathbf{c}$$

and vector addition is *associative* (see Fig. 7.4).

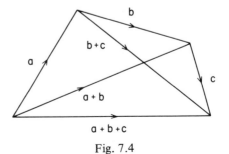

Fig. 7.4

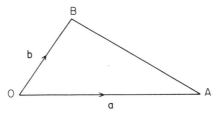

Fig. 7.5

For the subtraction of two vectors, we recall that the vector $-\mathbf{a}$ is of the same magnitude as $\mathbf{a}$ but of opposite direction. Thus, if $\overline{OA} = \mathbf{a}$, then $\overline{AO} = -\mathbf{a}$ and reference to Fig. 7.5 shows that

$$\overline{AB} = \overline{AO} + \overline{OB}$$
$$= -\mathbf{a} + \mathbf{b}$$
$$= \mathbf{b} - \mathbf{a}$$

which can be re-written in the form

$$\overline{AB} = \mathbf{b} + (-\mathbf{a})$$

and interpreted as the addition of $-\mathbf{a}$ to $\mathbf{b}$. In particular, if A and B coincide so that $\mathbf{a} = \mathbf{b}$, then

$$\mathbf{a} - \mathbf{b} = \mathbf{a} - \mathbf{a} = \mathbf{0},$$

$\mathbf{0}$ being a vector of zero magnitude, called a *zero* or *null vector*.

3. Multiplication of a vector by a scalar

Since the addition of n equal vectors $\mathbf{a}$, produces a vector having the same direction as $\mathbf{a}$ but with n times the magnitude of $\mathbf{a}$, it is appropriate to express the result in the form

$$\mathbf{a} + \mathbf{a} + \mathbf{a} + \cdots = n\mathbf{a}.$$

For *any* real number n, the vector $n\mathbf{a}$ is defined to be a vector of magnitude $|n\mathbf{a}|$, and direction equal to that of $\mathbf{a}$ if $n > 0$, and opposite to that of $\mathbf{a}$ if $n < 0$.

It follows immediately, that if m and n are real numbers, then

$$m(n\mathbf{a}) = (mn)\mathbf{a} = n(m\mathbf{a})$$

and

$$(m + n)\mathbf{a} = m\mathbf{a} + n\mathbf{a}.$$

Also, from a comparison of similar triangles OAC, OA$'$C$'$ (see Fig. 7.6) it can be seen that if $\overline{OA'} = n\mathbf{a}$, $\overline{OB'} = n\mathbf{b}$, then

$$n(\mathbf{a} + \mathbf{b}) = n\mathbf{a} + n\mathbf{b},$$

since OC$'$ = nOC.

F

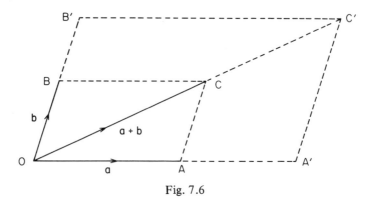

Fig. 7.6

4. Components of a vector

Let $Oxyz$ form a set of orthogonal Cartesian coordinates axes, and let A be a point in space having coordinates $x = a_1$, $y = a_2$, $z = a_3$ (see Fig. 7.7). Then A is uniquely determined by the *position vector* $\overline{OA}(=a)$.

The axes, as drawn, constitute a *right-handed system* and are indicative of the fact that a right-handed screw, having the axis along Ox, will advance along Ox when rotated through $90°$ from Oy to Oz. Similar notions apply to each of the cyclic variations of the axes, viz. a rotation through $90°$ from Oz to Ox will produce an advance along Oy, and a rotation through $90°$ from Ox to Oy will produce an advance along Oz.

Introducing unit vectors $\mathbf{i}, \mathbf{j}, \mathbf{k}$ along the axes Ox, Oy, Oz respectively, and noting that $\overline{OM} = a_1\mathbf{i}$, $\overline{MN} = a_2\mathbf{j}$, $\overline{NA} = a_3\mathbf{k}$, application of the triangle law of addition gives

$$\overline{ON} = \overline{OM} + \overline{MN}$$

$$= a_1\mathbf{i} + a_2\mathbf{j}$$

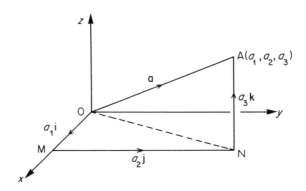

Fig. 7.7

and

$$\overline{OA} = \overline{ON} + \overline{NA}$$
$$= (a_1\mathbf{i} + a_2\mathbf{j}) + a_3\mathbf{k}$$
$$= a_1\mathbf{i} + a_2\mathbf{j} + a_3\mathbf{k}.$$

Thus

$$\mathbf{a} = a_1\mathbf{i} + a_2\mathbf{j} + a_3\mathbf{k} \tag{1}$$

and

$$a^2 = a_1^2 + a_2^2 + a_3^2 \tag{2}$$

from the theorem of Pythagoras, the positive square root being taken when determining a.

If b is expressed in component form, $\mathbf{b} = b_1\mathbf{i} + b_2\mathbf{j} + b_3\mathbf{k}$, it follows that

$$\mathbf{a} \pm \mathbf{b} = (a_1\mathbf{i} + a_2\mathbf{j} + a_3\mathbf{k}) \pm (b_1\mathbf{i} + b_2\mathbf{j} + b_3\mathbf{k})$$
$$= (a_1 \pm b_1)\mathbf{i} + (a_2 \pm b_2)\mathbf{j} + (a_3 \pm b_3)\mathbf{k}$$

from Section 3, and

$$|\mathbf{a} \pm \mathbf{b}| = +\sqrt{\{(a_1 \pm b_1)^2 + (a_2 \pm b_2)^2 + (a_3 \pm b_3)^2\}}. \tag{3}$$

Further, if α, β, γ represent the angles between $\mathbf{a}$ and $\mathbf{i}, \mathbf{j}, \mathbf{k}$ respectively, then

$$a_1 = a\cos\alpha, \quad a_2 = a\cos\beta, \quad a_3 = a\cos\gamma$$

where $\cos^2\alpha + \cos^2\beta + \cos^2\gamma = 1$ from equation (2) and the unit vector in the direction of $\mathbf{a}$ is

$$\hat{\mathbf{a}} = (a_1/a)\mathbf{i} + (a_2/a)\mathbf{j} + (a_3/a)\mathbf{k} \tag{4}$$
$$= \mathbf{i}\cos\alpha + \mathbf{j}\cos\beta + \mathbf{k}\cos\gamma,$$

$\cos\alpha, \cos\beta, \cos\gamma$ being the *direction cosines* of $\mathbf{a}$ (and of any other vector parallel to $\mathbf{a}$).

Example 1

Find the vector having direction opposite to $2\mathbf{i} - 3\mathbf{j} - 6\mathbf{k}$ and magnitude 14.

Let $\mathbf{a} = 2\mathbf{i} - 3\mathbf{j} - 6\mathbf{k}$. Then the magnitude of $\mathbf{a}$ is

$$a = +\sqrt{\{(2)^2 + (-3)^2 + (-6)^2\}} = +\sqrt{(49)} = 7$$

and the corresponding unit vector is

$$\hat{\mathbf{a}} = \mathbf{a}/a = \tfrac{1}{7}(2\mathbf{i} - 3\mathbf{j} - 6\mathbf{k}).$$

Therefore the vector having opposite direction to $\mathbf{a}$ and magnitude 14 is

$$-14\hat{\mathbf{a}} = -2(2\mathbf{i} - 3\mathbf{j} - 6\mathbf{k}) = -4\mathbf{i} + 6\mathbf{j} + 12\mathbf{k}.$$

154

Example 2

If $a = 3i + j - 2k$ and $b = i - 2j + k$, determine the magnitude and direction cosines of $a + b$.

From Section 4,

$$a + b = 4i - j - k$$
$$\therefore \ |a + b| = \sqrt{\{(4)^2 + (-1)^2 + (-1)^2\}} = +\sqrt{(18)} = 3\sqrt{2}$$

and the direction cosines of $a + b$ are $4/3\sqrt{2}, -1/3\sqrt{2}, -1/3\sqrt{2}$.

Example 3

Determine the equation of the straight line having direction cosines $2/\sqrt{(21)}$, $1/\sqrt{(21)}, -4/\sqrt{(21)}$ and passing through the point $(1, -2, 3)$.

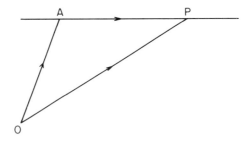

Fig. 7.8

Let A be the point $(1, -2, 3)$, $P \equiv P(x, y, z)$ any point on the line (see Fig. 7.8) and O the origin of coordinates.

Then, by the triangle law

$$\overline{OP} = \overline{OA} + \overline{AP}$$

where, $\overline{OP} = xi + yj + zk$ and $\overline{OA} = i - 2j + 3k$.

Now

$$\widehat{AP} = \frac{2}{\sqrt{(21)}} i + \frac{1}{\sqrt{(21)}} j - \frac{4}{\sqrt{(21)}} k$$

from equation (4), so that if we write $\overline{AP} = \sqrt{(21)}\ \widehat{AP}t$, where t is some parameter, then

$$\overline{AP} = (2i + j - 4k)t.$$

Thus,

$$xi + yj + zk = i - 2j + 3k + (2i + j - 4k)t.$$

Equating components gives the required equations, viz.

$$x = 1 + 2t, \quad y = -2 + t, \quad z = 3 - 4t$$

which may be expressed in the form

$$\frac{x-1}{2} = \frac{y+2}{1} = \frac{z-3}{-4}$$

on eliminating t.

5. Vector multiplication – the scalar product

There are two kinds of vector multiplication, those which generate scalar quantities and those which generate vector quantities. The former is referred to as the *scalar product* of two vectors and the latter the *vector product*.

The scalar product of two vectors **a** and **b** is defined as the product of the magnitudes of **a** and **b** and the cosine of the angle θ between them, the appropriate angle being established by localizing as in Fig. 7.3 of Section 2.

It is expressed in the form

$$\mathbf{a} \cdot \mathbf{b} = ab \cos \theta = \mathbf{b} \cdot \mathbf{a} \tag{5}$$

(read as **a** *dot* **b**), and is seen to be equal to the product of the length of one vector and the projected length of the other upon it. (See Fig. 7.9 in which $OA' = a \cos \theta$, $OB' = b \cos \theta$, and $\mathbf{a} \cdot \mathbf{b} = \overline{OA} \cdot \overline{OB'} = \overline{OB} \cdot \overline{OA'}$.)

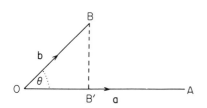 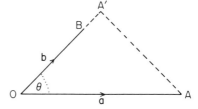

Fig. 7.9

The validity of the distributive law can be demonstrated by reference to Fig. 7.10. For,

$$(\mathbf{b} \cdot \mathbf{c}) \cdot \mathbf{a} = \overline{OD} \cdot \mathbf{a} = (OD')(OA)$$
$$= (OB' + B'D')(OA)$$
$$= (OB' + OC')(OA)$$
$$= (OB')(OA) + (OC')(OA)$$
$$= \mathbf{b} \cdot \mathbf{a} + \mathbf{c} \cdot \mathbf{a}$$

and similarly,

$$\mathbf{a} \cdot (\mathbf{b} + \mathbf{c}) = \mathbf{a} \cdot \mathbf{b} + \mathbf{a} \cdot \mathbf{c}$$

from the commutative property.

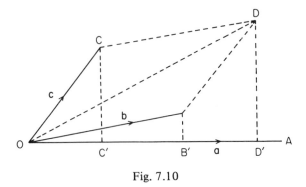

Fig. 7.10

The definition of a scalar product provides a convenient test of perpendicularity, for if **a** is perpendicular to **b**, then **a** · **b** = 0. Alternatively, if **a** and **b** are parallel, then **a** · **b** = ab. In particular, we note that for the unit vectors **i**, **j** and **k**,

$$\mathbf{i} \cdot \mathbf{j} = \mathbf{j} \cdot \mathbf{k} = \mathbf{k} \cdot \mathbf{i} = \mathbf{j} \cdot \mathbf{i} = \mathbf{k} \cdot \mathbf{j} = \mathbf{i} \cdot \mathbf{k} = 0,$$

and

$$\mathbf{i} \cdot \mathbf{i} = \mathbf{j} \cdot \mathbf{j} = \mathbf{k} \cdot \mathbf{k} = 1.$$

Applying these results to the vectors

$$\mathbf{a} = a_1 \mathbf{i} + a_2 \mathbf{j} + a_3 \mathbf{k}, \quad \mathbf{b} = b_1 \mathbf{i} + b_2 \mathbf{j} + b_3 \mathbf{k}$$

we find that

$$\mathbf{a} \cdot \mathbf{b} = (a_1 \mathbf{i} + a_2 \mathbf{j} + a_3 \mathbf{k}) \cdot (b_1 \mathbf{i} + b_2 \mathbf{j} + b_3 \mathbf{k})$$

$$= a_1 b_1 + a_2 b_2 + a_3 b_3 \tag{6}$$

which is the component form of the scalar product.

It follows at once, from equations (5) and (6), that

$$\cos \theta = \frac{\mathbf{a} \cdot \mathbf{b}}{|\mathbf{a}| |\mathbf{b}|} = \frac{a_1 b_1 + a_2 b_2 + a_3 b_3}{ab}. \tag{7}$$

Example 4

Determine the angle between the vectors **a** = **i** − 5**k** and **b** = −**i** + 2**j** − **k**.

The magnitudes of the given vectors can be derived as in Section 4, or, by using the result **a** · **a** = a^2, **b** · **b** = b^2. Using either technique we find $a = \sqrt{(26)}$, $b = \sqrt{6}$, and equation (7) gives

$$\cos \theta = \frac{(\mathbf{i} - 5\mathbf{k}) \cdot (-\mathbf{i} + 2\mathbf{j} - \mathbf{k})}{\sqrt{(26)}\sqrt{6}} = \frac{(1)(-1) + (0)(2) + (-5)(-1)}{2\sqrt{(39)}}$$

$$= \frac{2}{\sqrt{(39)}},$$

from which θ may be determined.

6. Vector multiplication – the vector product

With the angle θ defined as in Section 5, the vector product of two vectors **a** and **b** is defined to be that vector which has magnitude $ab \sin \theta$ and direction perpendicular to both **a** and **b**, and in the sense of a right-handed screw turn from **a** to **b**.

It is expressed in the form

$$\mathbf{a} \wedge \mathbf{b} = \hat{\mathbf{n}} \, ab \sin \theta \qquad (8)$$

(or **a** × **b**), and read as **a** *cross* **b**, $\hat{\mathbf{n}}$ being a unit vector having the prescribed direction (see Fig. 7.11).

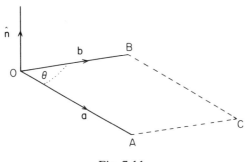

Fig. 7.11

It will be useful to note that the magnitude $ab \sin \theta$ is equal to the area of the parallelogram OACB in which $\overline{OA} = \mathbf{a}$ and $\overline{OB} = \mathbf{b}$, which in turn is equal to twice the area of the triangle OAB.

Since a rotation of **b** through θ to **a** produces, according to the above definition, a vector of direction opposite to that of **a** $\wedge$ **b**, it follows that

$$\mathbf{b} \wedge \mathbf{a} = -\mathbf{a} \wedge \mathbf{b},$$

and that vector products are non-commutative.

For the special case in which **a** and **b** are parallel, $\sin \theta = 0$ and **a** $\wedge$ **b** = **0**. Conversely, if $\mathbf{a} \neq \mathbf{0}$, $\mathbf{b} \neq \mathbf{0}$, then **a** $\wedge$ **b** = **0** provides an appropriate test for *parallelism*.

From the definition it follows that for any scalar quantity m,

$$m\mathbf{a} \wedge \mathbf{b} = \hat{\mathbf{n}} \, mab \sin \theta = \mathbf{a} \wedge m\mathbf{b}$$

and for the unit vectors **i**, **j** and **k**,

$$\mathbf{i} \wedge \mathbf{i} = \mathbf{j} \wedge \mathbf{j} = \mathbf{k} \wedge \mathbf{k} = \mathbf{0},$$

$$\mathbf{i} \wedge \mathbf{j} = -\mathbf{j} \wedge \mathbf{i} = \mathbf{k},$$

$$\mathbf{j} \wedge \mathbf{k} = -\mathbf{k} \wedge \mathbf{j} = \mathbf{i},$$

$$\mathbf{k} \wedge \mathbf{i} = -\mathbf{i} \wedge \mathbf{k} = \mathbf{j}. \qquad (9)$$

158.

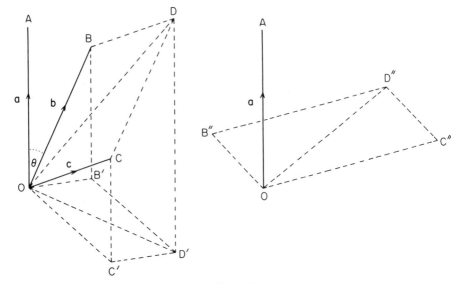

Fig. 7.12

To establish the distributive law

$$\mathbf{a} \wedge (\mathbf{b} + \mathbf{c}) = \mathbf{a} \wedge \mathbf{b} + \mathbf{a} \wedge \mathbf{c} \tag{10}$$

for any three vectors $\mathbf{a}$, $\mathbf{b}$ and $\mathbf{c}$, consider the projections of $\mathbf{b}$, $\mathbf{c}$ and $\mathbf{b} + \mathbf{c}$ onto a plane perpendicular to $\mathbf{a}$ (see Fig. 7.12).

With the notation of equation (8),

$$\mathbf{a} \wedge \mathbf{b} = \hat{\mathbf{n}}\, ab \sin \theta = \hat{\mathbf{n}}(a)(OB') = \mathbf{a} \wedge \overline{OB}'$$

and similarly

$$\mathbf{a} \wedge \mathbf{c} = \mathbf{a} \wedge \overline{OC}', \quad \mathbf{a} \wedge (\mathbf{b} + \mathbf{c}) = \mathbf{a} \wedge \overline{OD}'.$$

Now the vectors $\mathbf{a} \wedge \overline{OB}'$, $\mathbf{a} \wedge \overline{OC}'$ and $\mathbf{a} \wedge \overline{OD}'$ are perpendicular to $\mathbf{a}$ and $\overline{OB}'$, $\mathbf{a}$ and $\overline{OC}'$, $\mathbf{a}$ and $\overline{OD}'$ respectively, and in the sense of the same right-handed system. Consequently, the effect of forming these three vector products with $\mathbf{a}$ induces a $90°$ rotation about $\mathbf{a}$ of the parallelogram $OB'D'C'$ to the position $OB''D''C''$, in which

$$\overline{OB}'' = \mathbf{a} \wedge \overline{OB}', \quad \overline{OC}'' = \mathbf{a} \wedge \overline{OC}', \quad \overline{OD}'' = \mathbf{a} \wedge \overline{OD}'$$

where, $\overline{OD}'' = \overline{OB}'' + \overline{OC}''$, by the parallelogram law.

Thus,

$$\mathbf{a} \wedge \overline{OD}' = \mathbf{a} \wedge \overline{OB}' + \mathbf{a} \wedge \overline{OC}'$$

i.e.

$$\mathbf{a} \wedge (\mathbf{b} + \mathbf{c}) = \mathbf{a} \wedge \mathbf{b} + \mathbf{a} \wedge \mathbf{c}.$$

The component form of a vector product may now be obtained. For, if $a = a_1 i + a_2 j + a_3 k$ and $b = b_1 i + b_2 j + b_3 k$, then

$$a \wedge b = (a_1 i + a_2 j + a_3 k) \wedge (b_1 i + b_2 j + b_3 k)$$

$$= (a_2 b_3 - a_3 b_2)i + (a_3 b_1 - a_1 b_3)j + (a_1 b_2 - a_2 b_1)k$$

from equations (9).

In determinantal form the vector product becomes

$$a \wedge b = \begin{vmatrix} i & j & k \\ a_1 & a_2 & a_3 \\ b_1 & b_2 & b_3 \end{vmatrix} \tag{11}$$

Example 5

Evaluate $a \wedge b$ if $a = 3i + j - k$ and $b = 2i - j + 3k$.

Using the determinantal form,

$$a \wedge b = \begin{vmatrix} i & j & k \\ 3 & 1 & -1 \\ 2 & -1 & 3 \end{vmatrix} = 2i - 11j - 5k.$$

Example 6

Determine the area of the triangle having vertices at the points $A(1, 2, -1)$, $B(2,1, 1)$ and $C(1, -3, 2)$.

If O is the origin of coordinates, the position vectors of the points A, B, C are

$$\overline{OA} = i + 2j - k, \quad \overline{OB} = 2i + j + k, \quad \overline{OC} = i - 3j + 2k,$$

so that

$$\overline{AB} = \overline{OB} - \overline{OA} = i - j + 2k$$

and

$$\overline{AC} = \overline{OC} - \overline{OA} = -5j + 3k.$$

Thus

$$\overline{AB} \wedge \overline{AC} = \begin{vmatrix} i & j & k \\ 1 & -1 & 2 \\ 0 & -5 & 3 \end{vmatrix} = 7i - 3j - 5k$$

and the area of the triangle ABC is, from Section 6,

$$\tfrac{1}{2} | \overline{AB} \wedge \overline{AC} | = \tfrac{1}{2}\sqrt{\{(7)^2 + (-3)^2 + (-5)^2\}} = \tfrac{1}{2}\sqrt{(83)}.$$

Example 7

Prove that $(a + b + c) \wedge (a - b - c) = 2(b + c) \wedge a$.

We have

$$(a + b + c) \wedge (a - b - c) = a \wedge a - a \wedge b - a \wedge c + b \wedge a - b \wedge b - b \wedge c$$
$$+ c \wedge a - c \wedge b - c \wedge c$$
$$= -a \wedge b - a \wedge c + b \wedge a - b \wedge c + c \wedge a - c \wedge b$$
$$(\text{since } a \wedge a = b \wedge b = c \wedge c = 0)$$
$$= b \wedge a + c \wedge a + b \wedge a + c \wedge b + c \wedge a - c \wedge b$$
$$= 2(b \wedge a + c \wedge a)$$
$$= 2(b + c) \wedge a.$$

7. Scalar triple product

The scalar product of **a** with **b** ∧ **c** is called a *scalar triple product* and is denoted by **a** · (**b** ∧ **c**) or [**a**, **b**, **c**]. It may be interpreted geometrically, as the volume of the parallelepiped having **a**, **b** and **c** as adjoining edges (see Fig. 7.13).

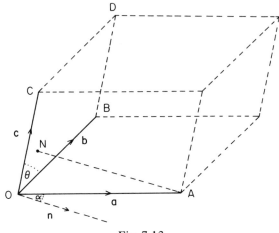

Fig. 7.13

With θ and $\hat{n}$ as shown, the vector **b** ∧ **c** has magnitude $bc \sin \theta$ and direction $\hat{n}$. Thus, if α represents the angle between **a** and $\hat{n}$, then

$$\mathbf{a} \cdot (\mathbf{b} \wedge \mathbf{c}) = (a)(bc \sin \theta)\cos \alpha$$
$$= (bc \sin \theta)(a \cos \alpha)$$
$$= (\text{area of parallelogram OBDC}) \times (\text{'height' AN of parallelepiped})$$
$$= \text{volume of parallelepiped}$$

establishing the result.

Similarly, we find that the volume may equally be represented by $\mathbf{b} \cdot (\mathbf{c} \wedge \mathbf{a})$ or $\mathbf{c} \cdot (\mathbf{a} \wedge \mathbf{b})$, and therefore

$$[\mathbf{a}, \mathbf{b}, \mathbf{c}] = [\mathbf{c}, \mathbf{a}, \mathbf{b}] = [\mathbf{b}, \mathbf{c}, \mathbf{a}]. \tag{12}$$

Also, since

$$\mathbf{c} \cdot (\mathbf{a} \wedge \mathbf{b}) = (\mathbf{a} \wedge \mathbf{b}) \cdot \mathbf{c}$$

from the commutative property, it follows that

$$\mathbf{a} \cdot (\mathbf{b} \wedge \mathbf{c}) = (\mathbf{a} \wedge \mathbf{b}) \cdot \mathbf{c}$$

showing that the value of the scalar triple product is unchanged if the dot and cross symbols are interchanged. However, if any two of the vectors are interchanged the sign of the triple product will change as a consequence of the non-commutative property of vector multiplication, referred to in Section 6.

Thus,

$$[\mathbf{b}, \mathbf{a}, \mathbf{c}] = -[\mathbf{a}, \mathbf{b}, \mathbf{c}]$$

and so on.

Furthermore, we note that the volume of the parallelepiped is zero when $\mathbf{a}$, $\mathbf{b}$ and $\mathbf{c}$ are coplanar, in which case $[\mathbf{a}, \mathbf{b}, \mathbf{c}] = 0$.

To derive the component form, let

$$\mathbf{a} = a_1\mathbf{i} + a_2\mathbf{j} + a_3\mathbf{k}, \quad \mathbf{b} = b_1\mathbf{i} + b_2\mathbf{j} + b_3\mathbf{k}, \quad \mathbf{c} = c_1\mathbf{i} + c_2\mathbf{j} + c_3\mathbf{k}.$$

Then,

$$\mathbf{a} \cdot (\mathbf{b} \wedge \mathbf{c}) = \mathbf{a} \cdot \begin{vmatrix} \mathbf{i} & \mathbf{j} & \mathbf{k} \\ b_1 & b_2 & b_3 \\ c_1 & c_2 & c_3 \end{vmatrix}$$

$$= (a_1\mathbf{i} + a_2\mathbf{j} + a_3\mathbf{k}) \cdot \{(b_2c_3 - b_3c_2)\mathbf{i} + (b_3c_1 - b_1c_3)\mathbf{j} + (b_1c_2 - b_2c_1)\mathbf{k}\}$$

$$= a_1(b_2c_3 - b_3c_2) + a_2(b_3c_1 - b_1c_3) + a_3(b_1c_2 - b_2c_1)$$

$$= \begin{vmatrix} a_1 & a_2 & a_3 \\ b_1 & b_2 & b_3 \\ c_1 & c_2 & c_3 \end{vmatrix} \tag{13}$$

which will be zero if any two of the vectors are equal or parallel.

Example 8

If $\overline{OA} = -2\mathbf{i} + 2\mathbf{j} + \mathbf{k}$, $\overline{OB} = 2\mathbf{i} - 2\mathbf{j} + 3\mathbf{k}$ and $\overline{OC} = 2\mathbf{i} + \mathbf{j} + \mathbf{k}$, determine the volume of the tetrahedron OABC.

Referring to Fig. 7.13, we observe that the area OAB of the base of the tetrahedron is equal to one half that of the base OADB of the parallelepiped. The volume of the tetrahedron, viz., $\frac{1}{3}$ (base area) × (height), is therefore

$V = \frac{1}{6}$ (volume of the parallelepiped)

$$= \frac{1}{6} \, [\overline{OA}, \overline{OB}, \overline{OC}] = \frac{1}{6} \begin{vmatrix} -2 & 2 & 1 \\ 2 & -2 & 3 \\ 2 & 1 & 1 \end{vmatrix} = 4 \text{ cubic units.}$$

Example 9

If $d = a + b + c$, prove that $[d, b, c]$, $[d, c, a]$ and $[d, a, b]$ have equal values.
    Forming the scalar product of both sides of the given equation with $b \wedge c$, gives

$d \cdot (b \wedge c) = (a + b + c) \cdot (b \wedge c)$

$\quad = a \cdot (b \wedge c) + b \cdot (b \wedge c) + c \cdot (b \wedge c)$

$\quad = a \cdot (b \wedge c)$

since a scalar triple product containing two identical vectors is zero.
    Similarly, the scalar products of d with $c \wedge a$ and $a \wedge b$ respectively, give

$d \cdot (c \wedge a) = b \cdot (c \wedge a)$

and

$d \cdot (a \wedge b) = c \cdot (a \wedge b).$

    Hence

$[d, b, c] = [d, c, a] = [d, a, b]$

using the cyclic property of scalar triple products (see equation (12)).

8. Vector triple product

The vector product of a with $b \wedge c$ is called a *vector triple product* and is denoted by $a \wedge (b \wedge c)$. We note that since $b \wedge c$ is perpendicular to both b and c, and $a \wedge (b \wedge c)$ is perpendicular to both $b \wedge c$ and a, therefore $a \wedge (b \wedge c)$ is coplanar with b and c.

    In component form

$$a \wedge (b \wedge c) = a \wedge \begin{vmatrix} i & j & k \\ b_1 & b_2 & b_3 \\ c_1 & c_2 & c_3 \end{vmatrix}$$

$\quad = (a_1 i + a_2 j + a_3 k) \wedge \{(b_2 c_3 - b_3 c_2)i + (b_3 c_1 - b_1 c_3)j$
$\quad \quad + (b_1 c_2 - b_2 c_1)k\}$

$\quad = i\{a_2(b_1 c_2 - b_2 c_1) + a_3(b_1 c_3 - b_3 c_1)\} + \text{terms in } j \text{ and } k$

$\quad = i\{b_1(a_2 c_2 + a_3 c_3) - c_1(a_2 b_2 + a_3 b_3)\} + \cdots$

$$= i\{b_1(a_1c_1 + a_2c_2 + a_3c_3) - c_1(a_1b_1 + a_2b_2 + a_3b_3)\} + \cdots$$

$$= i\{b_1(a \cdot c) - c_1(a \cdot b)\} + \cdots$$

$$= (b_1i + b_2j + b_3k)(a \cdot c) - (c_1i + c_2j + c_3k)(a \cdot b)$$

giving

$$a \wedge (b \wedge c) = b(a \cdot c) - c(a \cdot b) \tag{14}$$

Particular care should be taken when expanding vector triple products according to equation (14), the position of the brackets being of paramount importance. For example,

$$(a \wedge b) \wedge c = -c \wedge (a \wedge b)$$

$$= -a(c \cdot b) + b(c \cdot a) \tag{15}$$

$$\neq a \wedge (b \wedge c)$$

Example 10

If $a = i + 2j - k$, $b = 3i - j + k$ and $c = -2i + j + 4k$, evaluate $a \wedge (b \wedge c)$ and $(a \wedge b) \wedge c$.

Now

$$a \cdot b = (i + 2j - k) \cdot (3i - j + k) = 0,$$

$$a \cdot c = (i + 2j - k) \cdot (-2i + j + 4k) = -4,$$

and equation (14) gives

$$a \wedge (b \wedge c) = -4b = -12i + 4j - 4k.$$

Also

$$c \cdot b = (-2i + j + 4k) \cdot (3i - j + k) = -3$$

and equation (15) gives

$$(a \wedge b) \wedge c = 3a - 4b = -9i + 10j - 7k.$$

Example 11

Derive expansions for: (i) $(a \wedge b) \cdot (c \wedge d)$; and (ii) $(a \wedge b) \wedge (c \wedge d)$.

From Section 7 and using equation (14),

(i) $(a \wedge b) \cdot (c \wedge d) = a \cdot \{b \wedge (c \wedge d)\}$

$$= a \cdot \{c(b \cdot d) - d(b \cdot c)\}$$

$$= (a \cdot c)(b \cdot d) - (a \cdot d)(b \cdot c)$$

which is a scalar quantity, and

(ii) $(a \wedge b) \wedge (c \wedge d) = c\{(a \wedge b) \cdot d\} - d\{(a \wedge b) \cdot c\}$

$$= c[a, b, d] - d[a, b, c]$$

which is a vector quantity.

164

Example 12

Determine the angular momentum of a rigid body rotating about a fixed point.

Suppose that the body is rotating about a fixed point O with instantaneous angular velocity ω, about an instantaneous axis $\overline{ON}$ (see Fig. 7.14).

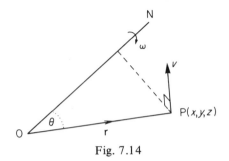

Fig. 7.14

If the direction $\overline{ON}$ is such that a right-handed screw advancing along $\overline{ON}$ rotates in the same sense as ω, then the velocity of a point P fixed in the body is

$$v = \omega r \sin \theta$$

where $\overline{OP} = \mathbf{r}$ and $N\hat{O}P = \theta$.

Denoting the unit vector in the direction of $\overline{ON}$ by $\hat{n}$, the corresponding vector form of the velocity may be expressed as

$$\mathbf{v} = \omega r(\hat{n} \wedge \hat{r}) = (\omega\hat{n}) \wedge (r\hat{r}) = \boldsymbol{\omega} \wedge \mathbf{r},$$

$\boldsymbol{\omega}$ being the *vector angular velocity*, and the *linear momentum* of a particle of mass m situated at the point $P \equiv P(x, y, z)$ is

$$m\mathbf{v} = m\boldsymbol{\omega} \wedge \mathbf{r}.$$

If we regard the body as a collection of such particles, the *vector angular momentum* of the whole system about O is defined to be

$$\mathbf{H} = \Sigma\, m\mathbf{r} \wedge (\boldsymbol{\omega} \wedge \mathbf{r})$$

the summation extending throughout the system.

Expanding the vector triple product and writing $\mathbf{r} = x\mathbf{i} + y\mathbf{j} + z\mathbf{k}$, $\boldsymbol{\omega} = \omega_x\mathbf{i} + \omega_y\mathbf{j} + \omega_z\mathbf{k}$, we find that

$$\mathbf{H} = \Sigma\, m\{\boldsymbol{\omega}(\mathbf{r} \cdot \mathbf{r}) - \mathbf{r}(\mathbf{r} \cdot \boldsymbol{\omega})\}$$
$$= \Sigma\, m\{\boldsymbol{\omega}r^2 - \mathbf{r}(\mathbf{r} \cdot \boldsymbol{\omega})\}$$
$$= \Sigma\, m\{\omega_x\mathbf{i} + \omega_y\mathbf{j} + \omega_z\mathbf{k})(x^2 + y^2 + z^2) - (x\mathbf{i} + y\mathbf{j} + z\mathbf{k})(x\omega_x + y\omega_y + z\omega_z)\}.$$

Thus, if $\mathbf{H} = H_x\mathbf{i} + H_y\mathbf{j} + H_z\mathbf{k}$, then

$$H_x = \Sigma\, m\{\omega_x(x^2 + y^2 + z^2) - x(x\omega_x + y\omega_y + z\omega_z)\}$$
$$= \omega_x \Sigma\, m(y^2 + z^2) - \omega_y \Sigma\, m(xy) - \omega_z \Sigma\, m(xz)$$
$$= A\omega_x - F\omega_y - E\omega_z$$

and similarly

$$H_y = B\omega_y - D\omega_z - F\omega_x, \quad H_z = C\omega_z - E\omega_x - D\omega_y,$$

the expressions

$$A = \Sigma m(y^2 + z^2), \quad B = \Sigma m(z^2 + x^2), \quad C = \Sigma m(x^2 + y^2)$$

being the moments of inertia of the system about the x-, y-, z-axes respectively. The expressions

$$D = \Sigma myz, \quad E = \Sigma mzx, \quad F = \Sigma mxy$$

are called the *products of inertia* of the system about the pairs of axes (Oy, Oz), (Oz, Ox), (Ox, Oy) respectively, the summation signs being replaced by integral signs when dealing with continuous distributions.

The results can conveniently be expressed in matrix form, to give

$$\begin{bmatrix} H_x \\ H_y \\ H_z \end{bmatrix} = \begin{bmatrix} A & -F & -E \\ -F & B & -D \\ -E & -D & C \end{bmatrix} \begin{bmatrix} \omega_x \\ \omega_y \\ \omega_z \end{bmatrix}$$

H_x, H_y, H_z representing the physical components of angular momentum about the x-, y-, z-axes respectively.

9. Differentiation of a vector

Suppose that a vector $\mathbf{r}$ is a function of some scalar parameter t, i.e. $\mathbf{r} = \mathbf{r}(t)$, and that as t varies $\mathbf{r}$ traces out some curve in space (see Fig. 7.15).

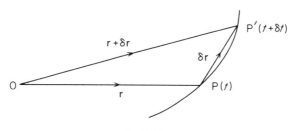

Fig. 7.15

If an increment δt in t produces a corresponding increment $\delta \mathbf{r}$ in $\mathbf{r}$, then $\mathbf{r} + \delta \mathbf{r} = \mathbf{r}(t + \delta t)$ and the change in $\mathbf{r}$ due to the change in t, is

$$\delta \mathbf{r} = \mathbf{r}(t + \delta t) - \mathbf{r}(t).$$

The existence of the derivative of the vector function $\mathbf{r}(t)$ is dependent upon the existence of the limit

$$\lim_{\delta t \to 0} \frac{\mathbf{r}(t + \delta t) - \mathbf{r}(t)}{\delta t}$$

and, if it exists, we write

$$\frac{d\mathbf{r}}{dt} = \lim_{\delta t \to 0} \frac{\delta \mathbf{r}}{\delta t} = \lim_{\delta t \to 0} \frac{\mathbf{r}(t + \delta t) - \mathbf{r}(t)}{\delta t}$$

in accordance with the notation of Chapter 3.

Of particular relevance is the case in which $\mathbf{r}$ is the position vector of a moving particle (see Example 12). If the particle is at P at time t and at P' at time $t + \delta t$, where $\overline{PP'} = \delta \mathbf{r}$ (Fig. 7.15), the value of the quotient approaches a limiting value as $\delta t \to 0$, which represents the velocity $\mathbf{v}$ of the particle at time t.

We write

$$\mathbf{v} = \lim_{\delta t \to 0} \left[\frac{\delta \mathbf{r}}{\delta t} \right] = \frac{d\mathbf{r}}{dt}.$$

In general $d\mathbf{r}/dt$ will be a function of t and may be differentiated with respect to t to give, in the case of the moving particle, the acceleration $\mathbf{a}$ at time t, viz.,

$$\mathbf{a} = \frac{d\mathbf{v}}{dt} = \frac{d^2\mathbf{r}}{dt^2}.$$

Since the definition of the derivative of a vector is similar to that of the derivative of a scalar, it follows from Chapter 3 that if $\mathbf{r} = r_1\mathbf{i} + r_2\mathbf{j} + r_3\mathbf{k}$, where r_1, r_2, r_3 are functions of t, then

$$\frac{d\mathbf{r}}{dt} = \frac{dr_1}{dt}\mathbf{i} + \frac{dr_2}{dt}\mathbf{j} + \frac{dr_3}{dt}\mathbf{k}$$

and similarly, if $\mathbf{s} = s_1\mathbf{i} + s_2\mathbf{j} + s_3\mathbf{k}$ is a differentiable vector function of t, then

$$\frac{d}{dt}(\mathbf{r} + \mathbf{s}) = \frac{d\mathbf{r}}{dt} + \frac{d\mathbf{s}}{dt}.$$

Differentiation of products may be effected by resorting to component form. For example, if ϕ is a scalar function of t, then

$$\frac{d}{dt}(\phi\mathbf{r}) = \frac{d}{dt}\{(\phi r_1)\mathbf{i} + (\phi r_2)\mathbf{j} + (\phi r_3)\mathbf{k}\}$$

$$= \left[\frac{d\phi}{dt}r_1 + \phi\frac{dr_1}{dt}\right]\mathbf{i} + \left[\frac{d\phi}{dt}r_2 + \phi\frac{dr_2}{dt}\right]\mathbf{j} + \left[\frac{d\phi}{dt}r_3 + \phi\frac{dr_3}{dt}\right]\mathbf{k}$$

$$= \frac{d\phi}{dt}(r_1\mathbf{i} + r_2\mathbf{j} + r_3\mathbf{k}) + \phi\left[\frac{dr_1}{dt}\mathbf{i} + \frac{dr_2}{dt}\mathbf{j} + \frac{dr_3}{dt}\mathbf{k}\right]$$

$$= \frac{d\phi}{dt}\mathbf{r} + \phi\frac{d\mathbf{r}}{dt}. \tag{16}$$

For scalar and vector products, we have

$$\frac{d}{dt}(\mathbf{r} \cdot \mathbf{s}) = \frac{d}{dt}(r_1 s_1 + r_2 s_2 + r_3 s_3)$$

$$= \left[\frac{dr_1}{dt} s_1 + r_1 \frac{ds_1}{dt}\right] + \left[\frac{dr_2}{dt} s_2 + r_2 \frac{ds_2}{dt}\right] + \left[\frac{dr_3}{dt} s_3 + r_3 \frac{ds_3}{dt}\right]$$

$$= \left[\frac{dr_1}{dt} s_1 + \frac{dr_2}{dt} s_2 + \frac{dr_3}{dt} s_3\right] + \left[r_1 \frac{ds_1}{dt} + r_2 \frac{ds_2}{dt} + r_3 \frac{ds_3}{dt}\right]$$

$$= \frac{d\mathbf{r}}{dt} \cdot \mathbf{s} + \mathbf{r} \cdot \frac{d\mathbf{s}}{dt} , \qquad (17)$$

and

$$\frac{d}{dt}(\mathbf{r} \wedge \mathbf{s}) = \frac{d}{dt}\begin{vmatrix} \mathbf{i} & \mathbf{j} & \mathbf{k} \\ r_1 & r_2 & r_3 \\ s_1 & s_2 & s_3 \end{vmatrix}$$

$$= \frac{d}{dt}\{(r_2 s_3 - r_3 s_2)\mathbf{i} + (r_3 s_1 - r_1 s_3)\mathbf{j} + (r_1 s_2 - r_2 s_1)\mathbf{k}$$

$$= \left[\frac{dr_2}{dt} s_3 + r_2 \frac{ds_3}{dt} - \frac{dr_3}{dt} s_2 - r_3 \frac{ds_2}{dt}\right]\mathbf{i} + \text{terms in } \mathbf{j} \text{ and } \mathbf{k}$$

$$= \left[\frac{dr_2}{dt} s_3 - \frac{dr_3}{dt} s_2\right]\mathbf{i} + \cdots + \left[r_2 \frac{ds_3}{dt} - r_3 \frac{ds_2}{dt}\right]\mathbf{i} + \cdots$$

$$= \begin{vmatrix} \mathbf{i} & \mathbf{j} & \mathbf{k} \\ \dfrac{dr_1}{dt} & \dfrac{dr_2}{dt} & \dfrac{dr_3}{dt} \\ s_1 & s_2 & s_3 \end{vmatrix} + \begin{vmatrix} \mathbf{i} & \mathbf{j} & \mathbf{k} \\ r_1 & r_2 & r_3 \\ \dfrac{ds_1}{dt} & \dfrac{ds_2}{dt} & \dfrac{ds_3}{dt} \end{vmatrix}$$

$$= \frac{d\mathbf{r}}{dt} \wedge \mathbf{s} + \mathbf{r} \wedge \frac{d\mathbf{s}}{dt} \qquad (18)$$

In fact, apart from the presence of the symbols representing vector multiplication, the order of which should be preserved, the rules for differentiating vectors are very similar to those for differentiating scalars.

Thus, we find

$$\frac{d}{dt}[\mathbf{r}, \mathbf{s}, \mathbf{w}] = [d\mathbf{r}/dt, \mathbf{s}, \mathbf{w}] + [\mathbf{r}, d\mathbf{s}/dt, \mathbf{w}] + [\mathbf{r}, \mathbf{s}, d\mathbf{w}/dt],$$

$$\frac{d}{dt}\{\mathbf{r} \wedge (\mathbf{s} \wedge \mathbf{w})\} = \frac{d\mathbf{r}}{dt} \wedge (\mathbf{s} \wedge \mathbf{w}) + \mathbf{r} \wedge \left[\frac{d\mathbf{s}}{dt} \wedge \mathbf{w}\right] + \mathbf{r} \wedge \left[\mathbf{s} \wedge \frac{d\mathbf{w}}{dt}\right],$$

and so on.

Example 13

Determine $d\mathbf{r}/dt$ and $d^2\mathbf{r}/dt^2$ given that $\mathbf{r} = 3\mathbf{i}\cos 4t + \mathbf{j}(t^2 + 5t - 1) + \mathbf{k}\log_e(2t+1)$.
 Differentiating the components of $\mathbf{r}$ with respect to t, we obtain

$$\frac{d\mathbf{r}}{dt} = -12\mathbf{i}\sin 4t + \mathbf{j}(2t+5) + \mathbf{k}\frac{2}{2t+1}$$

and

$$\frac{d^2\mathbf{r}}{dt^2} = -48\mathbf{i}\cos 4t + 2\mathbf{j} - \frac{4}{(2t+1)^2}\mathbf{k}.$$

Example 14

If $\mathbf{r} = \mathbf{r}(t)$ and $\mathbf{s} = \mathbf{s}(t)$, obtain the second derivatives of: (i) $\mathbf{r}\cdot\mathbf{s}$; and (ii) $\mathbf{r}\wedge\mathbf{s}$ with respect to t.
 Using primes to denote differentiation with respect to t,

(i) $\quad\dfrac{d}{dt}(\mathbf{r}\cdot\mathbf{s}) = \mathbf{r}'\cdot\mathbf{s} + \mathbf{r}\cdot\mathbf{s}', \quad$ from equation (17).

$$\therefore \quad \frac{d^2}{dt^2}(\mathbf{r}\cdot\mathbf{s}) = \frac{d}{dt}(\mathbf{r}'\cdot\mathbf{s}) + \frac{d}{dt}(\mathbf{r}\cdot\mathbf{s}')$$

$$= (\mathbf{r}''\cdot\mathbf{s} + \mathbf{r}'\cdot\mathbf{s}') + (\mathbf{r}'\cdot\mathbf{s}' + \mathbf{r}\cdot\mathbf{s}'')$$

$$= \mathbf{r}''\cdot\mathbf{s} + 2\mathbf{r}'\cdot\mathbf{s}' + \mathbf{r}\cdot\mathbf{s}''.$$

(ii) $\quad\dfrac{d}{dt}(\mathbf{r}\wedge\mathbf{s}) = \mathbf{r}'\wedge\mathbf{s} + \mathbf{r}\wedge\mathbf{s}', \quad$ from equation (18).

$$\therefore \quad \frac{d^2}{dt^2}(\mathbf{r}\wedge\mathbf{s}) = \frac{d}{dt}(\mathbf{r}'\wedge\mathbf{s}) + \frac{d}{dt}(\mathbf{r}\wedge\mathbf{s}')$$

$$= (\mathbf{r}''\wedge\mathbf{s} + \mathbf{r}'\wedge\mathbf{s}') + (\mathbf{r}'\wedge\mathbf{s}' + \mathbf{r}\wedge\mathbf{s}'')$$

$$= \mathbf{r}''\wedge\mathbf{s} + 2\mathbf{r}'\wedge\mathbf{s}' + \mathbf{r}\wedge\mathbf{s}''.$$

PROBLEMS

1. Find the magnitude of the vectors

(i) $\mathbf{i} + 3\mathbf{j} - 2\mathbf{k}$, (ii) $-3\mathbf{i} + \mathbf{j} + 4\mathbf{k}$, (iii) $2\mathbf{i} - 5\mathbf{k}$.

2. Determine the unit vectors for the following:

(i) $3\mathbf{i} + 6\mathbf{j} - 2\mathbf{k}$, (ii) $-4\mathbf{i} + \mathbf{j} - 8\mathbf{k}$, (iii) $2\mathbf{i} - 3\mathbf{j} + 6\mathbf{k}$.

3. Find a vector having direction opposite to $7\mathbf{i} + 4\mathbf{j} - 4\mathbf{k}$ and magnitude 18.

4. If $\mathbf{a} = \mathbf{i} - 2\mathbf{j} + \mathbf{k}$, $\mathbf{b} = -\mathbf{i} + \mathbf{j} - 3\mathbf{k}$, $\mathbf{c} = 2\mathbf{i} + \mathbf{j}$, determine the magnitude and direction cosines of $\mathbf{a} - \mathbf{b} - \mathbf{c}$.

5. Determine the equation of the straight line having direction cosines $(-7/9, 4/9, 4/9)$ and passing through the point $(4, 1, -2)$.

6. Determine the equation of the plane containing the point $(4, -1, 3)$, given that the normal to the plane is parallel to the vector $\mathbf{i} + 5\mathbf{j} - 3\mathbf{k}$.

7. If $\mathbf{a}, \mathbf{b}$ and $\mathbf{c}$ are defined as in Question 4, evaluate

(i) $\mathbf{a} \cdot \mathbf{b}$,   (ii) $\mathbf{b} \cdot \mathbf{c}$,   (iii) $(\mathbf{a} + \mathbf{b}) \cdot (\mathbf{a} + \mathbf{c})$.

8. If $\mathbf{a}, \mathbf{b}, \mathbf{c}$ are mutually orthogonal vectors having magnitudes 4, 1, 8 respectively, determine the magnitude of $\mathbf{a} + \mathbf{b} + \mathbf{c}$.

9. Show that the vectors $3\mathbf{i} + \mathbf{j} - 2\mathbf{k}$, $2\mathbf{i} + 4\mathbf{j} + 5\mathbf{k}$, are perpendicular. Determine a unit vector perpendicular to both vectors.

10. Evaluate

(i) $(2\mathbf{i} - 3\mathbf{k}) \wedge (\mathbf{i} + 7\mathbf{j} + 2\mathbf{k})$,   (ii) $(-3\mathbf{i} + \mathbf{j} - \mathbf{k}) \wedge (-\mathbf{i} + 4\mathbf{j} + 3\mathbf{k})$,

(iii) $(5\mathbf{i} - \mathbf{j} + \mathbf{k}) \wedge (-2\mathbf{i} + 7\mathbf{k})$.

11. Prove that

(i) $(3\mathbf{a} - 2\mathbf{b}) \wedge (2\mathbf{b} - \mathbf{a}) = 4\mathbf{b} \wedge \mathbf{a}$

(ii) $(2\mathbf{a} + \mathbf{b} - \mathbf{c}) \wedge (\mathbf{a} - 3\mathbf{b} + 2\mathbf{c}) = 5\mathbf{a} \wedge \mathbf{c} + 7\mathbf{b} \wedge \mathbf{a} + \mathbf{c} \wedge \mathbf{b}$.

12. If $\mathbf{a} = 2\mathbf{i} + 2\mathbf{j} - \mathbf{k}$, $\mathbf{b} = 2\mathbf{i} + 3\mathbf{j} - \mathbf{k}$, $\mathbf{c} = \mathbf{i} - \mathbf{j} + 3\mathbf{k}$, evaluate

(i) $\mathbf{a} \cdot (\mathbf{b} \wedge \mathbf{c})$,   (ii) $\mathbf{a} \wedge (\mathbf{b} \wedge \mathbf{c})$.

13. Prove that

(i) $(\mathbf{a} \wedge \mathbf{b})^2 + (\mathbf{a} \cdot \mathbf{b})^2 = a^2 b^2$,

(ii) $\mathbf{a} \wedge \{(\mathbf{a} \wedge \mathbf{b}) \wedge (\mathbf{c} \wedge \mathbf{a})\} = \mathbf{0}$,

(iii) $[(\mathbf{a} \wedge \mathbf{b}), (\mathbf{b} \wedge \mathbf{c}), (\mathbf{c} \wedge \mathbf{a})] = [\mathbf{a}, \mathbf{b}, \mathbf{c}]^2$.

14. If $\mathbf{a} \wedge \mathbf{x} = \mathbf{a} \wedge \mathbf{b}$ and $\mathbf{a} \cdot (\mathbf{x} - \mathbf{a} - \mathbf{b}) = 2a^2$, show that

$\mathbf{x} = 3\mathbf{a} + \mathbf{b}$.

15. A crystal has a unit cell, in the form of a cube, which consists of an atom C at the centre of the cube and four other atoms A_1, A_2, A_3, A_4 positioned on the edges. Rectangular coordinate axes $Oxyz$ are chosen with origin O at one corner of the cube, so that Ox, Oy and Oz are parallel to the edges. If the edge of the cube is of length 6 Å, and $\overline{OA}_1 = 4\mathbf{i}$, $\overline{OA}_2 = 6\mathbf{i} + 2\mathbf{j}$, $\overline{OA}_3 = \mathbf{i} + 6\mathbf{j} + 6\mathbf{k}$, $\overline{OA}_4 = 5\mathbf{j} + 6\mathbf{k}$, determine the distance of the A_2 atom from:

(i)   each of the A_1, A_3, A_4 atoms; and

(ii)   the C atom.

16. The electronic transition moment $\mathbf{M}$ due to four charges having position vectors $\mathbf{r}_1, \mathbf{r}_2, \mathbf{r}_3, \mathbf{r}_4$, corresponding to points $(3, 1, 1), (2, 1, 1), (0, 0, 2)$ and $(-1, -2, 2)$ respectively, is given by

$$\mathbf{M} = a(\mathbf{r}_1 - \mathbf{r}_4) + b(\mathbf{r}_2 - \mathbf{r}_3),$$

a and b being constants.

Show that

$$|\mathbf{M}|^2 = a^2 + 2ab + 5b^2.$$

17. A rigid body is rotating about a fixed point $(-4, 1, 2)$, with angular velocity ω about an instantaneous axis having the direction of the vector $2\mathbf{i} - \mathbf{j} + \mathbf{k}$.

Determine the magnitude and direction of the velocity of a point within the body having coordinates $(1, 2, 3)$.

18. If $\mathbf{r}(t) = (t^2 - 3)\mathbf{i} + (2t + 1)\mathbf{j} + (t^2 + t + 3)\mathbf{k}$, determine

(i) $\dfrac{d\mathbf{r}}{dt}$, (ii) $\dfrac{d}{dt}(\mathbf{r} \cdot \mathbf{r})$, (iii) $\dfrac{d}{dt}\left[\mathbf{r} \wedge \dfrac{d\mathbf{r}}{dt}\right]$.

19. The position of an electron at any time t is specified by the vector $\mathbf{r} = \mathbf{r}(t)$, where

$$e^t\mathbf{r} = \mathbf{i} \sin t + \mathbf{j} \cos t + \mathbf{k}.$$

Show that the ratio of the magnitude of its velocity to that of its acceleration is $\sqrt{3} : \sqrt{5}$.

CHAPTER 8

Ordinary Differential Equations

The *order* of a differential equation is defined to be the order of the highest derivative appearing in it. The *degree* of a differential equation is the degree of that highest derivative when the differential coefficients and dependent variable are free from all radicals and fractions.

Thus, the equations

$$\frac{dy}{dx} = \frac{\sqrt{y}}{x^2 + 2}, \quad \frac{d^2 y}{dx^2} = \left\{ 1 + \left(\frac{dy}{dx}\right)^2 \right\}^{3/2},$$

$$\frac{d^3 y}{dx^3} - 7 \left(\frac{dy}{dx}\right)^2 + y = x^2 \cos x,$$

are respectively of the first order, second degree; second order, second degree; and third order, first degree.

A differential equation is defined to be *linear* if it is linear in its dependent variable and all its derivatives.

The *general solution* or *complete primitive* of the nth order linear differential equation

$$\frac{d^n y}{dx^n} + f_1(x) \frac{d^{n-1} y}{dx^{n-1}} + f_2(x) \frac{d^{n-2} y}{dx^{n-2}} + \cdots + f_{n-1}(x) \frac{dy}{dx} + f_n(x)y = g(x),$$

where $f_1(x), f_2(x), \ldots, f_n(x)$, and $g(x)$, are functions of x only, contains n arbitrary constants. A *singular* or *particular* solution occurs when there are less than n arbitrary constants.

We now determine the solutions of several types of first order differential equations.

1. Variables separable

Consider the equation

$$f(x) \frac{dy}{dx} = g(y),$$

where $f(x)$ is a function of x only, and $g(y)$ is a function of y only.

Since the variables are separable we obtain the general solution by direct integration, giving

$$\int \frac{dy}{g(y)} = \int \frac{dx}{f(x)} + A,$$

where A is an arbitrary constant.

Example 1

Solve the differential equation

$$(x^2 - 1)\frac{dy}{dx} = 2y.$$

The solution is given by

$$\int \frac{dy}{2y} = \int \frac{dx}{x^2 - 1} + \text{constant}$$

$$= \frac{1}{2}\int\left(\frac{1}{x-1} - \frac{1}{x+1}\right)dx + \text{constant}$$

i.e.

$$\tfrac{1}{2}\log_e y = \tfrac{1}{2}\log_e(x - 1) - \tfrac{1}{2}\log_e(x + 1) + \text{constant}.$$

As each term is logarithmic it is convenient to express the constant as a logarithm. Hence we write

$$\tfrac{1}{2}\log_e y = \tfrac{1}{2}\log_e(x - 1) - \tfrac{1}{2}\log_e(x + 1) + \tfrac{1}{2}\log_e A,$$

where A is an arbitrary constant, giving

$$\log_e y = \log_e \frac{A(x - 1)}{(x + 1)}$$

and hence,

$$y = \frac{A(x - 1)}{x + 1}.$$

Example 2

Solve the differential equation

$$(3y^4 + 3y^2 + 1)\frac{dy}{dx} + 3(1 + y^2)\sqrt{(x + 4)} = 0.$$

Separating the variables we obtain

$$\int \frac{3y^4 + 3y^2 + 1}{y^2 + 1}\,dy + 3\int\sqrt{(x + 4)}\,dx = \text{constant}.$$

Thus

$$\int \left(3y^2 + \frac{1}{y^2 + 1} \right) dy + 2(x + 4)^{3/2} = \text{constant.}$$

Hence, the general solution is given by

$$y^3 + \tan^{-1} y + 2(x + 4)^{3/2} = \text{constant.}$$

Example 3

In the reversible chemical reaction $A \underset{k_2}{\overset{k_1}{\rightleftharpoons}} B + C$, the amount of A decomposed at time t is x, where

$$\frac{dx}{dt} = k_1 (a - x) - k_2 x^2,$$

a being the initial concentration of A.

If $2k_1 = ak_2$, show that

$$k_1 = \frac{1}{3t} \log_e \left(\frac{a + x}{a - 2x} \right).$$

The equation is

$$\frac{dx}{dt} = k_1 (a - x) - \frac{2k_1}{a} x^2$$

$$= \frac{k_1}{a} (a^2 - ax - 2x^2)$$

$$= \frac{k_1}{a} (a + x)(a - 2x).$$

Separating the variables now gives

$$\int \frac{k_1}{a} dt = \int \frac{dx}{(a + x)(a - 2x)} = \frac{1}{3a} \int \left(\frac{1}{a + x} + \frac{2}{a - 2x} \right) dx$$

i.e.

$$\frac{k_1 t}{a} = \frac{1}{3a} \log_e(a + x) - \frac{1}{3a} \log_e(a - 2x) + A.$$

When $t = 0, x = 0$

$$\therefore \quad A = \frac{1}{3a} \log_e(1) = 0.$$

$$\therefore \quad k_1 t = \frac{1}{3} \log_e \left(\frac{a + x}{a - 2x} \right)$$

2. Homogeneous equations

A function $f(kx, ky)$ of the two variables x and y, is said to be homogeneous and of degree n if it satisfies the relationship

$$f(kx, ky) = k^n f(x, y),$$

where k is some parameter.

For example, the function $f(x, y) = x^4 - 4x^3 y + 3y^4$ may be written as

$$f(x, y) = x^4 \left(1 - 4\frac{y}{x} + 3\frac{y^4}{x^4}\right)$$

$$= x^4 f(1, y/x).$$

Hence $f(x, y)$ is a homogeneous function of x and y, and is of the fourth degree. Let the differential equation

$$f(x, y)\frac{dy}{dx} = g(x, y)$$

be such that $f(x, y)$ and $g(x, y)$ are homogeneous functions of x and y, and both of the same degree n, say.

Taking out the factor x^n, the equation becomes

$$x^n f(1, y/x)\frac{dy}{dx} = x^n g(1, y/x)$$

and hence

$$\frac{dy}{dx} = \frac{g(1, y/x)}{f(1, y/x)} = F(y/x), \tag{1}$$

where $F(y/x)$ is a function of y/x.

Putting $y = vx$, where v is some function of x, and then differentiating with respect to x, gives

$$\frac{dy}{dx} = v + x\frac{dv}{dx}. \tag{2}$$

Equations (1) and (2) now give

$$v + x\frac{dv}{dx} = F(v)$$

i.e.

$$x\frac{dv}{dx} = F(v) - v$$

and hence

$$\int\frac{dv}{F(v) - v} = \int\frac{dx}{x} + \text{constant}.$$

The general solution is obtained by replacing v by y/x.

Example 4

Solve the equation

$$(x^2 + 2xy)\frac{dy}{dx} = xy + y^2,$$

and find the particular solution for which $y = 1$ when $x = 1$.
    Taking out a factor x^2 gives,

$$x^2(1 + 2y/x)\frac{dy}{dx} = x^2(y/x + y^2/x^2)$$

i.e.

$$\frac{dy}{dx} = \frac{y/x + y^2/x^2}{1 + 2y/x}. \tag{3}$$

Let $y = vx$, so that

$$\frac{dy}{dx} \equiv v + x\frac{dv}{dx} = \frac{v + v^2}{1 + 2v}, \quad \text{from (3)},$$

i.e.

$$x\frac{dv}{dx} = \frac{v + v^2}{1 + 2v} - v$$

$$= \frac{-v^2}{1 + 2v}.$$

Separating the variables we obtain

$$\int \frac{1 + 2v}{v^2} \, dv = - \int \frac{dx}{x} + \text{constant}$$

i.e.

$$\int (1/v^2 + 2/v) \, dv = -\log_e x + \text{constant}$$

$$\therefore \quad -\frac{1}{v} + 2\log_e v = -\log_e x + \text{constant}.$$

Replacing v by y/x gives,

$$-\frac{x}{y} + 2\log_e \frac{y}{x} = -\log_e x + \text{constant},$$

which is the general solution. As $y = 1$ when $x = 1$, substituting into the equation gives

$$-1 = \text{constant.}$$

Thus, the particular solution is given by

$$2 \log_e(y/x) + \log_e x = x/y - 1$$

i.e.

$$\log_e(y^2/x) = \frac{x - y}{y},$$

and hence

$$y^2 = x \exp\left(\frac{x - y}{y}\right).$$

Example 5

Obtain the general solution of the differential equation

$$\frac{dy}{dx} = \frac{9x^2 + 2xy + 3y^2}{2x(x + y)}. \tag{4}$$

Putting $y = vx$, equation (4) becomes

$$\frac{dy}{dx} \equiv v + x \frac{dv}{dx} = \frac{9 + 2v + 3v^2}{2(1 + v)}$$

i.e.

$$x \frac{dv}{dx} = \frac{9 + 2v + 3v^2}{2(v + 1)} - v$$

$$= \frac{v^2 + 9}{2(v + 1)}.$$

Separating the variables, the solution is given by

$$2 \int \frac{v + 1}{v^2 + 9} \, dv = \int \frac{dx}{x} + \text{constant}$$

i.e.

$$2 \int \left(\frac{v}{v^2 + 9} + \frac{1}{v^2 + 9}\right) dv = \log_e x + \text{constant}$$

$$\therefore \; \log_e(v^2 + 9) + \tfrac{2}{3} \tan^{-1}(v/3) = \log_e x + \text{constant.}$$

Replacing v by y/x, the general solution becomes

$$\log_e\left(\frac{y^2 + 9x^2}{x^2}\right) + \tfrac{2}{3} \tan^{-1}\left(\frac{y}{3x}\right) = \log_e x + \text{constant}$$

i.e.

$$\log_e\left(\frac{y^2 + 9x^2}{x^3}\right) + \tfrac{2}{3}\tan^{-1}\left(\frac{y}{3x}\right) = \text{constant}.$$

Example 6

A mixture of two liquids A and B, is boiling in a vessel, the respective volumes being x and y at time t.

The liquid A evaporates at a rate of α times the volume of A present, and the rate at which B evaporates is equal to the rate at which A evaporates minus β times the volume of B present.

If the initial volumes of A and B are x_0, y_0 respectively, show that

$$\beta\log_e\left(\frac{x}{x_0}\right) = \alpha\log_e\left(\frac{\alpha x + (\beta - \alpha)y}{\alpha x_0 + (\beta - \alpha)y_0}\right).$$

The equations representing the actions are

$$\frac{dx}{dt} = -\alpha x, \qquad \frac{dy}{dt} = \frac{dx}{dt} - \beta y = -(\alpha x + \beta y).$$

$$\therefore \quad \frac{dy}{dx} = \frac{dy/dt}{dx/dt} = \frac{\alpha x + \beta y}{\alpha x}.$$

Putting $y = vx$,

$$\therefore \quad \frac{dy}{dx} = v + x\frac{dv}{dx} = 1 + \frac{\beta}{\alpha}v,$$

whence,

$$\int \frac{dv}{1 + \left(\dfrac{\beta}{\alpha} - 1\right)v} = \int \frac{dx}{x} + \text{constant}$$

i.e.

$$\frac{1}{\left(\dfrac{\beta}{\alpha} - 1\right)}\log_e\left\{1 + \left(\frac{\beta}{\alpha} - 1\right)v\right\} = \log_e x + \text{constant}$$

i.e.

$$\frac{\alpha}{\beta - \alpha}\log_e\left(\frac{\alpha x + (\beta - \alpha)y}{x}\right) = \log_e x + \text{constant}.$$

Since $x = x_0, y = y_0$ at $t = 0$,

$$\therefore \quad \text{constant} = \frac{\alpha}{\beta - \alpha}\log_e\left(\frac{\alpha x_0 + (\beta - \alpha)y_0}{x_0}\right) - \log_e x_0$$

and

$$\frac{\alpha}{\beta - \alpha} \log_e\left(\frac{\alpha x + (\beta - \alpha)y}{x}\right)$$

$$= \log_e x + \frac{\alpha}{\beta - \alpha} \log_e\left(\frac{\alpha x_0 + (\beta - \alpha)y_0}{x_0}\right) - \log_e x_0$$

i.e.

$$\frac{\alpha}{\beta - \alpha} \log_e\left(\frac{\alpha x + (\beta - \alpha)y}{\alpha x_0 + (\beta - \alpha)y_0} \cdot \frac{x_0}{x}\right) = \log_e\left(\frac{x}{x_0}\right)$$

$$\therefore \alpha \log_e\left(\frac{\alpha x + (\beta - \alpha)y}{\alpha x_0 + (\beta - \alpha)y_0}\right) + \alpha \log_e\left(\frac{x_0}{x}\right) = (\beta - \alpha)\log_e\left(\frac{x}{x_0}\right)$$

i.e.

$$\alpha \log_e\left(\frac{\alpha x + (\beta - \alpha)y}{\alpha x_0 + (\beta - \alpha)y_0}\right) - \alpha \log_e\left(\frac{x}{x_0}\right) = (\beta - \alpha)\log_e\left(\frac{x}{x_0}\right)$$

$$\therefore \alpha \log_e\left(\frac{\alpha x + (\beta - \alpha)y}{\alpha x_0 + (\beta - \alpha)y_0}\right) = \beta \log_e\left(\frac{x}{x_0}\right).$$

3. Reducible equations

The differential equations to be considered now may be expressed in the form

$$\frac{dy}{dx} = \frac{a_1 x + b_1 y + c_1}{a_2 x + b_2 y + c_2} \tag{5}$$

where $a_1, b_1, c_1, a_2, b_2, c_2$ are constants.
This equation may be re-written as

$$\frac{dy}{dx} = \frac{a_1(x - \alpha) + b_1(y - \beta) + (a_1\alpha + b_1\beta + c_1)}{a_2(x - \alpha) + b_2(y - \beta) + (a_2\alpha + b_2\beta + c_2)}$$

where α and β are to be determined.
If α, β are now chosen such that

$$a_1\alpha + b_1\beta + c_1 = 0$$

and $\hspace{6cm}$ (6)

$$a_2\alpha + b_2\beta + c_2 = 0,$$

then this equation reduces to

$$\frac{dy}{dx} = \frac{a_1(x - \alpha) + b_1(y - \beta)}{a_2(x - \alpha) + b_2(y - \beta)}. \tag{7}$$

By now making the substitution $X = x - \alpha$, $Y = y - \beta$, equation (7) is transformed into

$$\frac{dY}{dX} = \frac{a_1 X + b_1 Y}{a_2 X + b_2 Y}$$

which is homogeneous in X and Y, and hence can be solved by putting $Y = vX$, as in Section 2.

Example 7

Solve the equation

$$\frac{dy}{dx} = \frac{7y - 3x + 1}{5x - y + 2}. \tag{8}$$

Let $x = X + \alpha$, and $y = Y + \beta$ where

$$-3\alpha + 7\beta + 1 = 0$$

and

$$5\alpha - \beta + 2 = 0,$$

giving

$$\alpha = -\tfrac{15}{32}, \quad \beta = -\tfrac{11}{32}.$$

Equation (8) then becomes

$$\frac{dy}{dx} \equiv \frac{dY}{dX} = \frac{7(Y - \tfrac{11}{32}) - 3(X - \tfrac{15}{32}) + 1}{5(X - \tfrac{15}{32}) - (Y - \tfrac{11}{32}) + 2}$$

i.e.

$$\frac{dY}{dX} = \frac{7Y - 3X}{5X - Y}. \tag{9}$$

Now put $Y = vX$, giving

$$\frac{dY}{dX} = v + X\frac{dv}{dX} = \frac{7v - 3}{5 - v}, \quad \text{from (9),}$$

i.e.

$$X\frac{dv}{dX} = \frac{7v - 3}{5 - v} - v$$

$$= \frac{v^2 + 2v - 3}{5 - v}.$$

Separating the variables we obtain,

$$\int \frac{5 - v}{v^2 + 2v - 3} \, dv = \int \left(\frac{1}{v - 1} - \frac{2}{v + 3} \right) dv = \int \frac{dX}{X} + \text{constant}$$

and so

$$\log_e(v - 1) - 2\log_e(v + 3) = \log_e X + \text{constant}.$$

Replacing v by Y/X gives,

$$\log_e\left(\frac{Y - X}{X}\right) - 2\log_e\left(\frac{Y + 3X}{X}\right) = \log_e X + \text{constant}$$

i.e.

$$\log_e(Y - X) - 2\log_e(Y + 3X) = \text{constant}.$$

$$\therefore \quad \frac{Y - X}{(Y + 3X)^2} = \text{constant}.$$

Now

$$X = x + \tfrac{15}{32}, \quad Y = y + \tfrac{11}{32}$$

so that,

$$Y - X = y - x - \tfrac{1}{8}$$

and

$$Y + 3X = y + 3x + \tfrac{7}{4}.$$

Hence, the general solution is

$$\frac{y - x - \tfrac{1}{8}}{(y + 3x + \tfrac{7}{4})^2} = \text{constant}$$

i.e.

$$8y - 8x - 1 = A(4y + 12x + 7)^2,$$

where A is an arbitrary constant.

Returning now to equation (6), we see that α and β are given by

$$\alpha = \frac{b_1 c_2 - b_2 c_1}{a_1 b_2 - a_2 b_1}, \quad \beta = \frac{c_1 a_2 - c_2 a_1}{a_1 b_2 - a_2 b_1},$$

so that the previous method for solving equation (5) can only be utilized provided $a_1 b_2 - a_2 b_1 \neq 0$.

Suppose then that $a_1 b_2 - a_2 b_1$ *is* equal to zero. Let

$$\frac{a_1}{b_1} = \frac{a_2}{b_2} = \lambda, \quad \text{say}.$$

Equation (5), i.e.

$$\frac{dy}{dx} = \frac{a_1 x + b_1 y + c_1}{a_2 x + b_2 y + c_2},$$

may then be written as

$$\frac{dy}{dx} = \frac{b_1\left(\dfrac{a_1}{b_1}x + y\right) + c_1}{b_2\left(\dfrac{a_2}{b_2}x + y\right) + c_2} = \frac{b_1(\lambda x + y) + c_1}{b_2(\lambda x + y) + c_2}. \tag{10}$$

Hence, putting $z = \lambda x + y$, equation (10) gives

$$\frac{dz}{dx} = \lambda + \frac{dy}{dx} = \lambda + \frac{b_1 z + c_1}{b_2 z + c_2}$$

$$= \text{a function of } z \text{ only,}$$

i.e. it is of the type 'variables separable' and can be solved in the usual way.

Example 8

Solve the differential equation

$$\frac{dy}{dx} = \frac{2x - 7y + 1}{6x - 21y - 1}. \tag{11}$$

Following the method used for equation (10), put $z = 2x - 7y$, then

$$\frac{dz}{dx} = 2 - 7\frac{dy}{dx} = 2 - 7\frac{z + 1}{3z - 1}, \quad \text{from (11)}$$

i.e.

$$\frac{dz}{dx} = -\frac{z + 9}{3z - 1}.$$

Separating the variables we obtain

$$\int \frac{3z - 1}{z + 9}\, dz = \int \left(3 - \frac{28}{z + 9}\right) dz = -\int dx + \text{constant}$$

i.e.

$$3z - 28 \log_e(z + 9) = -x + \text{constant}.$$

Thus, as $z = 2x - 7y$, the general solution of equation (11) is

$$7x - 21y - 28 \log_e(2x - 7y + 9) = \text{constant}$$

i.e.

$$x - 3y - 4 \log_e(2x - 7y + 9) = \text{constant}.$$

4. Linear equations of the first order

Differential equations belonging to this group may be expressed in the form

$$\frac{dy}{dx} + f(x)y = g(x),\tag{12}$$

where $f(x)$ and $g(x)$ are functions of x only.

Method 1

The method to be used is referred to as *variation of parameters*.
We first require the solution of the complementary equation

$$\frac{dy}{dx} + f(x)y = 0,$$

which is obtained by equating the right-hand side of equation (12) to zero.
Separating the variables gives

$$\int \frac{dy}{y} + \int f(x)\, dx = 0$$

i.e.

$$\log_e y = -\int f(x)\, dx + \text{constant},$$

i.e.

$$y = \exp\left(\text{constant} - \int f(x)\, dx\right)$$

and hence

$$y = A \exp\left(-\int f(x)\, dx\right),$$

where A is an arbitrary constant.
We now assume that the general solution of equation (12) is

$$y = z \exp\left(-\int f(x)\, dx\right),\tag{13}$$

where z is a function of x which is to be determined.
From equation (13),

$$z = y \exp\left(\int f(x)\, dx\right)$$

so that differentiating with respect to x gives

$$\frac{dz}{dx} = \frac{dy}{dx} \exp\left\{\int f(x) \, dx\right\} + y f(x) \exp\left(\int f(x) \, dx\right)$$

$$= g(x) \exp\left(\int f(x) \, dx\right), \quad \text{from (12).}$$

Hence, integrating with respect to x,

$$z = \int g(x) \exp\left(\int f(x) \, dx\right) dx + \text{constant.}$$

The general solution of equation (12) is now obtained by substituting this value for z into equation (13).

Example 9

Solve the differential equation

$$(x^2 + 1)^2 \frac{dy}{dx} + 2x(x^2 + 1)y = 1. \tag{14}$$

Writing this equation in the standard form gives

$$\frac{dy}{dx} + \frac{2x}{x^2 + 1} y = \frac{1}{(x^2 + 1)^2},$$

so that the complementary equation is

$$\frac{dy}{dx} + \frac{2xy}{x^2 + 1} = 0.$$

Separating the variables we obtain

$$\int \frac{dy}{y} + \int \frac{2x}{x^2 + 1} \, dx = \text{constant}$$

i.e.

$$\log_e y + \log_e(x^2 + 1) = \text{constant}$$

and

$$y(x^2 + 1) = \text{constant.}$$

$$\therefore y = A(x^2 + 1)^{-1},$$

where A is an arbitrary constant.

Let the general solution of equation (14) be

$$y = z(x^2 + 1)^{-1},$$

G

so that

$$z = y(x^2 + 1).$$

Differentiating with respect to x gives

$$\frac{dz}{dx} = (x^2 + 1)\frac{dy}{dx} + 2xy$$

$$= \frac{1}{x^2 + 1}, \quad \text{from (14)}$$

and hence

$$z = \int \frac{dx}{x^2 + 1} + \text{constant}$$

$$= \tan^{-1} x + B,$$

where B is an arbitrary constant.

The general solution of equation (14) is thus

$$y = (x^2 + 1)^{-1}(\tan^{-1} x + B).$$

Method 2

The solution of equation (12) may also be obtained in the following way.

Multiply equation (12) throughout by $h(x)$, a function of x only, which is to be determined. Then,

$$h(x)\frac{dy}{dx} + h(x)f(x)y = h(x)g(x). \tag{15}$$

The left-hand side may be replaced by

$$\frac{d}{dx}\{yh(x)\} \equiv h(x)\frac{dy}{dx} + y\frac{dh(x)}{dx},$$

provided $h(x)$ is chosen such that

$$\frac{dh(x)}{dx} = h(x)f(x). \tag{16}$$

Assuming this to be possible, equation (15) may be written as

$$\frac{d}{dx}\{yh(x)\} = h(x)g(x) \tag{17}$$

and the general solution of equation (12) can be expressed as

$$yh(x) = \int h(x)g(x)\,dx + \text{constant}. \tag{18}$$

In order to determine the functional form of $h(x)$, we return to equation (16).

Separating the variables gives

$$\int \frac{dh(x)}{h(x)} \equiv \log_e h(x) = \int f(x) \, dx + \text{constant}$$

so that

$$h(x) = A \, \exp\left(\int f(x) \, dx\right),$$

where A is an arbitrary constant.

It is clear that A may be omitted in the general solution given by (18), and so we may take $h(x)$ to be given by

$$h(x) = \exp\left(\int f(x) \, dx\right).$$

The function $h(x)$ is called the *integrating factor* of equation (12).

Example 10

Solve the differential equation

$$(x + 1) \frac{dy}{dx} - (x + 2)y = e^x(x + 1)^3 \log_e x,$$

and find the particular solution for which $y = 0$ when $x = 1$.

Arranging the equation in the standard form gives

$$\frac{dy}{dx} - \left(\frac{x + 2}{x + 1}\right)y = e^x(x + 1)^2 \log_e x \tag{19}$$

so that the integrating factor is

$$\exp\left(-\int \frac{x + 2}{x + 1} \, dx\right),$$

where the constant of integration is to be omitted. Now

$$\int \frac{x + 2}{x + 1} \, dx = \int\left(1 + \frac{1}{x + 1}\right) dx = x + \log_e(x + 1) + \text{constant}.$$

Thus the integrating factor is

$$\exp\{-x - \log_e(x + 1)\} = \frac{e^{-x}}{x + 1}.$$

Referring to equation (17) we see that the general solution of equation (19) is given by

$$\frac{d}{dx}\left\{\frac{ye^{-x}}{x + 1}\right\} = (x + 1)\log_e x$$

so that integration with respect to x gives

$$\frac{ye^{-x}}{x+1} = \int (x+1)\log_e x \; dx + \text{constant}$$

$$= (\tfrac{1}{2}x^2 + x)\log_e x - \int (\tfrac{1}{2}x + 1) \; dx + \text{constant}$$

$$= (\tfrac{1}{2}x^2 + x)\log_e x - \tfrac{1}{4}x^2 - x + \text{constant}.$$

Since $y = 0$ when $x = 1$, the constant must be $\tfrac{5}{4}$, and the particular solution is

$$\frac{ye^{-x}}{x+1} = (\tfrac{1}{2}x^2 + x)\log_e x - \tfrac{1}{4}x^2 - x + \tfrac{5}{4}$$

whence,

$$y = \frac{e^x}{4}(x+1)\{(2x^2 + 4x)\log_e x - x^2 - 4x + 5\}.$$

Example 11

A quantity of thorium A breaks down into thorium B at a rate of λ_1 times the number of atoms of thorium A present. The thorium B breaks down at a rate of λ_2 times the number of atoms of B present.

If x, y are the numbers of atoms of thorium A and thorium B at time t, find the differential equations satisfied by x and y.

If $x = n_0, y = 0$ at time $t = 0$, show that

$$\frac{dy}{dt} + \lambda_2 y = \lambda_1 n_0 \exp(-\lambda_1 t),$$

and solve this equation for y.

If at time $t + \delta t$ there are $x + \delta x$ atoms of A present, then at this time there will be

$$y + \delta y + \frac{dx}{dt}\delta t$$

atoms of B present.

Thus,

$$\lim_{\delta t \to 0} \frac{(x + \delta x) - x}{\delta t} = -\lambda_1 x, \qquad \lim_{\delta t \to 0} \frac{\left(y + \delta y + \dfrac{dx}{dt}\delta t\right) - y}{\delta t} = -\lambda_2 y$$

i.e.

$$\frac{dx}{dt} = -\lambda_1 x, \qquad \frac{dy}{dt} + \frac{dx}{dt} = -\lambda_2 y.$$

(i) If

$$\frac{dx}{dt} = -\lambda_1 x, \quad \text{then } \log_e x = -\lambda_1 t + \text{constant},$$

and since $x = n_0$ at $t = 0$, the constant must be equal to $\log_e n_0$, giving

$$x = n_0 \exp(-\lambda_1 t).$$

(ii) The equation for y becomes

$$\frac{dy}{dt} + \lambda_2 y = \lambda_1 x = \lambda_1 n_0 \exp(-\lambda_1 t).$$

The integrating factor is thus

$$\exp\left(\int \lambda_2 \, dt\right) = \exp(\lambda_2 t)$$

$$\therefore \quad y \exp(\lambda_2 t) = \lambda_1 n_0 \int \exp(\lambda_2 - \lambda_1)t \, dt$$

$$= \frac{\lambda_1 n_0}{\lambda_2 - \lambda_1} \exp(\lambda_2 - \lambda_1)t + \text{constant}.$$

At $t = 0, y = 0$

$$\therefore \quad \text{constant} = -\frac{\lambda_1 n_0}{\lambda_2 - \lambda_1}.$$

Whence

$$y = \frac{\lambda_1 n_0}{\lambda_2 - \lambda_1} \{\exp(-\lambda_1 t) - \exp(-\lambda_2 t)\}.$$

5. Bernoulli's equation

This type of differential equation, which is reducible to a linear equation of the first order, may be expressed in the form

$$\frac{dy}{dx} + f(x)y = g(x)y^n \tag{20}$$

where n is a constant.

Dividing throughout by y^n, the equation becomes

$$y^{-n} \frac{dy}{dx} + f(x)y^{1-n} = g(x), \tag{21}$$

which may be simplified by making the substitution

$$z = y^{1-n}.$$

Differentiating with respect to x gives

$$\frac{dz}{dx} = (1 - n)y^{-n} \frac{dy}{dx}$$

so that equation (21) reduces to

$$\frac{dz}{dx} + (1 - n)f(x)z = (1 - n)g(x) \tag{22}$$

i.e. a linear differential equation of the first order, which may be solved by using one of the methods of Section 4.

Example 12

Solve the differential equation

$$2(x^2 + 7x - 8)\frac{dy}{dx} + (6x + 21)y = 3(x + 8)^2 y^{5/3}. \tag{23}$$

Re-arranging in the Bernoullian form gives

$$\frac{dy}{dx} + \frac{6x + 21}{2(x^2 + 7x - 8)} y = \frac{3(x + 8)^2}{2(x^2 + 7x - 8)} y^{5/3}$$

i.e.

$$\frac{dy}{dx} + \frac{6x + 21}{2(x^2 + 7x - 8)} y = \frac{3(x + 8)}{2(x - 1)} y^{5/3}.$$

Dividing throughout by $y^{5/3}$ we obtain

$$y^{-5/3}\frac{dy}{dx} + \frac{6x + 21}{2(x^2 + 7x - 8)} y^{-2/3} = \frac{3(x + 8)}{2(x - 1)}. \tag{24}$$

Let

$$z = y^{-2/3}, \quad \therefore \frac{dz}{dx} = -\frac{2}{3} y^{-5/3}\frac{dy}{dx}$$

and equation (24) becomes

$$\frac{dz}{dx} - \frac{(2x + 7)}{x^2 + 7x - 8} z = -\frac{(x + 8)}{x - 1}. \tag{25}$$

The integrating factor is

$$\exp\left(-\int \frac{2x + 7)}{x^2 + 7x - 8} \, dx\right) = \exp\{-\log_e(x^2 + 7x - 8)\}$$

$$= (x^2 + 7x - 8)^{-1},$$

so that the solution of equation (25) is

$$\frac{d}{dx}\left\{\frac{z}{x^2 + 7x - 8}\right\} = \frac{-(x + 8)}{(x - 1)(x^2 + 7x - 8)} = \frac{-1}{(x - 1)^2}.$$

Integrating with respect to x gives

$$\frac{z}{x^2 + 7x - 8} = \frac{1}{x - 1} + A,$$

where A is an arbitrary constant.

Substituting for z, the general solution of equation (23) is obtained in the form

$$\frac{y^{-2/3}}{x^2 + 7x - 8} = \frac{1}{x - 1} + A = \frac{1 + A(x - 1)}{x - 1}$$

i.e.

$$y^{-2/3} = (x + 8)\{1 + A(x - 1)\}.$$

6. Linear differential equations with constant coefficients

The general form of the differential equations now to be considered is

$$\frac{d^n y}{dx^n} + a_1 \frac{d^{n-1} y}{dx^{n-1}} + a_2 \frac{d^{n-2} y}{dx^{n-2}} + \cdots + a_{n-1} \frac{dy}{dx} + a_n y = g(x) \tag{26}$$

where the $a_r (r = 1, 2, \ldots, n)$ are constants, and $g(x)$ is a function of x only.

Differential equations of this type may be solved by the use of the *Laplace transform*, which not only gives the complete solution but also takes into account any initial conditions.

If $u(x)$ is a function of x only, then the *integral transform* $\bar{u}(s)$ of the function $u(x)$ in the range (a, b) is defined as

$$\bar{u}(s) = \int_a^b u(x) k(s, x) \, dx, \tag{27}$$

where $k(s, x)$, the *kernel* of the transform, is a known function of s and x.

The Laplace transform of a function is a special case of (27), and is such that $k(s, x) = \exp(-sx)$ and the limits are zero and infinity.

Hence the Laplace transform of the function $g(x)$ occurring on the right-hand side of equation (26) is given by

$$\mathcal{L}\{g(x)\} \equiv \bar{g}(s) = \int_0^\infty g(x) \exp(-sx) \, dx. \tag{28}$$

Using this definition we now obtain the transforms of various functions.
Case (i), $g(x) = 1$.

$$\bar{g}(s) = \int_0^\infty \exp(-sx) \, dx = -\left. \frac{\exp(-sx)}{s} \right|_0^\infty = \frac{1}{s},$$

$$\therefore \ \mathcal{L}\{1\} = \frac{1}{s} \quad \text{for } s > 0.$$

Case (ii), $g(x) = x^n$, (n integral).
Let

$$\bar{g}(s) = I_n = \int_0^\infty x^n \exp(-sx) \, dx$$

$$= -\frac{x^n}{s}\exp(-sx)\Big|_0^\infty + \frac{n}{s}\int_0^\infty x^{n-1}\exp(-sx)\,dx$$

$$\therefore I_n = \frac{n}{s}I_{n-1}, \quad \text{assuming } s > 0$$

i.e.

$$I_n = \frac{n}{s}\cdot\frac{n-1}{s}\cdot I_{n-2} = \frac{n(n-1)(n-2)\cdots 2\cdot 1}{s^n}\cdot I_0 = \frac{n!}{s^n}I_0.$$

But,

$$I_0 = \int_0^\infty \exp(-sx)\,dx = \frac{1}{s}, \quad \text{from (i)},$$

hence,

$$I_n = \frac{n!}{s^{n+1}}$$

and so

$$\mathcal{L}\{x^n\} = \frac{n!}{s^{n+1}}.$$

Case (iii), $g(x) = \exp(ax)$.

$$\bar{g}(s) = \int_0^\infty \exp\{-(s-a)x\}\,dx = -\frac{\exp\{-(s-a)x\}}{s-a}\Big|_0^\infty$$

$$= \frac{1}{s-a}, \quad \text{assuming that } s > a.$$

$$\therefore \mathcal{L}\{\exp(ax)\} = \frac{1}{s-a}.$$

Case (iv), $g(x) = \sin(bx)$.

$$\bar{g}(s) = \int_0^\infty \sin(bx)\exp(-sx)\,dx$$

$$= -\frac{1}{s}\sin(bx)\exp(-sx)\Big|_0^\infty + \frac{b}{s}\int_0^\infty \cos(bx)\exp(-sx)\,dx$$

$$= -\frac{b}{s^2}\cos(bx)\exp(-sx)\Big|_0^\infty + \frac{b^2}{s^2}\int_0^\infty \sin(bx)\exp(-sx)\,dx$$

i.e.

$$\bar{g}(s) = \frac{b}{s^2} + \frac{b^2}{s^2}\bar{g}(s), \quad \text{provided } s > 0,$$

so that

$$\bar{g}(s) = \frac{b}{s^2 + b^2}$$

and so

$$\mathcal{L}\{\sin(bx)\} = \frac{b}{s^2 + b^2}.$$

Case (v), $g(x) = \cosh(bx)$.

$$\mathcal{L}\{\cosh(bx)\} = \mathcal{L}\left(\frac{\exp(bx) + \exp(-bx)}{2}\right)$$

$$= \frac{1}{2(s - b)} + \frac{1}{2(s + b)}, \quad \text{if } s > b$$

i.e.

$$\mathcal{L}\{\cosh(bx)\} = \frac{s}{s^2 - b^2}.$$

A 'shift' theorem

If $\bar{g}(s)$ is the transform of $g(x)$, then $\bar{g}(s + a)$ is the transform of $\exp(-ax)g(x)$, since

$$\mathcal{L}\{\exp(-ax)g(x)\} = \int_0^\infty \exp(-ax)g(x)\exp(-sx)\,dx$$

$$= \int_0^\infty g(x)\exp\{-(s + a)x\}\,dx$$

$$= \bar{g}(s + a),$$

by comparison with (28).

Thus, we obtain:

Case (vi), $g(x) = \exp(ax)\sin(bx)$.

$$\mathcal{L}\{\sin(bx)\} = \frac{b}{s^2 + b^2}, \quad \therefore \ \mathcal{L}\{\exp(ax)\sin(bx)\} = \frac{b}{(s - a)^2 + b^2}.$$

Case (vii), $g(x) = \exp(ax)\cosh(bx)$

$$\mathcal{L}\{\cosh(bx)\} = \frac{s}{s^2 - b^2}, \quad \therefore \ \mathcal{L}\{\exp(ax)\cosh(bx)\} = \frac{s - a}{(s - a)^2 - b^2}.$$

Case (viii), $g(x) = \exp(ax)x^n$.

$$\mathcal{L}\{x^n\} = \frac{n!}{s^{n+1}}, \quad \therefore \ \mathcal{L}\{\exp(ax)x^n\} = \frac{n!}{(s - a)^{n+1}}.$$

The Laplace transforms of various functions are given in Table 8.1.

Table 8.1

Function	Laplace transform	
1	$\dfrac{1}{s}$	$s > 0$
x^n	$\dfrac{n!}{s^{n+1}}$	$s > 0, n$ integral
$\exp(ax)$	$\dfrac{1}{s - a}$	$s > a$
$\sin(bx)$	$\dfrac{b}{s^2 + b^2}$	$s > 0$
$\cos(bx)$	$\dfrac{s}{s^2 + b^2}$	$s > 0$
$\sinh(bx)$	$\dfrac{b}{s^2 - b^2}$	$s > b$
$\cosh(bx)$	$\dfrac{s}{s^2 - b^2}$	$s > b$
$\exp(ax)\cos(bx)$	$\dfrac{s - a}{(s - a)^2 + b^2}$	$s > a$
$\exp(ax)\sinh(bx)$	$\dfrac{b}{(s - a)^2 - b^2}$	$s > a + b$
$\exp(ax)x^n$	$\dfrac{n!}{(s - a)^{n+1}}$	$s > a, n$ integral

In order to obtain solutions of equations of the type given by (26) it will be necessary to *invert* a function of s into its corresponding function of x. The process involved is best illustrated by means of examples.

Example 13

Invert the expression

$$\frac{4}{s^2 + 9}.$$

Let

$$\bar{g}(s) = \frac{4}{s^2 + 9} = \frac{4}{3} \cdot \frac{3}{s^2 + 3^2}.$$

Referring to Table 8.1 we see that the corresponding function of x is

$$g(x) = \tfrac{4}{3} \sin 3x.$$

Example 14

Invert the expression

$$\frac{2s - 3}{s^2 + 4s + 20}.$$

Let

$$\bar{g}(s) = \frac{2s - 3}{s^2 + 4s + 20} = \frac{2s - 3}{(s + 2)^2 + 16}$$

i.e.

$$\bar{g}(s) = \frac{2(s + 2) - 7}{(s + 2)^2 + 16} = 2 \cdot \frac{s + 2}{(s + 2)^2 + 4^2} - \frac{7}{4} \cdot \frac{4}{(s + 2)^2 + 4^2}$$

Hence, Table 8.1 gives

$$g(x) = 2 \exp(-2x)\cos 4x - \tfrac{7}{4} \exp(-2x)\sin 4x.$$

Example 15

Invert the expression

$$\frac{5(s - 13)}{(2s - 1)(s^2 - 6s + 34)}.$$

Expressing the function in terms of partial fractions gives

$$\bar{g}(s) = \frac{s - 3}{s^2 - 6s + 34} - \frac{2}{2s - 1}$$

$$= \frac{s - 3}{(s - 3)^2 + 5^2} - \frac{1}{s - \tfrac{1}{2}}.$$

Whence,

$$g(x) = \exp(3x)\cos 5x - \exp(\tfrac{1}{2}x).$$

The Laplace transform of $d^n y/dx^n$

We now consider the left-hand side of equation (26), viz.

$$\frac{d^n y}{dx^n} + a_1 \frac{d^{n-1}y}{dx^{n-1}} + a_2 \frac{d^{n-2}y}{dx^{n-2}} + \cdots + a_{n-1} \frac{dy}{dx} + a_n y,$$

whose transform is also required.

Integrating by parts, we see that

$$\mathcal{L}\left(\frac{dy}{dx}\right) = \int_0^\infty \frac{dy}{dx} \exp(-sx)\,dx = y \exp(-sx)\Big|_0^\infty + s \int_0^\infty y \exp(-sx)\,dx$$

Assuming that $\lim_{x \to \infty} \{y \exp(-sx)\} = 0$, and writing $\mathcal{L}(y) = \bar{y}(s)$, we obtain

$$\mathcal{L}\left(\frac{dy}{dx}\right) = -y(0) + s\bar{y}(s), \tag{29}$$

where $y(0)$ denotes the value of y when $x = 0$.
   Similarly,

$$\mathcal{L}\left(\frac{d^2 y}{dx^2}\right) = \int_0^\infty \frac{d^2 y}{dx^2} \exp(-sx)\, dx = \frac{dy}{dx} \exp(-sx)\bigg|_0^\infty$$

$$+ s \int_0^\infty \frac{dy}{dx} \exp(-sx)\, dx.$$

Assuming that

$$\lim_{x \to \infty} \left(\frac{dy}{dx} \exp(-sx)\right) = 0,$$

and writing y' for dy/dx, we obtain

$$\mathcal{L}\left(\frac{d^2 y}{dx^2}\right) = -y'(0) + s\{-y(0) + s\bar{y}(s)\}, \quad \text{from (29)}$$

where $y'(0)$ denotes the values of dy/dx when $x = 0$, i.e.

$$\mathcal{L}\left(\frac{d^2 y}{dx^2}\right) = -y'(0) - sy(0) + s^2 \bar{y}(s).$$

Thus, assuming that $\lim_{x \to \infty} \{y^{(r-1)} \exp(-sx)\} = 0$, where $y^{(r)}$ denotes $d^r y/dx^r$, we obtain the general result

$$\mathcal{L}\left(\frac{d^n y}{dx^n}\right) = s^n \bar{y}(s) - s^{n-1}y(0) - s^{n-2}y'(0) - s^{n-3}y''(0) - \cdots - y^{(n-1)}(0), \tag{30}$$

where $y^{(r)}(0)$ denotes the value of $d^r y/dx^r$ when $x = 0$.
   We now have sufficient information to enable the solution of equation (26) to be determined.

Example 16

Solve the equation

$$\frac{d^2 y}{dx^2} - \frac{dy}{dx} - 2y = \exp(3x),$$

given that $y = dy/dx = 0$ when $x = 0$.
   Let $\mathcal{L}(y) = \bar{y}$, then for the left-hand side

$$\mathcal{L}\left(\frac{dy}{dx}\right) = s\bar{y}(s) - y(0) = s\bar{y}$$

$$\mathcal{L}\left(\frac{d^2 y}{dx^2}\right) = s^2 \bar{y}(s) - sy(0) - y'(0) = s^2 \bar{y},$$

and for the right-hand side

$$\mathcal{L}\{\exp(3x)\} = \frac{1}{s-3}.$$

The transform of the equation is thus

$$s^2 \bar{y} - s\bar{y} - 2\bar{y} = \frac{1}{s-3}$$

i.e.

$$\bar{y}(s^2 - s - 2) = \frac{1}{s-3},$$

so that

$$\bar{y} = \frac{1}{(s-3)(s^2 - s - 2)} = \frac{1}{(s-3)(s-2)(s+1)}$$

$$= \frac{\frac{1}{4}}{s-3} - \frac{\frac{1}{3}}{s-2} + \frac{\frac{1}{12}}{s+1}$$

and hence

$$y = \tfrac{1}{4}\exp(3x) - \tfrac{1}{3}\exp(2x) + \tfrac{1}{12}\exp(-x).$$

Example 17

Solve the equation

$$\frac{d^2 y}{dx^2} + 2\frac{dy}{dx} + 5y = \sin 3x,$$

given that $y = 1$ and $dy/dx = -2$ when $x = 0$.

Let $\mathcal{L}(y) = \bar{y}$, then

$$\mathcal{L}\left(\frac{dy}{dx}\right) = s\bar{y}(s) - y(0) = s\bar{y} - 1$$

$$\mathcal{L}\left(\frac{d^2 y}{dx^2}\right) = s^2 \bar{y}(s) - sy(0) - y'(0) = s^2 \bar{y} - s + 2.$$

For the right-hand side

$$\mathcal{L}\{\sin(3x)\} = \frac{3}{s^2 + 9}.$$

Hence the transform of the equation to be solved is

$$(s^2 \bar{y} - s + 2) + 2(s\bar{y} - 1) + 5\bar{y} = \frac{3}{s^2 + 9}$$

i.e.

$$\bar{y}(s^2 + 2s + 5) - s = \frac{3}{s^2 + 9}$$

$$\therefore \bar{y} = \frac{3}{(s^2 + 9)(s^2 + 2s + 5)} + \frac{s}{(s^2 + 2s + 5)}.$$

The first term on the right-hand side of this equation is

$$\frac{3(s + 4)}{26(s^2 + 2s + 5)} - \frac{3(s + 2)}{26(s^2 + 9)}$$

so that

$$26\bar{y} = \frac{29s + 12}{(s^2 + 2s + 5)} - \frac{3(s + 2)}{s^2 + 9}$$

$$= \frac{29(s + 1)}{(s + 1)^2 + 2^2} - \frac{17}{(s + 1)^2 + 2^2} - \frac{3s}{s^2 + 3^2} - \frac{6}{s^2 + 3^2}.$$

Referring to Table 8.1, we obtain the general solution in the form

$$26y = 29 \exp(-x) \cos 2x - \tfrac{17}{2} \exp(-x) \sin 2x - 3 \cos 3x - 2 \sin 3x.$$

In the cases where the values of

$$y, \frac{dy}{dx}, \frac{d^2 y}{dx^2}, \ldots, \frac{d^n y}{dx^n}$$

are not known for $x = 0$, the general solution may be obtained by assuming that when $x = 0$,

$$y = c_1, \frac{dy}{dx} = c_2, \frac{d^2 y}{dx^2} = c_3, \ldots, \frac{d^n y}{dx^n} = c_{n+1},$$

where the c_r $(r = 1, 2, \ldots, n + 1)$ are constants.

Example 18

Solve the equation

$$\frac{d^2 y}{dx^2} - 4y = x.$$

Let $\mathcal{L}(y) = \bar{y}(s)$, and suppose that $y = c_1$ and $dy/dx = c_2$ when $x = 0$. Then,

$$\mathcal{L}\left(\frac{d^2 y}{dx^2}\right) = s^2 \bar{y}(s) - sy(0) - y'(0) = s^2 \bar{y} - c_1 s - c_2.$$

Also,

$$\mathcal{L}(x) = \frac{1}{s^2}.$$

Thus the differential equation is transformed into

$$(s^2 \bar{y} - c_1 s - c_2) - 4\bar{y} = \frac{1}{s^2}$$

i.e.

$$\bar{y}(s^2 - 4) = \frac{1}{s^2} + c_1 s + c_2$$

$$\therefore \ \bar{y} = \frac{1}{s^2(s^2 - 4)} + \frac{c_1 s + c_2}{s^2 - 4}$$

$$= \frac{1}{4(s^2 - 4)} - \frac{1}{4s^2} + \frac{c_1 s + c_2}{s^2 - 4}$$

i.e.

$$\bar{y} = \frac{c_1 s}{s^2 - 4} + \frac{c_2 + \frac{1}{4}}{s^2 - 4} - \frac{1}{4s^2} \ .$$

Hence,

$$y = c_1 \cosh 2x + \tfrac{1}{2}(c_2 + \tfrac{1}{4}) \sinh 2x - \tfrac{1}{4}x$$

or,

$$y = A_1 \cosh 2x + A_2 \sinh 2x - \tfrac{1}{4}x,$$

where A_1 and A_2 are arbitrary constants.

Example 19

Solve the simultaneous differential equations

$$9\frac{dx}{dt} + 4\frac{dy}{dt} - 2x - 7y = 36t \tag{31}$$

$$\frac{dx}{dt} + 2x + y = 12, \tag{32}$$

if $x = y = 0$ when $t = 0$.

Let

$$\mathcal{L}(x) = \bar{x}, \qquad \mathcal{L}(y) = \bar{y}.$$

Then,

$$\mathcal{L}\left(\frac{dx}{dt}\right) = s\bar{x}(s) - x(0) = s\bar{x}; \qquad \mathcal{L}\left(\frac{dy}{dt}\right) = s\bar{y}(s) - y(0) = s\bar{y}.$$

Also,

$$\mathcal{L}(36t) = \frac{36}{s^2} \quad \text{and} \quad \mathcal{L}(12) = \frac{12}{s} \ .$$

Thus, the transforms of equations (31) and (32) are

$$\bar{x}(9s - 2) + \bar{y}(4s - 7) = \frac{36}{s^2}$$
$$\bar{x}(s + 2) + \bar{y} = \frac{12}{s} \quad\quad\quad (33)$$

Solving for $\bar{x}$ gives,

$$\bar{x}(s^2 - 2s - 3) = \frac{3(4s - 7)}{s} - \frac{9}{s^2} = \frac{12s^2 - 21s - 9}{s^2}$$

$$\therefore \bar{x} = \frac{12s^2 - 21s - 9}{s^2(s^2 - 2s - 3)} = \frac{1}{s - 3} - \frac{6}{s + 1} + \frac{3}{s^2} + \frac{5}{s} \quad\quad (34)$$

so that

$$x = \exp(3t) - 6 \exp(-t) + 3t + 5.$$

From (33),

$$\bar{y} = \frac{12}{s} - (s + 2)\bar{x}$$

$$= \frac{12}{s} - \frac{5(s + 2)}{s} - \frac{3(s + 2)}{s^2} + \frac{6(s + 2)}{s + 1} - \frac{(s + 2)}{s - 3},$$

from (34), i.e.

$$\bar{y} = \frac{6}{s + 1} - \frac{5}{s - 3} - \frac{6}{s^2} - \frac{1}{s}$$

whence

$$y = 6 \exp(-t) - 5 \exp(3t) - 6t - 1.$$

Example 20

In the theory of chain reactions, the concentration $n(x)$ of active centres in a layer of gas, width δx, between the walls of a vessel distance c apart, is given by

$$D\frac{d^2 n}{dx^2} - an + b = 0$$

where D is the coefficient of diffusion, and a and b constants.

If $n = 0$ when $x = 0$ and when $x = c$, show that

$$an = b\left(1 - \frac{\cosh\{\alpha(x - c/2)\}}{\cosh(\alpha c/2)}\right)$$

where $\alpha = \sqrt{(a/D)}$.

Let $\mathcal{L}(n) = \bar{n}$, then

$$\mathcal{L}\left(\frac{d^2 n}{dx^2}\right) = s^2 \bar{n}(s) - sn(0) - n'(0) = s^2 \bar{n} - A,$$

where A is to be determined.

The transform of the given equation is thus

$$D(s^2\bar{n} - A) - a\bar{n} = -\frac{b}{s}.$$

$$\therefore \bar{n} = \frac{AD}{Ds^2 - a} - \frac{b}{s(Ds^2 - a)}$$

$$= \frac{A}{s^2 - a/D} + \frac{b}{a}\cdot\frac{1}{s} - \frac{b}{a}\cdot\frac{s}{s^2 - a/D}$$

$$\therefore n = A\sqrt{\left(\frac{D}{a}\right)}\sinh\left\{\sqrt{\left(\frac{a}{D}\right)}x\right\} + \frac{b}{a} - \frac{b}{a}\cosh\left\{\sqrt{\left(\frac{a}{D}\right)}x\right\}$$

$$= \frac{A}{\alpha}\sinh(\alpha x) + \frac{b}{a}\{1 - \cosh(\alpha x)\}.$$

Since $n = 0$ when $x = c$, we have

$$\frac{A}{\alpha} = \frac{b}{a}\cdot\frac{\cosh(\alpha c) - 1}{\sinh(\alpha c)} = \frac{b}{a}\cdot\frac{2\sinh^2(\alpha c/2)}{2\sinh(\alpha c/2)\cosh(\alpha c/2)}$$

$$= \frac{b}{\alpha}\cdot\tanh\frac{\alpha c}{2}.$$

Hence,

$$n = \frac{b}{a}\left(1 - \cosh(\alpha x) + \sinh(\alpha x)\frac{\sinh(\alpha c/2)}{\cosh(\alpha c/2)}\right)$$

$$= \frac{b}{a}\left(1 - \frac{\cosh(\alpha x)\cosh(\alpha c/2) - \sinh(\alpha x)\sinh(\alpha c/2)}{\cosh(\alpha c/2)}\right)$$

i.e.

$$an = b\left(1 - \frac{\cosh\{\alpha(x - c/2)\}}{\cosh(\alpha c/2)}\right).$$

Finally, we consider a further type of differential equation which is solved by introducing a change of variable.

7. Non-linear differential equations

One such equation has already been discussed, namely, Bernoulli's equation, in which we made the substitution

$$z = y^{1-n}.$$

Other non-linear equations containing terms such as y^2, $(dy/dx)^2$ may be solved by putting $dy/dx = p$, so that

$$\frac{d^2y}{dx^2} = \frac{dp}{dx}.$$

Alternatively, as

$$\frac{dp}{dy} = \frac{d}{dy}\left(\frac{dy}{dx}\right) = \frac{d}{dx}\left(\frac{dy}{dx}\right)\frac{dx}{dy} = \frac{d^2 y}{dx^2} \cdot \frac{1}{p},$$

a solution may be obtained by replacing $d^2 y/dx^2$ by $p\, dp/dy$.

By so doing, the non-linear equation reduces to a linear equation and can be solved by one of the previous methods.

Example 21

Solve the equation

$$\frac{d^2 y}{dx^2} + \left(\frac{dy}{dx}\right)^2 + 1 = 0.$$

Let $p = dy/dx$, and so $dp/dx = d^2 y/dx^2$ and the equation becomes

$$\frac{dp}{dx} + p^2 + 1 = 0$$

i.e. 'variables separable'. Thus

$$\int \frac{dp}{p^2 + 1} + \int dx = A_1$$

where A_1 is an arbitrary constant. Therefore

$$\tan^{-1} p + x = A_1$$

i.e.

$$p \equiv \frac{dy}{dx} = \tan(A_1 - x).$$

Hence,

$$y = A_2 - \log_e \sec(A_1 - x)$$

where A_2 is an arbitrary constant.

Example 22

Solve the equation

$$2y\frac{d^2 y}{dx^2} - 3\left(\frac{dy}{dx}\right)^2 = y^2,$$

given that $dy/dx = 1$ when $y = 1$.

Let

$$p = \frac{dy}{dx}, \quad \text{then} \quad p\frac{dp}{dy} = \frac{d^2 y}{dx^2},$$

and the equation reduces to

$$2py \frac{dp}{dy} - 3p^2 = y^2$$

i.e.

$$\frac{dp}{dy} = \frac{3p^2 + y^2}{2py},$$

(35)

which is homogeneous in p and y. Following the method of Section 2, let $p = vy$, where v is some function of y. Then,

$$\frac{dp}{dy} \equiv v + y \frac{dv}{dy} = \frac{3v^2 + 1}{2v}, \quad \text{from (35)}$$

so that

$$y \frac{dv}{dy} = \frac{3v^2 + 1}{2v} - v = \frac{v^2 + 1}{2v}.$$

Separating the variables gives

$$\int \frac{2v}{v^2 + 1} \, dv = \int \frac{dy}{y} + \text{constant},$$

i.e.

$$\log_e(v^2 + 1) = \log_e(A_1 y),$$

where A_1 is an arbitrary constant.
Hence,

$$v^2 + 1 \equiv p^2/y^2 + 1 = A_1 y$$

i.e.

$$p^2 = y^2(A_1 y - 1).$$

Thus,

$$\frac{dy}{dx} = \pm y\sqrt{(A_1 y - 1)}$$

and since $dy/dx = 1$ when $y = 1$, thus $A_1 = 2$, and the positive root must be taken.
Therefore

$$\frac{dy}{dx} = y\sqrt{(2y - 1)}.$$

Separating the variables we obtain

$$\frac{dy}{y\sqrt{(2y - 1)}} = \int dx + \text{constant}.$$

(36)

For the integral on the left-hand side let

$$2y = \sec^2 \theta, \quad \therefore \frac{dy}{d\theta} = \sec^2 \theta \tan \theta$$

and thus

$$\int \frac{dy}{y\sqrt{(2y-1)}} = \int \frac{2 \sec^2 \theta \tan \theta \, d\theta}{\sec^2 \theta \tan \theta} = 2 \int d\theta$$

$$= 2\theta = 2 \sec^{-1}\sqrt{(2y)}.$$

Hence, equation (36) gives

$$2 \sec^{-1} \sqrt{(2y)} = x + A_2$$

where A_2 is an arbitrary constant, and so

$$y = \tfrac{1}{2} \sec^2 \left(\frac{x + A_2}{2} \right).$$

Example 23

A liquid of density ρ contains a spherical bubble of gas, the radius of the bubble being a and the pressure of the gas being p_0.

If the gas expands adiabatically, the radius r of the bubble at time t, is given by

$$2\rho r \frac{d^2 r}{dt^2} + 3\rho \left(\frac{dr}{dt} \right)^2 = 2p_0 \left(\frac{a}{r} \right)^{3\gamma},$$

where γ is the ratio of the specific heats of gas at constant pressure and volume respectively.

Show that

$$pr^3 \left(\frac{dr}{dt} \right)^2 = \frac{2p_0 a^3}{3(\gamma - 1)} \left\{ 1 - \left(\frac{a}{r} \right)^{3(\gamma - 1)} \right\}$$

and that when $\gamma = \tfrac{4}{3}$, r is given by the equation

$$3r^2 + 4ar + 8a^2 = 15at \left(\frac{ap_0}{2\rho(r-a)} \right)^{1/2}.$$

Let

$$u = \frac{dr}{dt}, \quad \text{then} \quad u \frac{du}{dr} = \frac{d^2 r}{dt^2}$$

and the equation becomes

$$2\rho r u \frac{du}{dr} + 3\rho u^2 = 2p_0 \left(\frac{a}{r} \right)^{3\gamma},$$

which is of the Bernoulli type.

The substitution

$$y = u^2, \quad \frac{dy}{dr} = 2u \frac{du}{dr}$$

now gives

$$\rho r \frac{dy}{dr} + 3\rho y = 2p_0 \left(\frac{a}{r}\right)^{3\gamma},$$

i.e.

$$\frac{dy}{dr} + \left(\frac{3}{r}\right) y = \frac{2p_0 \, a^{3\gamma}}{\rho r^{3\gamma + 1}}.$$

The integrating factor is $\exp(3 \int dr/r) = r^3$

$$\therefore \; yr^3 = \frac{2p_0 a^{3\gamma}}{\rho} \int r^{(2 - 3\gamma)} \, dr$$

$$= \frac{2p_0 a^{3\gamma}}{\rho} \cdot \frac{r^{3(1-\gamma)}}{3(1 - \gamma)} + A.$$

Since $y \equiv (dr/dt)^2 = 0$ when $r = a$, thus $A = 2p_0 \, a^3/\{3\rho(\gamma - 1)\}$ and hence,

$$yr^3 = \frac{2p_0}{3\rho(\gamma - 1)} \left\{ a^3 - r^3 \left(\frac{a}{r}\right)^{3\gamma} \right\},$$

i.e.

$$\rho r^3 \left(\frac{dr}{dt}\right)^2 = \frac{2p_0 a^3}{3(\gamma - 1)} \left\{ 1 - \left(\frac{a}{r}\right)^{3(\gamma - 1)} \right\}.$$

When $\gamma = \frac{4}{3}$, the equation simplifies to

$$\int \frac{r^2}{\sqrt{(r - a)}} \, dr = \left(\frac{2p_0 a^3}{\rho}\right)^{1/2} t + B. \tag{37}$$

In the integrand, let $r - a = x^2$, so that $dr/dx = 2x$ and

$$\int \frac{r^2}{\sqrt{(r - a)}} \, dr = 2 \int (x^4 + 2ax^2 + a^2) \, dx$$

$$= 2\left(\frac{x^5}{5} + 2a\frac{x^3}{3} + a^2 x\right)$$

$$= \frac{2x}{15} (3x^4 + 10ax^2 + 15a^2)$$

$$= \frac{2\sqrt{(r - a)}}{15} \{3(r - a)^2 + 10a(r - a) + 15a^2\}$$

$$= \frac{2\sqrt{(r - a)}}{15} (3r^2 + 4ar + 8a^2).$$

Subtituting into equation (37) now gives

$$\sqrt{(r-a)} \cdot \{3r^2 + 4ar + 8a^2\} = 15at\left(\frac{ap_0}{2\rho}\right)^{1/2},$$

B being zero, since $r = a$ at $t = 0$.

8. Solution in series

Consider the differential equation

$$\frac{d^2y}{dx^2} - \frac{dy}{dx} - 2y = 0, \tag{38}$$

the general solution of which is

$$y = A_1 e^{-x} + A_2 e^{2x},$$

A_1 and A_2 being arbitrary constants (see Example 16).

Since expansion of the exponential functions enables the solution to be expressed in terms of infinite series, viz.,

$$y = A_1\left(1 - \frac{x}{1!} + \frac{x^2}{2!} - \cdots\right) + A_2\left(1 + \frac{(2x)}{1!} + \frac{(2x)^2}{2!} + \cdots\right),$$

this suggests that a *trial* solution of the form

$$y = a_0 + a_1x + a_2x^2 + \cdots + a_rx^r + \cdots = \sum_{r=0}^{\infty} a_rx^r,$$

where the constants a_r are to be determined, may provide an alternative means of solving the equation and subsequently a technique for solving equations of a more complex form.

Assuming, therefore, that differentiation of the series term by term is permissible, we may write

$$\frac{dy}{dx} = a_1 + 2a_2x + 3a_3x^2 + 4a_4x^3 + \cdots + ra_rx^{r-1} + \cdots$$

and

$$\frac{d^2y}{dx^2} = 2a_2 + 6a_3x + 12a_4x^2 + \cdots + r(r-1)a_rx^{r-2} + \cdots.$$

Substitution into equation (38) then gives

$$(2a_2 + 6a_3x + 12a_4x^2 + \cdots) - (a_1 + 2a_2x + 3a_3x^2 + \cdots)$$
$$- 2(a_0 + a_1x + a_2x^2 + \cdots) = 0$$

or, re-arranging,

$$(2a_2 - a_1 - 2a_0) + x(6a_3 - 2a_2 - 2a_1) + x^2(12a_4 - 3a_3 - 2a_2) + \cdots = 0,$$

an equation which must hold for *all* values of x.

Equating to zero the coefficient of each power of x, gives

$$2a_2 - a_1 - 2a_0 = 0,$$

$$6a_3 - 2a_2 - 2a_1 = 0,$$

$$12a_4 - 3a_3 - 2a_2 = 0,$$

$$\vdots \qquad \vdots$$

and so on.

Although there are more unknowns than equations, it is possible to express any one unknown in terms of two others. For example, $a_2, a_3, a_4, \ldots$, may each be expressed in terms of a_0 and a_1 to give

$$a_2 = \tfrac{1}{2}(2a_0 + a_1),$$

$$a_3 = \tfrac{1}{6}(2a_1 + 2a_2) = \tfrac{1}{6}(2a_0 + 3a_1),$$

$$a_4 = \tfrac{1}{12}(2a_2 + 3a_3) = \tfrac{1}{12}(3a_0 + \tfrac{5}{2}a_1),$$

and so on.

Thus,

$$y = a_0 + a_1 x + a_2 x^2 + a_3 x^3 + a_4 x^4 + \cdots$$

$$= a_0 + a_1 x + \tfrac{1}{2}(2a_0 + a_1)x^2 + \tfrac{1}{6}(2a_0 + 3a_1)x^3 + \tfrac{1}{12}(3a_0 + \tfrac{5}{2}a_1)x^4 + \cdots$$

$$= a_0 + a_1 \frac{x}{1!} + (2a_0 + a_1)\frac{x^2}{2!} + (2a_0 + 3a_1)\frac{x^3}{3!} + (6a_0 + 5a_1)\frac{x^4}{4!} + \cdots.$$

The original form of the solution may be obtained substituting $a_0 = A_1 + A_2$, $a_1 = 2A_2 - A_1$, to give

$$y = A_1 + A_2 + (2A_2 - A_1)\frac{x}{1!} + (A_1 + 4A_2)\frac{x^2}{2!} + (-A_1 + 8A_2)\frac{x^3}{3!}$$

$$+ (A_1 + 16A_2)\frac{x^4}{4!} + \cdots$$

$$= A_1\left(1 - \frac{x}{1!} + \frac{x^2}{2!} - \frac{x^3}{3!} + \frac{x^4}{4!} - \cdots\right) +$$

$$+ A_2\left(1 + \frac{(2x)}{1!} + \frac{(2x)^2}{2!} + \frac{(2x)^3}{3!} + \frac{(2x)^4}{4!} + \cdots\right),$$

as before.

9. The method of Frobenius

When dealing with equations of the form

$$f(x)\frac{\mathrm{d}^2 y}{\mathrm{d}x^2} + g(x)\frac{\mathrm{d}y}{\mathrm{d}x} + h(x)y = 0,$$

where $f(x)$, $g(x)$, $h(x)$ are functions of x only—of which equation (38) is a special case—it is usual to adopt a more general trial solution of the form

$$y = x^c(a_0 + a_1 x + a_2 x^2 + \cdots + a_r x^r + \cdots) = \sum_{r=0}^{\infty} a_r x^{c+r} \tag{39}$$

where c is a constant to be determined later.

Before proceeding with this technique, it is important that we recognize the necessity to determine the range of values of x for which the series is valid.

Normally this would require extensive knowledge of a wide variety of *tests of convergence*. However, the validity of the series arising in the examples discussed here, may be examined by resorting to one test only, namely, *d'Alembert's ratio test*, which may be stated as follows:

If $\sum_{r=0}^{\infty} u_r$ is a series of terms such that

$$\lim_{r \to \infty} \left| \frac{u_{r+1}}{u_r} \right| = l,$$

then the series is said to be absolutely convergent if $l < 1$ (in which case it is also convergent, i.e. the sum to infinity is a finite quantity), and divergent if $l > 1$.

Alternative tests need to be consulted if $l = 1$.

Example 24

Solve the equation

$$3x \frac{d^2 y}{dx^2} + (1 - x) \frac{dy}{dx} + y = 0.$$

The procedure is as before, but for convenience we retain the summation notation and write

$$y = \sum_{r=0}^{\infty} a_r x^{c+r}, \qquad \frac{dy}{dx} = \sum_{r=0}^{\infty} a_r(c + r)x^{c+r-1},$$

$$\frac{d^2 y}{dx^2} = \sum_{r=0}^{\infty} a_r(c + r)(c + r - 1)x^{c+r-2}. \tag{40}$$

Substituting these expressions into the given equation gives

$$3x \sum_{r=0}^{\infty} a_r(c + r)(c + r - 1)x^{c+r-2} + (1 - x) \sum_{r=0}^{\infty} a_r(c + r)x^{c+r-1}$$

$$+ \sum_{r=0}^{\infty} a_r x^{c+r} = 0$$

i.e.

$$3 \sum_{r=0}^{\infty} a_r(c + r)(c + r - 1)x^{c+r-1} + \sum_{r=0}^{\infty} a_r(c + r)x^{c+r-1}$$

$$- \sum_{r=0}^{\infty} a_r(c+r)x^{c+r} + \sum_{r=0}^{\infty} a_r x^{c+r} = 0.$$

Combining similar terms, we find

$$\sum_{r=0}^{\infty} a_r(c+r)(3c+3r-2)x^{c+r-1} - \sum_{r=0}^{\infty} a_r(c+r-1)x^{c+r} = 0.$$

Removing the term corresponding to $r = 0$ from the first summation, the equation becomes

$$a_0(c)(3c-2)x^{c-1} + \sum_{r=1}^{\infty} a_r(c+r)(3c+3r-2)x^{c+r-1}$$

$$- \sum_{r=0}^{\infty} a_r(c+r-1)x^{c+r} = 0.$$

Replacing r by $r + 1$ in the first summation gives

$$a_0(c)(3c-2)x^{c-1} + \sum_{r=0}^{\infty} a_{r+1}(c+r+1)(3c+3r+1)x^{c+r}$$

$$- \sum_{r=0}^{\infty} a_r(c+r-1)x^{c+r} = 0.$$

Equating to zero the coefficient of x^{c-1} (the lowest power of x) gives

$$a_0(c)(3c-2) = 0,$$

a quadratic equation in c, called the *indicial equation*. Assuming, without loss of generality, that a_0 is non-zero, the solution in this case is $c = 0$ or $c = 2/3$.

The remaining constants, a_r, follow immediately on equating to zero the coefficients of x^{c+r}, to give

$$a_{r+1}(c+r+1)(3c+3r+1) - a_r(c+r-1) = 0$$

i.e.

$$a_{r+1} = \frac{c+r-1}{(c+r+1)(3c+3r+1)} a_r, \quad r = 0, 1, 2, \ldots \tag{41}$$

Case (i) $c = 0$:

With $c = 0$, equation (41) becomes

$$a_{r+1} = \frac{r-1}{(r+1)(3r+1)} a_r,$$

so that $a_1 = -a_0$ and $a_2 = a_3 = \cdots = 0$.

Thus,

$$y = a_0 + a_1 x, \quad \text{from equation (39)},$$

$$= a_0(1 - x)$$

$$= A_1(1 - x)$$

if we put $a_0 = A_1$.

Case (ii) c = 2/3:

With $c = 2/3$, equation (41) becomes

$$a_{r+1} = \frac{3r - 1}{3(3r + 5)(r + 1)} a_r,$$

so that

$$a_1 = \frac{-1}{3.5.1} a_0, \qquad a_2 = \frac{2}{3.8.2} a_1,$$

$$a_3 = \frac{5}{3.11.3} a_2, \qquad a_4 = \frac{8}{3.14.4} a_3,$$

and so on. Expressing each constant in terms of a_0 leads to

$$y = A_2 x^{2/3} \left(1 - \frac{2}{3(5.2)(1!)} x - \frac{2}{3^2(8.5)(2!)} x^2 - \frac{2}{3^3(11.8)(3!)} x^3 - \cdots \right)$$

if we put $a_0 = A_2$.

Applying d'Alembert's ratio test, we see that

$$\lim_{r \to \infty} \left| \frac{u_{r+1}}{u_r} \right| = \lim_{r \to \infty} \left| \frac{a_{r+1} x^{r+1}}{a_r x^r} \right|$$

$$= \lim_{r \to \infty} \left| \frac{a_{r+1} x}{a_r} \right|$$

$$= \lim_{r \to \infty} \left| \frac{(3r - 1)x}{3(3r + 5)(r + 1)} \right|$$

$$= 0,$$

showing that the series is convergent for all values of x.

It can be shown that the general solution is the sum of the two independent solutions.

Difficulties occur when the roots of the indicial equation are equal or differ by an integer, for then equation (39) yields only one of the two independent solutions expected. However, we concern ourselves here with simple cases only as a means of introducing a few of the *special functions* that may be of use to the reader, viz., Bessel functions, Hermite and Legendre polynomials.

The following examples illustrate the difficulties that can arise.

Example 25

Find a series solution to *Bessel's equation* of order n, viz.,

$$x^2 \frac{d^2 y}{dx^2} + x \frac{dy}{dx} + (x^2 - n^2)y = 0.$$

Inserting the series expansions of equation (40) into the given equation, we find

$$\sum_{r=0}^{\infty} a_r(c+r)(c+r-1)x^{c+r} + \sum_{r=0}^{\infty} a_r(c+r)x^{c+r} + \sum_{r=0}^{\infty} a_r x^{c+r+2}$$

$$- n^2 \sum_{r=0}^{\infty} a_r x^{c+r} = 0,$$

or, combining terms in x^{c+r},

$$\sum_{r=0}^{\infty} a_r(c+r+n)(c+r-n)x^{c+r} + \sum_{r=0}^{\infty} a_r x^{c+r+2} = 0. \tag{42}$$

Removing terms corresponding to $r = 0, 1$, the first summation can be written in the form

$$a_0(c+n)(c-n)x^c + a_1(c+n+1)(c-n+1)x^{c+1}$$

$$+ \sum_{r=2}^{\infty} a_r(c+r+n)(c+r-n)x^{c+r}$$

wherein

$$\sum_{r=2}^{\infty} a_r(c+r+n)(c+r-n)x^{c+r} \equiv \sum_{r=0}^{\infty} a_{r+2}(c+r+n+2)(c+r-n+2)x^{c+r+2},$$

so that equation (42) becomes

$$a_0(c+n)(c-n)x^c + a_1(c+n+1)(c-n+1)x^{c+1}$$

$$+ \sum_{r=0}^{\infty} \{a_{r+2}(c+r+n+2)(c+r-n+2) + a_r\}x^{c+r+2} = 0.$$

Equating coefficients to zero gives

$$a_0(c+n)(c-n) = 0, \tag{43}$$

$$a_1(c+n+1)(c-n+1) = 0, \tag{44}$$

$$a_{r+2}(c+r+n+2)(c+r-n+2) + a_r = 0. \tag{45}$$

Assuming $a_0 \neq 0$, equation (43) shows that $c = \pm n$, leading to two independent solutions if n is non-integral.

Suppose, however, that n *is* integral and consider first the case $c = +n$ (n positive). From equation (44) we see that $a_1 = 0$ and from equation (45)

$$a_{r+2} = -\frac{a_r}{(r+2n+2)(r+2)}.$$

Thus, $a_1 = a_3 = a_5 = \cdots = 0$, and

$$a_2 = -\frac{a_0}{(2n+2)(2)} = -\frac{a_0}{2^2(n+1)(1)},$$

$$a_4 = -\frac{a_2}{(2n+4)(4)} = -\frac{a_2}{2^2(n+2)(2)} = \frac{a_0}{2^4(n+2)(n+1)(2!)},$$

$$a_6 = -\frac{a_4}{(2n+6)(6)} = -\frac{a_4}{2^2(n+3)(3)} = -\frac{a_0}{2^6(n+3)(n+2)(n+1)(3!)},$$

and so on, giving

$$y = a_0 x^n \left(1 - \frac{(x/2)^2}{1!(n+1)} + \frac{(x/2)^4}{2!(n+1)(n+2)} - \frac{(x/2)^6}{3!(n+1)(n+2)(n+3)} + \cdots\right).$$

The *Bessel function*, $J_n(x)$, of the *first kind* of order n, is obtained by writing $a_0 = 1/(2^n.n!)$ to give

$$J_n(x) = \frac{(x/2)^n}{n!}\left(1 - \frac{(x/2)^2}{1!(n+1)} + \frac{(x/2)^4}{2!(n+1)(n+2)} - \cdots\right), \tag{46}$$

which is convergent for all values of x.

In the case $c = -n$, equation (45) becomes

$$a_{r+2}(r+2)(r-2n+2) + a_r = 0,$$

giving

$$a_{r+2} = -\frac{a_r}{(r+2)(r-2n+2)},$$

yielding coefficients of infinite value when $n = 1, 2, \ldots$, and no second solution. For $n = 0$, equation (46) still holds, but in this case the indicial equation has equal roots ($c = 0$, twice), and again no second solution is found.

However, by making use of the known solutions prior to insertion of the value found for c, it can be shown that second solutions take the form $(\partial y/\partial c)_{c=n}$, the appropriate value of c being inserted *after* differentiation.

Example 26

Obtain a power series solution to *Hermite's differential equation*

$$\frac{d^2 y}{dx^2} - 2x\frac{dy}{dx} + 2ny = 0, \quad n \text{ integral}, \geqslant 0.$$

Utilizing the expansions of equation (40) we find that

$$\sum_{r=0}^{\infty} a_r(c+r)(c+r-1)x^{c+r-2} - 2\sum_{r=0}^{\infty} a_r(c+r)x^{c+r} + 2n\sum_{r=0}^{\infty} a_r x^{c+r} = 0. \tag{47}$$

The first summation may be re-arranged by removing the terms corresponding to $r = 0, 1$, to give

$$\sum_{r=0}^{\infty} a_r(c+r)(c+r-1)x^{c+r-2}$$

$$= a_0(c)(c-1)x^{c-2} + a_1(c+1)(c)x^{c-1} + \sum_{r=2}^{\infty} a_r(c+r)(c+r-1)x^{c+r-2}$$

$$= a_0(c)(c-1)x^{c-2} + a_1(c+1)(c)x^{c-1} + \sum_{r=0}^{\infty} a_{r+2}(c+r+2)(c+r+1)x^{c+r}.$$

Substituting into equation (47) and combining terms in a_r, gives

$$a_0(c)(c-1)x^{c-2} + a_1(c+1)(c)x^{c-1} + \sum_{r=0}^{\infty} a_{r+2}(c+r+2)(c+r+1)x^{c+r}$$

$$-2\sum_{r=0}^{\infty} a_r(c+r-n)x^{c+r} = 0.$$

Thus, equating coefficients to zero,

$$a_0(c)(c-1) = 0,$$

giving $c = 0$ or 1, if we assume $a_0 \neq 0$, and

$$a_1(c+1)(c) = 0, \tag{48}$$

$$a_{r+2}(c+r+2)(c+r+1) = 2a_r(c+r-n). \tag{49}$$

If $c = 0$, equation (48) shows that a_1 is indeterminate, so that in this case we have no choice but to express the remaining coefficients in terms of both a_0 and a_1.

From equation (49),

$$a_{r+2}(r+2)(r+1) = 2a_r(r-n)$$

and therefore

$$a_{r+2} = -\frac{2(n-r)}{(r+2)(r+1)} a_r,$$

giving

$$a_2 = -\frac{2n}{2!}a_0, \qquad\qquad a_3 = -\frac{2(n-1)}{3!}a_1,$$

$$a_4 = -\frac{2(n-2)}{4.3}a_2, \qquad a_5 = -\frac{2(n-3)}{5.4}a_3,$$

$$= \frac{2^2 n(n-2)}{4!}a_0, \qquad\qquad = \frac{2^2(n-1)(n-3)}{5!}a_1,$$

and so on.

Thus

$$y = a_0\left(1 - \frac{2n}{2!}x^2 + \frac{2^2 n(n-2)}{4!}x^4 - \cdots\right)$$

$$+ a_1\left(x - \frac{2(n-1)}{3!}x^3 + \frac{2^2(n-1)(n-3)}{5!}x^5 - \cdots\right), \tag{50}$$

both series being convergent for all values of x.

It is left as an exercise for the reader, to show that no new independent solution arises when $c = 1$; the solution corresponding to $c = 1$ is simply a multiple of the second series arising in equation (50).

If the initial conditions are chosen so that $a_0 = 1$, $a_1 = 0$ for n even, and $a_0 = 0$, $a_1 = 1$ for n odd, then the infinite series of equation (50) degenerate into finite polynomials. Normalizing each expression (by multiplying throughout by an appropriate constant) so that the coefficient of the highest power is $+2^n$, gives rise to the *Hermite polynomials* $H_n(x)$ *of order n.*

For example, for $n = 0, 1, 2, 3, 4, 5$, we find:

$$H_0(x) = 1, \qquad\qquad H_1(x) = 2x,$$

$$H_2(x) = 4x^2 - 2, \qquad\qquad H_3(x) = 8x^3 - 12x,$$

$$H_4(x) = 16x^4 - 48x^2 + 12, \qquad H_5(x) = 32x^5 - 160x^3 + 120x.$$

Example 27

Find a series solution to *Legendre's equation* of order n, viz.,

$$(1 - x^2)\frac{d^2 y}{dx^2} - 2x\frac{dy}{dx} + n(n + 1)y = 0.$$

In this case, insertion of equation (40) gives

$$(1 - x^2) \sum_{r=0}^{\infty} a_r(c + r)(c + r - 1)x^{c+r-2} - 2x \sum_{r=0}^{\infty} a_r(c + r)x^{c+r-1}$$

$$+ n(n + 1) \sum_{r=0}^{\infty} a_r x^{c+r} = 0,$$

or, re-arranging as in the previous examples,

$$\sum_{r=0}^{\infty} a_r(c + r)(c + r - 1)x^{c+r-2} - \sum_{r=0}^{\infty} a_r(c + r + n + 1)(c + r - n)x^{c+r} = 0.$$

The left-hand summation is equivalent to

$$a_0(c)(c - 1)x^{c-2} + a_1(c + 1)(c)x^{c-1} + \sum_{r=0}^{\infty} a_{r+2}(c + r + 2)(c + r + 1)x^{c+r}$$

so that the complete equation becomes

$$a_0(c)(c - 1)x^{c-2} + a_1(c + 1)(c)x^{c-1}$$

$$+ \sum_{r=0}^{\infty} \{a_{r+2}(c + r + 2)(c + r + 1) - a_r(c + r + n + 1)(c + r - n)\}x^{c+r}$$

Thus,

$$a_0(c)(c - 1) = 0$$

giving $c = 0$ or 1, if we assume $a_0 \neq 0$,

$$a_1(c+1)(c) = 0 \tag{51}$$

and

$$a_{r+2}(c+r+2)(c+r+1) = a_r(c+r+n+1)(c+r-n). \tag{52}$$

If $c = 0$, a_1 is indeterminate and we proceed as in the previous example.

From equation (52),

$$a_{r+2} = \frac{-(n-r)(n+r+1)}{(r+2)(r+1)} a_r,$$

giving

$$a_2 = -\frac{n(n+1)}{2!} a_0, \qquad\qquad a_3 = -\frac{(n-1)(n+2)}{3!} a_1,$$

$$a_4 = -\frac{(n-2)(n+3)}{4.3} a_2, \qquad\qquad a_5 = -\frac{(n-3)(n+4)}{5.4} a_3,$$

$$= \frac{n(n-2)(n+1)(n+3)}{4!} a_0, \qquad\qquad = \frac{(n-1)(n-3)(n+2)(n+4)}{5!} a_1,$$

and similarly for the higher coefficients.

Thus,

$$y = a_0 \left(1 - \frac{n(n+1)}{2!} x^2 + \frac{n(n-2)(n+1)(n+3)}{4!} x^4 - \cdots \right)$$

$$+ a_1 \left(x - \frac{(n-1)(n+2)}{3!} x^3 + \frac{(n-1)(n-3)(n+2)(n+4)}{5!} x^5 - \cdots \right),$$

the series being convergent for $|x| < 1$. No new independent solution arises when $c = 1$.

If n is a positive integer, the *Legendre polynomials* $P_n(x)$ may be obtained by choosing $a_1 = 0$ for n even, $a_0 = 0$ for n odd, and $a_0 (n$ even) or $a_1 (n$ odd) such that $P_n(1) = 1$.

The Legendre polynomials corresponding to $n = 0, 1, 2, 3, 4, 5$ are thus

$$P_0(x) = 1, \qquad\qquad P_1(x) = x,$$

$$P_2(x) = \tfrac{1}{2}(3x^2 - 1), \qquad\qquad P_3(x) = \tfrac{1}{2}(5x^3 - 3x),$$

$$P_4(x) = \tfrac{1}{8}(35x^4 - 30x^2 + 3), \qquad P_5(x) = \tfrac{1}{8}(63x^5 - 70x^3 + 15x).$$

PROBLEMS

1. Solve the following differential equations:

(i) $\dfrac{dy}{dx} = \dfrac{2x + 3y}{x - y}$, (ii) $\dfrac{dy}{dx} = \dfrac{3x^2 - 2xy + y^2}{xy + y^2}$,

(iii) $\dfrac{dy}{dx} = \dfrac{3x + y - 5}{x + y - 1}$, (iv) $\dfrac{dy}{dx} = \dfrac{x - 2y + 3}{3x - 6y - 1}$,

(v) $(2x^2 - x - 1)\dfrac{dy}{dx} - 3(x + 1)y = (x - 1)^3(2x + 1)$,

(vi) $\cos^2 x \dfrac{dy}{dx} - 3y \cos^2 x \cot x = \sin^3 x$,

(vii) $\dfrac{dy}{dx} + \dfrac{8xy}{1 + x} = 4(1 + x)y^{3/4} \exp(-2x)$.

2. Find the Laplace transforms of the following functions:

(i) x^5,   (ii) $\exp(-3x)$,   (iii) $\sin 4x$,

(iv) $\exp(-x/2)\cos 2x$,   (v) $x^3 e^{2x}$.

3. Invert the following transforms:

(i) $\dfrac{8 - s}{s^2 + 4}$,   (ii) $\dfrac{s^2 - 4s + 5}{(s - 3)(s - 1)(s + 1)}$,   (iii) $\dfrac{4s}{4s^2 - 8s + 5}$

(iv) $\dfrac{s^4 - 5s^3 + 3s^2 - 4s + 2}{s^3(s - 1)^2}$,   (v) $\dfrac{s^2 + 27s + 9}{s^2(s^2 + 9)}$.

4. Find from first principles, the Laplace transforms of,

$x \sin 2x$, $e^x \sin 3x$.

5. Solve the equation

$$2\dfrac{d^2 y}{dx^2} - 3\dfrac{dy}{dx} + y = x,$$

given that $y = dy/dx = 0$ when $x = 0$.

6. Use the substitution $x = e^t$ to convert the equation

$$x^3 \dfrac{d^3 y}{dx^3} + 3x^2 \dfrac{d^2 y}{dx^2} - 3x \dfrac{dy}{dx} = 0,$$

into a linear equation with constant coefficients, and solve for y given that

$$y = \dfrac{dy}{dx} = 0, \quad \dfrac{d^2 y}{dx^2} = 1 \quad \text{when} \quad x = 1.$$

7. Solve the simultaneous differential equations

$$\dfrac{dx}{dt} + 5\dfrac{dy}{dt} + x + y = t, \quad \dfrac{dx}{dt} + 4\dfrac{dy}{dt} - y = 4 \quad (x = y = 0 \text{ when } t = 0).$$

8. Derive the Laplace transform of the function $t \sin(wt)$ where w is a positive constant.
Hence solve the equation

$$\dfrac{d^2 x}{dt^2} + 4x = \cos 2t,$$

given that $x = 3$, $dx/dt = 2$ when $t = 0$.

9. Solve the equation

$$\frac{d^2 y}{dx^2} = \frac{1}{d^2 x/dy^2}.$$

10. When a catalyst was added to N cm^3 of hydrogen peroxide, it was found that 15.4 cm^3 of oxygen were produced in 5 minutes and 25.0 cm^3 in 10 minutes. Assuming the reaction to obey the law

$$\frac{dx}{dt} = k(N - x),$$

where x is the amount of oxygen produced after time t minutes, and k is the velocity constant, determine the amount of oxygen produced after 15 minutes.

11. In a chemical reaction, a chemical A forms a chemical B which in turn forms a chemical C. If x, y, z are the concentrations of A, B, C respectively at time t, then

$$\frac{dx}{dt} = -\lambda_1 x, \quad \frac{dy}{dt} = \lambda_1 x - \lambda_2 y, \quad \frac{dz}{dt} = \lambda_2 y,$$

where λ_1, λ_2 are constants. Given that $x = a$, $y = z = 0$, when $t = 0$, obtain expressions for x, y, z in terms of t.

If $\lambda_1 = \frac{2}{3}$ and $\lambda_2 = \frac{8}{3}$, show that the maximum value of y is $(2^{-8/3})a$.

12. In the irreversible ter-molecular reaction $A + B + C \rightarrow D$, the initial concentration of A is 7 g mol l^{-1}, and the concentration of both B and C is 5 g mol l^{-1}. If x g mol l^{-1} of D are produced after a time t minutes, where

$$\frac{dx}{dt} = k(7 - x)(5 - x)^2,$$

k being the velocity constant for the reaction, show that

$$kt = \frac{1}{4} \log_e \left(\frac{1 - x/5}{1 - x/7}\right) + \frac{x}{10(5 - x)}.$$

If 1 g mol l^{-1} of D is produced after 1 minute, find how long it will take to produce 3 g mol l^{-1} of D.

13. Heat is applied to one end of a metal rod of radius a. If the temperature of the rod is T at a distance x from the heated end, the lateral heat loss is given by the equation,

$$\frac{d^2 T}{dx^2} - \frac{2ET}{ka} = 0,$$

where E is the heat transfer coefficient and k is the thermal conductivity, both being constants for the rod concerned.

If $T = T_0$ when $x = 0$, and $T = 0$ when x is very large, show that

$$T = T_0 \exp\left\{-\sqrt{\left(\frac{2E}{ka}\right)}x\right\}.$$

14. The coordinates of an electron moving in the xy-plane about its nucleus, are given by

$$2\frac{dx}{dt} + y = 3, \quad \frac{dy}{dt} - 2x = 4,$$

at any time t.

H

Solve the equations, given that $x = 0$ and $y = 3$ when $t = 0$. By eliminating t, show that the equation of the path of the electron is the ellipse

$$4x^2 + y^2 + 16x - 6y + 9 = 0.$$

15. In the successive reversible reaction $A \rightleftharpoons B \rightleftharpoons C$, the amount of A transformed into B at time t is x, where

$$\frac{d^2x}{dt^2} + 3\frac{dx}{dt} + 2x = a,$$

a being the initial concentration of A. If $x = 0$ and $dx/dt = 2a/3$ at $t = 0$, obtain x in terms of a and t.

16. In the theory of vibrating diatomic molecules, the following equation occurs

$$\mu\frac{d^2x}{dt^2} + 2w\frac{dx}{dt} + kx = 0,$$

where μ, w, k are constants, and x is the distance of a molecule from its rest position at time t.

Determine the general solution and show that in the case $w = 0$,

$$x = A \sin\left\{\sqrt{\left(\frac{k}{\mu}\right)}t\right\},$$

where A is an arbitrary constant.

17. A liquid of density ρ contains a sperical bubble of gas of mean radius a, the pressure of the gas being P when the radius is a.

Assuming the gas to obey the law $pv = \text{constant}$, the radius r of the bubble at time t, is given by

$$\rho r\frac{d^2r}{dt^2} + \frac{3\rho}{2}\left(\frac{dr}{dt}\right)^2 + P = p.$$

Show that the period of oscillation of the surface of the bubble is approximately

$$2\pi a\left(\frac{\rho}{3P}\right)^{\frac{1}{2}}.$$

18. Solve the following differential equations by the method of Frobenius:

(i) $3x\dfrac{d^2y}{dx^2} + 2\dfrac{dy}{dx} + y = 0,$ (ii) $(1 - x^2)\dfrac{d^2y}{dx^2} - 2x\dfrac{dy}{dx} + 12y = 0,$

(iii) $x(2 + x^2)\dfrac{d^2y}{dx^2} - \dfrac{dy}{dx} - 6xy = 0.$

19. Show that

$$P_n(x) = \frac{1}{2^n n!}\frac{d^n}{dx^n}(x^2 - 1)^n$$

satisfies Legendre's equation of order n. Use the formula to verify the results of example 27. (This generating function for the Legendre polynomials is known as *Rodrigues' formula*.)

20. Use the method of solution in series to show that

$$y = 1 - nx + \frac{n(n-1)}{(2!)^2} x^2 - \frac{n(n-1)(n-2)}{(3!)^2} x^3 + \cdots$$

is a solution of *Laguerre's differential equation*

$$x \frac{d^2 y}{dx^2} + (1 - x) \frac{dy}{dx} + ny = 0.$$

CHAPTER 9

Partial Differentiation

Consider the function

$$\phi = x^2 y^3 - 2y^2 + \sin x \cos y,$$

of the two independent variables x and y.

If we differentiate ϕ with respect to x, treating y as a constant, we obtain the expression

$$2xy^3 + \cos x \cos y$$

which is called the *partial derivative* of ϕ with respect to x, and is denoted by $\partial\phi/\partial x$ or ϕ_x.

Thus,

$$\frac{\partial\phi}{\partial x} = 2xy^3 + \cos x \cos y.$$

Similarly, if we differentiate ϕ with respect to y, treating x as a constant, we obtain

$$3x^2 y^2 - 4y - \sin x \sin y$$

which is called the partial derivative of ϕ with respect to y, and is denoted by $\partial\phi/\partial y$ or ϕ_y.

Thus,

$$\frac{\partial\phi}{\partial y} = 3x^2 y^2 - 4y - \sin x \sin y.$$

If we now differentiate $\partial\phi/\partial x$ with respect to x keeping y constant, we obtain the expression

$$2y^3 - \sin x \cos y$$

which is called the partial derivative of $\partial\phi/\partial x$ with respect to x, and is denoted by $\partial^2\phi/\partial x^2$ or ϕ_{xx}, so that

$$\frac{\partial^2\phi}{\partial x^2} = 2y^3 - \sin x \cos y.$$

Similarly, differentiating $\partial\phi/\partial y$ *partially* with respect to y gives

$$\frac{\partial^2\phi}{\partial y^2} = 6x^2y - 4 - \sin x \cos y \equiv \phi_{yy}.$$

If we differentiate $\partial\phi/\partial x$ with respect to y keeping x constant, we obtain

$$6xy^2 - \cos x \sin y$$

which is denoted by $\partial^2\phi/\partial y\partial x$ or ϕ_{xy}. Hence

$$\frac{\partial^2\phi}{\partial y\partial x} = 6xy^2 - \cos x \sin y.$$

Similarly the partial derivative of $\partial\phi/\partial y$ with respect to x is

$$\frac{\partial^2\phi}{\partial x\partial y} = 6xy^2 - \cos x \sin y.$$

Note that for the above function

$$\frac{\partial^2\phi}{\partial x\partial y} = \frac{\partial^2\phi}{\partial y\partial x}.$$

This is always true for functions which are continuous within the region in which we are differentiating. This condition will be assumed in the ensuing theory, so that we may obtain particular partial derivatives by differentiating in any order

Thus, differentiating $\partial^2\phi/\partial x\partial y$ partially with respect to y gives

$$\frac{\partial^3\phi}{\partial y\partial x\partial y} \equiv \frac{\partial^3\phi}{\partial x\partial y^2} = 12xy - \cos x \cos y,$$

and so on.

1. Partial derivative as a limit

In Chapter 3, we saw that the derivative of the function $y = f(x)$ with respect to x, is defined as

$$\frac{dy}{dx} = \lim_{\delta x \to 0}\left(\frac{\delta y}{\delta x}\right) = \lim_{\delta x \to 0}\frac{f(x + \delta x) - f(x)}{\delta x}.$$

Partial derivatives may be defined in a similar way.

Let $\phi = \phi(x, y)$ be any function of x and y, and let an increment δx in x produce a corresponding increment $\delta\phi$ in ϕ, y being kept constant.

Then,

$$\phi = \phi(x, y)$$

$$\phi + \delta\phi = \phi(x + \delta x, y)$$

and

$$\delta\phi = \phi(x + \delta x, y) - \phi(x, y).$$

Thus for the function

$$\phi = x^2 y^3 - 2y^2 + \sin x \cos y$$

$$\delta\phi = (x + \delta x)^2 y^3 - 2y^2 + \sin(x + \delta x)\cos y - x^2 y^3 + 2y^2 - \sin x \cos y$$

$$= 2xy^3 \delta x + y^3 (\delta x)^2 + \cos y\{\sin(x + \delta x) - \sin x\}$$

$$= 2xy^3 \delta x + y^3 (\delta x)^2 + 2\cos y \cos(x + \tfrac{1}{2}\delta x)\sin(\tfrac{1}{2} \delta x).$$

$$\therefore \frac{\delta\phi}{\delta x} = 2xy^3 + \cos y \cos(x + \tfrac{1}{2} \delta x)\frac{\sin(\tfrac{1}{2} \delta x)}{(\tfrac{1}{2} \delta x)} y^3 \delta x$$

$$\therefore \lim_{\delta x \to 0} \left(\frac{\delta\phi}{\delta x}\right) = 2xy^3 + \cos y \cos x = \frac{\partial\phi}{\partial x}.$$

We thus define

$$\frac{\partial\phi}{\partial x} = \lim_{\delta x \to 0}\left(\frac{\delta\phi}{\delta x}\right) = \lim_{\delta x \to 0}\frac{\phi(x + \delta x, y) - \phi(x, y)}{\delta x}$$

and similarly,

$$\frac{\partial\phi}{\partial y} = \lim_{\delta y \to 0}\left(\frac{\delta\phi}{\delta y}\right) = \lim_{\delta y \to 0}\frac{\phi(x, y + \delta y) - \phi(x, y)}{\delta y}$$

From this definition, it follows that the rules for obtaining a partial derivative of a function of several variables are the same as the rules for obtaining the derivative of a function of one variable.

In general if $\phi = \phi(x, y, z, \ldots)$ is a function of several independent variables $x, y, z, \ldots$, then the partial derivative of the function with respect to one of these variables, is obtained by differentiating the function with respect to that variable, treating all other variables as constants.

2. Function of a function

Suppose that $\phi = \phi(t)$, where t is some function of $x, y, z, \ldots$, i.e.

$$t = t(x, y, z, \ldots)$$

Then,

$$\frac{\partial\phi}{\partial x} = \frac{\partial\phi(t)}{\partial x}$$

$$= \frac{\partial}{\partial t}\{\phi(t)\}\frac{\partial t}{\partial x}.$$

Since $\phi(t)$ is a function of the single variable t, $\partial\phi/\partial t$ is superfluous and the expression

$$\frac{\partial}{\partial t}\{\phi(t)\}$$

may be replaced by $d\phi/dt$, or simply $\phi'(t)$, where the dash denotes differentiation with respect to t.

Hence,

$$\frac{\partial \phi}{\partial x} = \frac{d\phi}{dt} \frac{\partial t}{\partial x}, \quad \frac{\partial \phi}{\partial y} = \frac{d\phi}{dt} \frac{\partial t}{\partial y}, \quad \frac{\partial \phi}{\partial z} = \frac{d\phi}{dt} \frac{\partial t}{\partial z},$$

and so on.

Example 1

If $z = 3x^4 - x^3 y + 2x^2 y^2 - 7y^4$, show that

$$x^2 \frac{\partial^2 z}{\partial x^2} + y^2 \frac{\partial^2 z}{\partial y^2} + 2xy \frac{\partial^2 z}{\partial x \partial y} = 12z.$$

Differentiating z partially with respect to x gives

$$\frac{\partial z}{\partial x} = 12x^3 - 3x^2 y + 4xy^2,$$

and differentiating $\partial z/\partial x$ partially with respect to x gives

$$\frac{\partial^2 z}{\partial x^2} = 36x^2 - 6xy + 4y^2.$$

Similarly,

$$\frac{\partial z}{\partial y} = -x^3 + 4x^2 y - 28y^3,$$

$$\frac{\partial^2 z}{\partial y^2} = 4x^2 - 84y^2,$$

and

$$\frac{\partial^2 z}{\partial x \partial y} = 8xy - 3x^2.$$

$$\therefore x^2 \frac{\partial^2 z}{\partial x^2} + y^2 \frac{\partial^2 z}{\partial y^2} + 2xy \frac{\partial^2 z}{\partial x \partial y} = 36x^4 - 12x^3 y + 24x^2 y^2 - 84y^4$$

$$= 12z.$$

Example 2

If $z = \sin^{-1}(x/y)$, show that

$$x \frac{\partial z}{\partial x} + y \frac{\partial z}{\partial y} = 0.$$

We have,

$$\frac{\partial z}{\partial x} = \frac{1}{\sqrt{(1 - x^2/y^2)}} \cdot \frac{1}{y} = \frac{1}{\sqrt{(y^2 - x^2)}},$$

and

$$\frac{\partial z}{\partial y} = \frac{1}{\sqrt{(1 - x^2/y^2)}} \cdot \frac{-x}{y^2} = \frac{-x}{y\sqrt{(y^2 - x^2)}},$$

so that

$$x \frac{\partial z}{\partial x} + y \frac{\partial z}{\partial y} = \frac{x}{\sqrt{(y^2 - x^2)}} - \frac{x}{\sqrt{(y^2 - x^2)}} = 0.$$

Example 3

If $\phi(x, y) = (x^3 + y^3)/(x - y)$, show that

$$x \frac{\partial \phi}{\partial y} + y \frac{\partial \phi}{\partial x} = \frac{(x + y)^3}{x - y}.$$

Differentiating ϕ partially with respect to x gives

$$\frac{\partial \phi}{\partial x} = \frac{3x^2(x - y) - (x^3 + y^3)}{(x - y)^2} = \frac{2x^3 - 3x^2 y - y^3}{(x - y)^2}.$$

Differentiating ϕ partially with respect to y gives

$$\frac{\partial \phi}{\partial y} = \frac{3y^2(x - y) + (x^3 + y^3)}{(x - y)^2} = \frac{3y^2 x - 2y^3 + x^3}{(x - y)^2}.$$

Hence,

$$x \frac{\partial \phi}{\partial y} + y \frac{\partial \phi}{\partial x} = \frac{x^4 - 2y^3 x + 2x^3 y - y^4}{(x - y)^2}$$

$$= \frac{(x - y)(x^3 + 3x^2 y + 3xy^2 + y^3)}{(x - y)^2}$$

i.e.

$$x \frac{\partial \phi}{\partial y} + y \frac{\partial \phi}{\partial x} = \frac{(x + y)^3}{x - y}.$$

Example 4

Given that $z = x \log_e (1 + x^2 y^2) + \tan^{-1}(xy)$, and $x \neq 0, y \neq 0$, show that

$$x^2 \frac{\partial^2 z}{\partial x^2} + y^2 \frac{\partial^2 z}{\partial y^2} = 0, \quad \text{when } y = 2.$$

Differentiating z partially with respect to x we obtain,

$$\frac{\partial z}{\partial x} = \log_e(1 + x^2 y^2) + x \cdot \frac{2xy^2}{1 + x^2 y^2} + \frac{y}{1 + x^2 y^2}$$

$$= \log_e(1 + x^2 y^2) + \frac{2x^2 y^2 + y}{1 + x^2 y^2}.$$

$$\therefore \frac{\partial^2 z}{\partial x^2} = \frac{2xy^2}{1 + x^2 y^2} + \frac{4xy^2(1 + x^2 y^2) - (2x^2 y^2 + y)2xy^2}{(1 + x^2 y^2)^2}$$

$$= \frac{2xy^2}{1 + x^2 y^2} + \frac{2xy^2(2 - y)}{(1 + x^2 y^2)^2}$$

$$= \frac{2xy^2(x^2 y^2 - y + 3)}{(1 + x^2 y^2)^2}.$$

Similarly,

$$\frac{\partial z}{\partial y} = x \cdot \frac{2x^2 y}{1 + x^2 y^2} + \frac{x}{1 + x^2 y^2} = \frac{2x^3 y + x}{1 + x^2 y^2}.$$

$$\therefore \frac{\partial^2 z}{\partial y^2} = \frac{2x^3 (1 + x^2 y^2) - (2x^3 y + x)2x^2 y}{(1 + x^2 y^2)^2}$$

$$= \frac{-2x^3 (x^2 y^2 + y - 1)}{(1 + x^2 y^2)^2}$$

$$\therefore x^2 \frac{\partial^2 z}{\partial x^2} + y^2 \frac{\partial^2 z}{\partial y^2} = \frac{2x^3 y^2 (4 - 2y)}{(1 + x^2 y^2)^2} = 0 \quad \text{when} \quad y = 2.$$

Example 5

If $u = \exp(xyz)f(xy/z)$, where $f(xy/z)$ is an arbitrary function of xy/z, show that

$$x \frac{\partial u}{\partial x} + z \frac{\partial u}{\partial z} = y \frac{\partial u}{\partial y} + z \frac{\partial u}{\partial z} = 2xyzu,$$

and that

$$x^2 \frac{\partial^2 u}{\partial x^2} + z^2 \frac{\partial^2 u}{\partial z^2} + 2xz \frac{\partial^2 u}{\partial x \partial z} = 2xyz(2xyz + 1)u.$$

Since $f(xy/z)$ is a function of a function, therefore

$$\frac{\partial}{\partial x} f\left(\frac{xy}{z}\right) = f'\left(\frac{xy}{z}\right) \times \frac{\partial\left(\frac{xy}{z}\right)}{\partial x}$$

$$= \frac{y}{z} f'\left(\frac{xy}{z}\right).$$

Similarly,

$$\frac{\partial}{\partial y} f\left(\frac{xy}{z}\right) = \frac{x}{z} f'\left(\frac{xy}{z}\right), \quad \text{and} \quad \frac{\partial}{\partial z} f\left(\frac{xy}{z}\right) = \frac{-xy}{z^2} f'\left(\frac{xy}{z}\right).$$

Hence,

$$\frac{\partial u}{\partial x} = yz \exp(xyz)f\left(\frac{xy}{z}\right) + \frac{y}{z} \exp(xyz)f'\left(\frac{xy}{z}\right),$$

$$\frac{\partial u}{\partial y} = xz \exp(xyz)f\left(\frac{xy}{z}\right) + \frac{x}{z} \exp(xyz)f'\left(\frac{xy}{z}\right),$$

and

$$\frac{\partial u}{\partial z} = xy \exp(xyz)f\left(\frac{xy}{z}\right) - \frac{xy}{z^2} \exp(xyz)f'\left(\frac{xy}{z}\right).$$

$$\therefore x \frac{\partial u}{\partial x} + z \frac{\partial u}{\partial z} = y \frac{\partial u}{\partial y} + z \frac{\partial u}{\partial z} = 2xyz \exp(xyz)f\left(\frac{xy}{z}\right)$$

$$= 2xyzu.$$

i.e.

$$\left(x\frac{\partial}{\partial x} + z\frac{\partial}{\partial z} \right) u = 2xyzu.$$

Writing

$$D \equiv x\frac{\partial}{\partial x} + z\frac{\partial}{\partial z}$$

we have

$$D\left(x\frac{\partial u}{\partial x} + z\frac{\partial u}{\partial z} \right) = D(2xyzu)$$

i.e.

$$\left(x\frac{\partial}{\partial x} + z\frac{\partial}{\partial z} \right)\left(x\frac{\partial u}{\partial x} + z\frac{\partial u}{\partial z} \right) = \left(x\frac{\partial}{\partial x} + z\frac{\partial}{\partial z} \right)(2xyzu)$$

i.e.

$$x\frac{\partial}{\partial x}\left(x\frac{\partial u}{\partial x} \right) + x\frac{\partial}{\partial x}\left(z\frac{\partial u}{\partial z} \right) + z\frac{\partial}{\partial z}\left(x\frac{\partial u}{\partial x} \right) + z\frac{\partial}{\partial z} z\frac{\partial u}{\partial z}$$

$$= x\frac{\partial}{\partial x}(2xyzu) + z\frac{\partial}{\partial z}(2xyzu)$$

i.e.

$$x\frac{\partial u}{\partial x} + x^2\frac{\partial^2 u}{\partial x^2} + xz\frac{\partial^2 u}{\partial x\partial z} + zx\frac{\partial^2 u}{\partial z\partial x} + z\frac{\partial u}{\partial z} + z^2\frac{\partial^2 u}{\partial z^2}$$

$$= x\left(2yzu + 2xyz\frac{\partial u}{\partial x} \right) + z\left(2xyu + 2xyz\frac{\partial u}{\partial z} \right).$$

$$\therefore \quad x^2\frac{\partial^2 u}{\partial x^2} + z^2\frac{\partial^2 u}{\partial z^2} + 2xz\frac{\partial^2 u}{\partial x\partial z} + \left(x\frac{\partial u}{\partial x} + z\frac{\partial u}{\partial z} \right)$$

$$= 2xyz\left(x\frac{\partial u}{\partial x} + z\frac{\partial u}{\partial z} \right) + 4xyzu,$$

and hence,

$$x^2\frac{\partial^2 u}{\partial x^2} + z^2\frac{\partial^2 u}{\partial z^2} + 2xz\frac{\partial^2 u}{\partial x\partial z} = 4x^2 y^2 z^2 u + 2xyzu$$

$$= 2xyz(2xyz + 1)u.$$

Example 6

For a gas at constant temperature, the rate of increase of internal energy E with volume, is

$$T\left(\frac{\partial p}{\partial T} \right)_v - p.$$

If the gas obeys Van der Waal's equation

$$\left(p + \frac{a}{v^2}\right)(v - b) = RT,$$

where a, b, R are constants, show that the increase in internal energy, as the volume increases from v_1 to v_2 is

$$\frac{a}{v_1 v_2}(v_2 - v_1).$$

The expression $(\partial p / \partial T)_v$ is used to signify the partial derivative of p with respect to T, keeping v constant.

Thus, differentiating Van der Waal's equation with respect to T gives

$$(v - b)\left(\frac{\partial p}{\partial T}\right)_v = R.$$

$$\therefore \ T\left(\frac{\partial p}{\partial T}\right)_v = \frac{RT}{v - b} = p + \frac{a}{v^2}.$$

Now the rate of increase of internal energy with volume is dE/dv, where

$$\frac{dE}{dv} = \frac{a}{v^2}.$$

$$\therefore \ E = a \int_{v_1}^{v_2} \frac{dv}{v^2} = a\left(-\frac{1}{v}\right)\bigg|_{v_1}^{v_2} = \frac{a(v_2 - v_1)}{v_1 v_2}.$$

3. Taylor's theorem

We have seen that partial differentiation of a function of several variables implies the variation of one variable subject to the other variables being held constant.

We now consider the effect of simultaneous changes in x and y.

Let $\phi = \phi(x, y)$ be a function of the two independent variables x and y, and let increments $\delta x, \delta y$ in x and y respectively, produce a corresponding increment $\delta \phi$ in ϕ. Then

$$\phi = \phi(x, y)$$
$$\phi + \delta\phi = \phi(x + \delta x, y + \delta y) \tag{1}$$

and thus

$$\delta\phi = \phi(x + \delta x, y + \delta y) - \phi(x, y). \tag{2}$$

. In Chapter 5, we saw that one form of the corresponding Taylor series for the function $f(x + c)$, is

$$f(x + c) = f(x) + \frac{c}{1!}f'(x) + \frac{c^2}{2!}f''(x) + \cdots + \frac{c^n}{n!}f^{(n)}(x) + \cdots$$

where dashes denote differentiation with respect to x.

Thus, if in the function $\phi(x + \delta x, y) y$ is held constant, we may treat this function as a function of the single variable x; the corresponding Taylor series is then

$$\phi(x + \delta x, y) = \phi(x, y) + \frac{\delta x}{1!} \frac{\partial \phi(x, y)}{\partial x} + \frac{(\delta x)^2}{2!} \frac{\partial^2 \phi(x, y)}{\partial x^2} + \cdots$$

$$+ \frac{(\delta x)^n}{n!} \frac{\partial^n \phi(x, y)}{\partial x^n} + \cdots . \tag{3}$$

Similarly, if $x + \delta x$ is held constant in the function $\phi(x + \delta x, y + \delta y)$, we may then treat the function as a function of the single variable y. The corresponding Taylor series is then,

$$\phi(x + \delta x, y + \delta y)$$

$$= \phi(x + \delta x, y) + \frac{\delta y}{1!} \frac{\partial \phi(x + \delta x, y)}{\partial y}$$

$$+ \frac{(\delta y)^2}{2!} \frac{\partial^2 \phi(x + \delta x, y)}{\partial y^2} + \cdots + \frac{(\delta y)^n}{n!} \frac{\partial^{(n)} \phi(x + \delta x, y)}{\partial y^n} + \cdots . \tag{4}$$

Differentiating equation (3) partially and successively with respect to y, and substituting into equation (4), we obtain

$$\phi(x + \delta x, y + \delta y)$$

$$= \left(\phi(x, y) + \frac{\delta x}{1!} \frac{\partial \phi(x, y)}{\partial x} + \frac{(\delta x)^2}{2!} \frac{\partial^2 \phi(x, y)}{\partial x^2} + \frac{(\delta x)^3}{3!} \frac{\partial^3 \phi(x, y)}{\partial x^3} + \cdots \right)$$

$$+ \frac{\delta y}{1!} \left(\frac{\partial \phi(x, y)}{\partial y} + \frac{\delta x}{1!} \frac{\partial^2 \phi(x, y)}{\partial x \, \partial y} + \frac{(\delta x)^2}{2!} \frac{\partial^3 \phi(x, y)}{\partial x^2 \, \partial y} + \cdots \right)$$

$$+ \frac{(\delta y)^2}{2!} \left(\frac{\partial^2 \phi(x, y)}{\partial y^2} + \frac{\delta x}{1!} \frac{\partial^3 \phi(x, y)}{\partial x \, \partial y^2} + \cdots \right) + \cdots$$

i.e.

$$\phi(x + \delta x, y + \delta y) = \phi(x, y) + \frac{1}{1!} \left(\delta x \frac{\partial \phi}{\partial x} + \delta y \frac{\partial \phi}{\partial y} \right)$$

$$+ \frac{1}{2!} \left((\delta x)^2 \frac{\partial^2 \phi}{\partial x^2} + 2 \delta x \, \delta y \frac{\partial^2 \phi}{\partial x \, \partial y} + (\delta y)^2 \frac{\partial^2 \phi}{\partial y^2} \right) + \text{other terms}, \tag{5}$$

where

$$\frac{\partial \phi}{\partial x} \equiv \frac{\partial \phi(x, y)}{\partial x}, \quad \frac{\partial \phi}{\partial y} \equiv \frac{\partial \phi(x, y)}{\partial y}, \text{ etc.}$$

Thus, writing

$$\delta x \frac{\partial \phi}{\partial x} + \delta y \frac{\partial \phi}{\partial y} \equiv \left(\delta x \frac{\partial}{\partial x} + \delta y \frac{\partial}{\partial y} \right) \phi,$$

equation (5) may be written:

$$\phi(x + \delta x, y + \delta y)$$

$$= \phi(x, y) + \frac{1}{1!}\left(\delta x \frac{\partial}{\partial x} + \delta y \frac{\partial}{\partial y}\right)\phi$$

$$+ \frac{1}{2!}\left(\delta x \frac{\partial}{\partial x} + \delta y \frac{\partial}{\partial y}\right)^2 \phi + \frac{1}{3!}\left(\delta x \frac{\partial}{\partial x} + \delta y \frac{\partial}{\partial y}\right)^3 \phi + \cdots$$

$$+ \frac{1}{n!}\left(\delta x \frac{\partial}{\partial x} + \delta y \frac{\partial}{\partial y}\right)^n \phi + \cdots \qquad (6)$$

which is the Taylor series for the function $\phi = \phi(x, y)$. It is assumed that the function is continuous and possesses continuous partial derivatives within the region considered.

4. Small errors

If δx and δy represent small errors in x and y then equation (2) gives the corresponding error in ϕ, i.e.

$$\delta\phi \simeq \frac{\partial\phi}{\partial x}\delta x + \frac{\partial\phi}{\partial y}\delta y, \quad \text{to the first order.}$$

More generally, if $\phi = \phi(x, y, z, \ldots)$ and errors $\delta x, \delta y, \delta z, \ldots$, occur in ϕ, then the corresponding error in ϕ is

$$\delta\phi \simeq \frac{\partial\phi}{\partial x}\delta x + \frac{\partial\phi}{\partial y}\delta y + \frac{\partial\phi}{\partial z}\delta z + \cdots . \qquad (7)$$

Example 7

If z is given in terms of x and y by the relation

$$e^z = \sec x \cos y,$$

and errors of magnitudes μ and $-\mu$ are made in estimating x and y, when x and y are found to be $\pi/3$ and $\pi/6$ respectively, show that the corresponding error in z is approximately $4\mu/\sqrt{3}$.

Re-writing the given equation in the form

$$z = \log_e \sec x + \log_e \cos y$$

we obtain

$$\frac{\partial z}{\partial x} = \frac{\sec x \tan x}{\sec x} = \tan x = \sqrt{3}, \quad \text{when } x = \pi/3,$$

and

$$\frac{\partial z}{\partial y} = -\frac{\sin y}{\cos y} = -\tan y = -\frac{1}{\sqrt{3}}, \quad \text{when } x = \pi/6.$$

If the errors in x, y and z are $\delta x, \delta y, \delta z$ respectively then equation (7) gives

$$\delta z \simeq \frac{\partial z}{\partial x}\delta x + \frac{\partial z}{\partial y}\delta y,$$

i.e.

$$\delta z \simeq (\sqrt{3})(\mu) + \left(-\frac{1}{\sqrt{3}}\right)(-\mu)$$

i.e.

$$\delta z \simeq \frac{4\mu}{\sqrt{3}}.$$

Example 8

The area Δ of a triangle whose sides are a, b, c respectively, is to be calculated from the formula

$$\Delta^2 = s(s-a)(s-b)(s-c) \tag{8}$$

where $2s = a + b + c$. The lengths of the sides are measured and found to be $5k$, $5k$ and $2k$ units respectively. If a and b are too large by amounts α and β respectively, and c is too small by an amount γ, show that the error in the calculated area of the triangle is approximately

$$\frac{k}{4\sqrt{6}}(5\alpha + 5\beta - 23\gamma).$$

From equation (8),

$$2 \log_e \Delta = \log_e s + \log_e(s-a) + \log_e(s-b) + \log_e(s-c)$$

so that if $\delta\Delta$, δs, δa, δb and δc are the respective errors in Δ, s, a, b, and c, the application of equation (7) then gives,

$$\delta(2 \log_e \Delta) \equiv \frac{2\,\delta\Delta}{\Delta} \simeq \frac{\delta s}{s} + \frac{\delta s - \delta a}{s-a} + \frac{\delta s - \delta b}{s-b} + \frac{\delta s - \delta c}{s-c}. \tag{9}$$

Now,

$$2s = a + b + c$$

$$\therefore \ 2\delta s = \delta a + \delta b + \delta c$$

i.e.

$$2s = 5k + 5k + 2k = 12k, \quad \therefore \ s = 6k$$

and

$$2\delta s = \alpha + \beta - \gamma.$$

Equation (9) now becomes,

$$\frac{2\,\delta\Delta}{\Delta} \simeq \frac{\alpha + \beta - \gamma}{12k} + \frac{\beta - \alpha - \gamma}{2k} + \frac{\alpha - \beta - \gamma}{2k} + \frac{\alpha + \beta + \gamma}{8k}$$

i.e.

$$\frac{2\,\delta\Delta}{\Delta} \simeq \frac{5\alpha + 5\beta - 23\gamma}{24k}.$$

But

$$\Delta^2 = (6k)(k)(k)(4k) = 24k^4$$
$$\therefore \quad \Delta = 2\sqrt{(6)}k^2$$

and hence

$$\delta\Delta \simeq \frac{(5\alpha + 5\beta - 23\gamma)}{48k} \cdot 2\sqrt{(6)}k^2 = \frac{k}{4\sqrt{6}}(5\alpha + 5\beta - 23\gamma),$$

the estimated area thus being too large by this amount.

Example 9

If m_1 grams of solute of molecular weight M_1 are dissolved in m_2 grams of solvent of molecular weight M_2, then assuming that the vapour pressures p_2 and p_1 of solvent and solution can be measured, the molecular weight of the solute is given by

$$M_1 = \frac{m_1 M_2}{m_2} \times \frac{p_2}{p_2 - p_1}.$$

Write down an expression for the error produced in M_1 due to errors in the measurements of m_1, m_2, M_2, p_1, p_2.

If $p_1 = kp_2$ where k is a constant not equal to 1, and the errors in m_1, m_2, M_1 are $\epsilon m_1, \epsilon m_2, 3\epsilon M_1$ respectively, show that

$$\delta M_1 \simeq 3M_1 \epsilon.$$

Taking logarithms

$$\therefore \quad \log_e M_1 = \log_e m_1 + \log_e M_2 + \log_e p_2 - \log_e m_2 - \log_e (p_2 - p_1).$$

Let the respective errors in m_1, m_2, M_2, p_1, p_2 be $\delta m_1, \delta m_2, \delta M_2, \delta p_1$ and δp_2. Then

$$\frac{\delta M_1}{M_1} \simeq \frac{\delta m_1}{m_1} + \frac{\delta M_2}{M_2} + \frac{\delta p_2}{p_2} - \frac{\delta m_2}{m_2} - \frac{\delta p_2 - \delta p_1}{p_2 - p_1}.$$

Now $p_1 = kp_2$, giving $\delta p_1 = k\delta p_2$, so that

$$\frac{\delta p_2 - \delta p_1}{p_2 - p_1} = \frac{\delta p_2}{p_2}.$$

Hence,

$$\frac{\delta M_1}{M_1} \simeq \frac{\delta m_1}{m_1} + \frac{\delta M_2}{M_2} - \frac{\delta m_2}{m_2} = \epsilon + 3\epsilon - \epsilon = 3\epsilon.$$

5. Total derivative

It is useful in calculus problems to replace the increments $\delta x, \delta y, \delta z, \ldots$, by their *differentials* $dx, dy, dz, \ldots$, which are defined by

$$d\phi = \frac{\partial \phi}{\partial x} dx + \frac{\partial \phi}{\partial y} dy + \frac{\partial \phi}{\partial z} dz + \cdots. \tag{10}$$

This expression represents the *total differential* of ϕ. As an example, if $x, y, z, \ldots$, are all functions of some other variable t, then

$$\frac{d\phi}{dt} = \frac{\partial\phi}{\partial x}\frac{dx}{dt} + \frac{\partial\phi}{\partial y}\frac{dy}{dt} + \frac{\partial\phi}{\partial z}\frac{dz}{dt} + \cdots,$$

and is referred to as the *total derivative* of ϕ with respect to t.

The derivatives $d\phi/dt$, dx/dt, dy/dt, dz/dt, $\ldots$, are used here since $\phi, x, y, z, \ldots$, are now all functions of a *single* variable t.

If, on the other hand, $x, y, z, \ldots$, are all functions of several variables $u, v, w, \ldots$, then

$$\frac{\partial\phi}{\partial u} = \frac{\partial\phi}{\partial x}\frac{\partial x}{\partial u} + \frac{\partial\phi}{\partial y}\frac{\partial y}{\partial u} + \frac{\partial\phi}{\partial z}\frac{\partial z}{\partial u} + \cdots,$$

$$\frac{\partial\phi}{\partial v} = \frac{\partial\phi}{\partial x}\frac{\partial x}{\partial v} + \frac{\partial\phi}{\partial y}\frac{\partial y}{\partial v} + \frac{\partial\phi}{\partial z}\frac{\partial z}{\partial v} + \cdots,$$

$$(11)$$

and so on. The partial derivatives $\partial\phi/\partial u$, $\partial x/\partial u$, $\partial y/\partial u$, $\ldots$, being used in this case since $\phi, x, y, z, \ldots$, are all functions of *several* variables $u, v, w, \ldots$.

Example 10

If $z = z(u, v)$ where $u = \log_e(x/y)$ and $v = x^2 + y^2$, show that

$$x^2\frac{\partial^2 z}{\partial x^2} + 2xy\frac{\partial^2 z}{\partial x\,\partial y} + y^2\frac{\partial^2 z}{\partial y^2} = 4v^2\frac{\partial^2 z}{\partial v^2} + 2v\frac{\partial z}{\partial v}.$$

As in equation (11), we obtain

$$\frac{\partial z}{\partial x} = \frac{\partial z}{\partial u}\frac{\partial u}{\partial x} + \frac{\partial z}{\partial v}\frac{\partial v}{\partial x}, \qquad \frac{\partial z}{\partial y} = \frac{\partial z}{\partial u}\frac{\partial u}{\partial y} + \frac{\partial z}{\partial v}\frac{\partial v}{\partial y}$$

$$= \frac{\partial z}{\partial u}\cdot\frac{1}{x} + \frac{\partial z}{\partial v}\cdot 2x \qquad\quad = \frac{\partial z}{\partial u}\left(-\frac{1}{y}\right) + \frac{\partial z}{\partial v}\cdot 2y.$$

Hence,

$$x\frac{\partial z}{\partial x} + y\frac{\partial z}{\partial y} = 2(x^2 + y^2)\frac{\partial z}{\partial v} = 2v\frac{\partial z}{\partial v} \tag{12}$$

i.e.

$$\left(x\frac{\partial}{\partial x} + y\frac{\partial}{\partial y}\right) \equiv 2v\frac{\partial}{\partial v}$$

so that

$$\left(x\frac{\partial}{\partial x} + y\frac{\partial}{\partial y}\right)\left(x\frac{\partial z}{\partial x} + y\frac{\partial z}{\partial y}\right) = \left(2v\frac{\partial}{\partial v}\right)\left(2v\frac{\partial z}{\partial v}\right)$$

i.e.

$$x^2\frac{\partial^2 z}{\partial x^2} + 2xy\frac{\partial^2 z}{\partial x\,\partial y} + y^2\frac{\partial^2 z}{\partial y^2} + x\frac{\partial z}{\partial x} + y\frac{\partial z}{\partial y} = 4v^2\frac{\partial^2 z}{\partial v^2} + 4v\frac{\partial z}{\partial v}$$

i.e.

$$x^2 \frac{\partial^2 z}{\partial x^2} + 2xy \frac{\partial^2 z}{\partial x\, \partial y} + y^2 \frac{\partial^2 z}{\partial y^2} = 4v^2 \frac{\partial^2 z}{\partial v^2} + 2v \frac{\partial z}{\partial v}, \quad \text{from (12).}$$

Example 11

If $v = v(x, y)$ where $x = r \cos \theta$ and $y = r \sin \theta$ prove that

$$r^2 \frac{\partial^2 v}{\partial r^2} + r \frac{\partial v}{\partial r} + \frac{\partial^2 v}{\partial \theta^2} = (x^2 + y^2)\left(\frac{\partial^2 v}{\partial x^2} + \frac{\partial^2 v}{\partial y^2}\right).$$

Show also, that when $v = \sin \theta \, \log_e r$, the above equation reduces to

$$r^2 \frac{\partial^2 v}{\partial r^2} + r \frac{\partial v}{\partial r} + \frac{\partial^2 v}{\partial \theta^2} + v = 0.$$

Proof:

$$\frac{\partial v}{\partial r} = \frac{\partial v}{\partial x} \frac{\partial x}{\partial r} + \frac{\partial v}{\partial y} \frac{\partial y}{\partial r} = \frac{\partial v}{\partial x} \cos \theta + \frac{\partial v}{\partial y} \sin \theta.$$

$$= \frac{\partial v}{\partial x}\left(\frac{x}{r}\right) + \frac{\partial v}{\partial y}\left(\frac{y}{r}\right)$$

$$\therefore \quad r \frac{\partial v}{\partial r} = x \frac{\partial v}{\partial x} + y \frac{\partial v}{\partial y}.$$

Hence,

$$r \frac{\partial}{\partial r}\left(r \frac{\partial v}{\partial r}\right) = \left(x \frac{\partial}{\partial x} + y \frac{\partial}{\partial y}\right)\left(x \frac{\partial v}{\partial x} + y \frac{\partial v}{\partial y}\right)$$

i.e.

$$r^2 \frac{\partial^2 v}{\partial r^2} + r \frac{\partial v}{\partial r} = x^2 \frac{\partial^2 v}{\partial x^2} + 2xy \frac{\partial^2 v}{\partial x\, \partial y} + y^2 \frac{\partial^2 v}{\partial y^2} + x \frac{\partial v}{\partial x} + y \frac{\partial v}{\partial y}. \quad (13)$$

Also,

$$\frac{\partial v}{\partial \theta} = \frac{\partial v}{\partial x} \frac{\partial x}{\partial \theta} + \frac{\partial v}{\partial y} \frac{\partial y}{\partial \theta} = \frac{\partial v}{\partial x}(-r \sin \theta) + \frac{\partial v}{\partial y}(r \cos \theta)$$

i.e.

$$\frac{\partial v}{\partial \theta} = x \frac{\partial v}{\partial y} - y \frac{\partial v}{\partial x}$$

so that

$$\frac{\partial}{\partial \theta}\left(\frac{\partial v}{\partial \theta}\right) = \left(x \frac{\partial}{\partial y} - y \frac{\partial}{\partial x}\right)\left(x \frac{\partial v}{\partial y} - y \frac{\partial v}{\partial x}\right)$$

i.e.

$$\frac{\partial^2 v}{\partial \theta^2} = x^2 \frac{\partial^2 v}{\partial y^2} - 2xy \frac{\partial^2 v}{\partial x \partial y} + y^2 \frac{\partial^2 v}{\partial x^2} - x \frac{\partial v}{\partial x} - y \frac{\partial v}{\partial y}. \quad (14)$$

Equations (13) and (14) now give

$$r^2 \frac{\partial^2 v}{\partial r^2} + r\frac{\partial v}{\partial r} + \frac{\partial^2 v}{\partial \theta^2} = (x^2 + y^2)\left(\frac{\partial^2 v}{\partial x^2} + \frac{\partial^2 v}{\partial y^2}\right). \tag{15}$$

In the case when $v = \sin \theta \log_e r$, we have

$$\frac{\partial v}{\partial \theta} = \cos \theta \log_e r, \qquad \frac{\partial v}{\partial r} = \frac{\sin \theta}{r}.$$

$$\frac{\partial^2 v}{\partial \theta^2} = -\sin \theta \log_e r, \qquad \frac{\partial^2 v}{\partial r^2} = \frac{-\sin \theta}{r^2}.$$

$$\therefore \ r^2 \frac{\partial^2 v}{\partial r^2} + r\frac{\partial v}{\partial r} + \frac{\partial^2 v}{\partial \theta^2} = -\sin \theta \log_e r = -v. \qquad \text{Hence result.}$$

Note: Since $x^2 + y^2 = r^2$, equation (15) may also be written in the form

$$\frac{\partial^2 v}{\partial r^2} + \frac{1}{r}\frac{\partial v}{\partial r} + \frac{1}{r^2}\frac{\partial^2 v}{\partial \theta^2} = \frac{\partial^2 v}{\partial x^2} + \frac{\partial^2 v}{\partial y^2}.$$

The equations

$$\frac{\partial^2 v}{\partial x^2} + \frac{\partial^2 v}{\partial y^2} = 0,$$

$$\frac{\partial^2 v}{\partial r^2} + \frac{1}{r}\frac{\partial v}{\partial r} + \frac{1}{r^2}\frac{\partial^2 v}{\partial \theta^2} = 0,$$

represent Laplace's equation in two dimensions, in Cartesian coordinates and polar coordinates, respectively.

Example 12

If $x^5 - 4x^2 y^2 + y^3 - xy + 3x + 2y + 2 = 0$, determine the value of dy/dx at the point $(1,2)$.

We note that if $u = f(x, y)$, then

$$\frac{du}{dx} = \frac{\partial f}{\partial x} + \frac{\partial f}{\partial y}\frac{dy}{dx}$$

so that if $f(x, y) = 0$, then

$$\frac{\partial f}{\partial x} + \frac{\partial f}{\partial y}\frac{dy}{dx} = 0,$$

and hence

$$\frac{dy}{dx} = -\frac{\partial f/\partial x}{\partial f/\partial y}.$$

For the given case

$$f(x, y) = x^5 - 4x^2y^2 + y^3 - xy + 3x + 2y + 2$$

$$\therefore \quad \frac{\partial f}{\partial x} = 5x^4 - 8xy^2 - y + 3 = -26, \quad \text{at the point } (1, 2),$$

and

$$\frac{\partial f}{\partial y} = -8x^2y + 3y^2 - x + 2 = -3, \quad \text{at the point } (1, 2).$$

Hence $dy/dx = -\frac{26}{3}$ at the given point.

Example 13

The change of entropy dS in a system, corresponding to a change dE in internal energy and a change dT in temperature is given by

$$dS = \frac{1}{T} dE + \frac{1}{3} \frac{\psi}{T} dV,$$

dS being a perfect differential, and ψ being a function of T only.
 If $E = \psi V$, show that $\psi = AT^4$, where A is a constant.
 The expression

$$\frac{\partial \phi}{\partial x} dx + \frac{\partial \phi}{\partial y} dy + \frac{\partial \phi}{\partial z} dz + \cdots$$

on the right-hand side of equation (10) is called a *perfect differential*.
 Thus,

$$P(x, y)dx + Q(x, y)\, dy$$

is a perfect differential provided ϕ can be found such that

$$P(x, y) = \frac{\partial \phi}{\partial x}, \quad Q(x, y) = \frac{\partial \phi}{\partial y},$$

which means

$$\frac{\partial P(x, y)}{\partial y} = \frac{\partial Q(x, y)}{\partial x},$$

assuming that P and Q have continuous partial derivatives of the first order.
 Now

$$E = \psi V$$

$$\therefore \quad dE = \frac{\partial E}{\partial \psi} d\psi + \frac{\partial E}{\partial V} dV$$

$$= V d\psi + \psi dV,$$

so that

$$dS = \frac{V}{T}\,d\psi + \frac{4}{3}\frac{\psi}{T}\,dV$$

$$= \left(\frac{V}{T}\frac{d\psi}{dT}\right)dT + \frac{4\psi}{3T}\,dV,$$

since ψ is a function of T only.

Thus, as dS is a perfect differential, we may write

$$\frac{\partial}{\partial V}\left(\frac{V}{T}\frac{d\psi}{dT}\right) = \frac{\partial}{\partial T}\left(\frac{4\psi}{3T}\right)$$

i.e.

$$\frac{1}{T}\frac{d\psi}{dT} = \frac{4}{3}\left(\frac{1}{T}\frac{d\psi}{dT} - \frac{\psi}{T^2}\right)$$

$$\therefore \frac{d\psi}{dT} = 4\frac{\psi}{T}.$$

Separating the variables now gives

$$\int \frac{d\psi}{\psi} = 4\int \frac{dT}{T}$$

i.e.

$$\log_e \psi = 4 \log_e T + \text{constant}$$

and hence

$$\psi = AT^4.$$

6. Stationary values

For the function $\phi(x, y)$ to have a maximum value when $x = a, y = b$, we must have

$$\phi(a + \delta x, b + \delta y) - \phi(a, b) < 0 \qquad (16)$$

for all small values $\delta x, \delta y$.

For the function to have a minimum value when $x = a, y = b$, we must have

$$\phi(a + \delta x, b + \delta y) - \phi(a, b) > 0, \qquad (17)$$

for all small values $\delta x, \delta y$.

Equations (6) and (16) imply that:

$$\left(\delta x \frac{\partial}{\partial x} + \delta y \frac{\partial}{\partial y}\right)\phi + \frac{1}{2!}\left(\delta x \frac{\partial}{\partial x} + \delta y \frac{\partial}{\partial y}\right)^2 \phi + \cdots < 0, \qquad (18)$$

for all $\delta x, \delta y$, the expression being evaluated at $x = a, y = b$. For small values of $\delta x,$

δy, the predominant term in equation (18) is,

$$\frac{\partial \phi}{\partial x} \delta x + \frac{\partial \phi}{\partial y} \delta y.$$

Since its sign is dependent upon the signs of δx and δy, the left-hand side of equation (18) will not be true for all δx, δy, unless

$$\frac{\partial \phi}{\partial x} \delta x + \frac{\partial \phi}{\partial y} \delta y = 0, \quad \text{at} \quad x = a, y = b,$$

which implies that

$$\left(\frac{\partial \phi}{\partial x}\right)_{\substack{x=a \\ y=b}} = 0 \quad \text{and} \quad \left(\frac{\partial \phi}{\partial y}\right)_{\substack{x=a \\ y=b}} = 0.$$

Similarly, for equation (17) to hold, we must also have

$$\left(\frac{\partial \phi}{\partial x}\right)_{\substack{x=a \\ y=b}} = 0 \quad \text{and} \quad \left(\frac{\partial \phi}{\partial y}\right)_{\substack{x=a \\ y=b}} = 0.$$

Thus, the function $\phi = \phi(x, y)$ will possess a *stationary value* at $x = a, y = b$ provided

$$\left(\frac{\partial \phi}{\partial x}\right)_{\substack{x=a \\ y=b}} = 0, \quad \left(\frac{\partial \phi}{\partial y}\right)_{\substack{x=a \\ y=b}} = 0.$$

In general, for a function $\phi = \phi(x, y, z, \ldots)$ of several variables $x, y, z, \ldots$, a stationary value will occur at $x = a, y = b, z = c, \ldots$, provided

$$\frac{\partial \phi}{\partial x} = \frac{\partial \phi}{\partial y} = \frac{\partial \phi}{\partial z} = \cdots = 0.$$

when $x = a, y = b, z = c, \ldots$.

It is beyond the scope of this book to determine the conditions for the stationary value to be a maximum or a minimum. It is sufficient here to mention that in order to determine such conditions, it is necessary to consider the sign of the term:

$$\frac{\partial^2 \phi}{\partial x^2} (\delta x)^2 + 2 \frac{\partial^2 \phi}{\partial x \, \partial y} (\delta x)(\delta y) + \frac{\partial^2 \phi}{\partial y^2} (\delta y)^2$$

occurring in equation (6).

We state the conditions for the case of a function of two variables, namely: the function $\phi = \phi(x, y)$ possesses a maximum value at $x = a, y = b$ provided

$$\frac{\partial \phi}{\partial x} = \frac{\partial \phi}{\partial y} = 0, \quad \frac{\partial^2 \phi}{\partial x^2} < 0,$$

$$\left(\frac{\partial^2 \phi}{\partial x \, \partial y}\right)^2 - \left(\frac{\partial^2 \phi}{\partial x^2}\right)\left(\frac{\partial^2 \phi}{\partial y^2}\right) < 0,$$

and a minimum value, provided

$$\frac{\partial \phi}{\partial x} = \frac{\partial \phi}{\partial y} = 0, \quad \frac{\partial^2 \phi}{\partial x^2} > 0,$$

and

$$\left(\frac{\partial^2 \phi}{\partial x \, \partial y}\right)^2 - \left(\frac{\partial^2 \phi}{\partial x^2}\right)\left(\frac{\partial^2 \phi}{\partial y^2}\right) < 0,$$

all derivatives being evaluated at $x = a, x = b$.

In the case when

$$\left(\frac{\partial^2 \phi}{\partial x \, \partial y}\right)^2 - \left(\frac{\partial^2 \phi}{\partial x^2}\right)\left(\frac{\partial^2 \phi}{\partial y^2}\right) > 0,$$

the point $x = a, y = b$ is called a *saddle point* or *minimax*. If the expression is zero further analysis is required.

Example 14

Determine the nature of the stationary points of the function

$$\phi(x, y) = x^2 y - 5x^2 - y^2 - 6x + 3y + 1.$$

For stationary values,

$$\frac{\partial \phi}{\partial x} \equiv 2xy - 10x - 6 = 0, \tag{19}$$

$$\frac{\partial \phi}{\partial y} \equiv x^2 - 2y + 3 = 0. \tag{20}$$

From equation (20),

$$2y = x^2 + 3. \tag{21}$$

Substituting into equation (19) gives

$$x(x^2 + 3) - 10x - 6 = 0$$

i.e.

$$x^3 - 7x - 6 = 0$$

i.e.

$$(x + 1)(x + 2)(x - 3) = 0 \quad \therefore \ x = -1, -2, \text{ or } +3.$$

The stationary points are thus $(-1, 2), (-2, \frac{7}{2}), (3, 6), \ldots$ from equation (21).

Now

$$\frac{\partial^2 \phi}{\partial x^2} = 2y - 10, \quad \frac{\partial^2 \phi}{\partial y^2} = -2, \quad \frac{\partial^2 \phi}{\partial x \, \partial y} = 2x.$$

$$\therefore \left(\frac{\partial^2 \phi}{\partial x \, \partial y}\right)^2 - \left(\frac{\partial^2 \phi}{\partial x^2}\right)\left(\frac{\partial^2 \phi}{\partial y^2}\right) = 4x^2 + 2(2y - 10) = 4(x^2 + y - 5). \tag{22}$$

At the point $(-1, 2)$:
   The right-hand side of equation (22) = -8, i.e. < 0, and $\partial^2 \phi / \partial x^2 = -6$, i.e. < 0
   $\therefore$ $(-1, 2)$ gives a maximum.
At the point $(-2, \frac{7}{2})$:
   The right-hand side of equation (22) = $10 > 0$.
   $\therefore$ $(-2, \frac{7}{2})$ is a saddle point.
At the point $(3, 6)$:
   The right-hand side of equation (22) = $40 > 0$, so that $(3, 6)$ is also a saddle-point.

Example 15

Determine the nature of the stationary points of the function

$$\phi(x, y) = 3x^2y^2 + 9x^2y + 2y^3 + 6x^2 - 12y.$$

The first order partial derivatives of $\phi(x, y)$ are

$$\frac{\partial \phi}{\partial x} = 6xy^2 + 18xy + 12x, \qquad \frac{\partial \phi}{\partial y} = 6x^2y + 9x^2 + 6y^2 - 12.$$

Stationary values are thus given by

$$6xy^2 + 18xy + 12x = 0 \quad \text{and} \quad 6x^2y + 9x^2 + 6y^2 - 12 = 0$$

i.e.

$$6x(y + 1)(y + 2) = 0 \quad \text{and} \quad 2x^2y + 3x^2 + 2y^2 - 4 = 0.$$

i.e.

$$x = 0, \qquad y = \pm\sqrt{2},$$
$$x = \pm\sqrt{2}, \quad y = -1,$$
$$x = \pm 2, \qquad y = -2.$$

The second order partial derivatives of $\phi(x, y)$ are

$$\frac{\partial^2 \phi}{\partial x^2} = 6y^2 + 18y + 12, \qquad \frac{\partial^2 \phi}{\partial y^2} = 6x^2 + 12y,$$

$$\frac{\partial^2 \phi}{\partial x \, \partial y} = 12xy + 18x.$$

Let

$$\left(\frac{\partial^2 \phi}{\partial x \, \partial y} \right)^2 - \left(\frac{\partial^2 \phi}{\partial x^2} \right)\left(\frac{\partial^2 \phi}{\partial y^2} \right) \equiv E.$$

At the point $(0, \sqrt{2})$:
   E is negative, $\partial^2 \phi / \partial x^2$ is positive, and so the stationary value is a minimum.
At the point $(0, -\sqrt{2})$:
   E is negative, $\partial^2 \phi / \partial x^2$ is negative, and so the stationary value is a maximum.
   For the remaining four points E is positive, thus giving saddle-points.

Example 16

In the determination of the variation of the molar heat capacity of hydrogen with temperature, at one atmosphere pressure, in calories per degree centigrade, the following results were obtained:

Temperature, T_i (°C)	200	400	600	800	1000	1200	1400
m.h.c., h_i	7.01	7.15	7.30	7.40	7.49	7.77	7.80

Assuming that there is a linear relationship between h and T of the form

$$h = aT + b,$$

obtain the best values for a and b, by minimizing the expression

$$\sum_{i=1}^{7} (h_i - aT_i - b)^2.$$

In order that the given expressions may be a minimum its first order partial derivatives with respect to a and b respectively, must both be zero, i.e.,

$$\sum_{i=1}^{7} T_i(h_i - aT_i - b) = 0,$$

and

$$\sum_{i=1}^{7} (h_i - aT_i - b) = 0.$$

i.e.

$$\sum_{i=1}^{7} T_i h_i - a \sum_{i=1}^{7} T_i^2 - b \sum_{i=1}^{7} T_i = 0$$

and

$$\sum_{i=1}^{7} h_i - a \sum_{i=1}^{7} T_i - 7b = 0.$$

From the results given, we obtain

$$\sum_{i=1}^{7} T_i h_i = 42,296. \quad \sum_{i=1}^{7} T_i^2 = 5,600,000$$

$$\sum_{i=1}^{7} T_i = 5,600. \quad \sum_{i=1}^{7} h_i = 51.92.$$

so that the equations reduce to

$$56,000a + 56b - 422.96 = 0$$
$$5,600a + 7b - 51.92 = 0,$$

whence,

$$a = 0.00068 \quad b = 6.87$$

giving

$$h = 0.00068T + 6.87 \quad \text{calories/}°\text{C}.$$

Further theory is to be found in Chapter 19.

PROBLEMS

1. Evaluate $\partial z/\partial x$, $\partial^2 z/\partial x\, \partial y$, $\partial^2 z/\partial y^2$ when

(i) $z = x^5 y + 2x^3 y^2 + y^3 + 3xy + x + 3$,   (ii) $z = \sin(x + y)\sin(x - y)$,

(iii) $z = (x + y)\tan xy$.

2. If

$$u = \frac{x^2 + 2xy - y^2}{x + y},$$

show that

$$\frac{\partial^2 u}{\partial x\, \partial y} = \frac{4xy}{(x + y)^3}.$$

3. If $x = \alpha_1$ is a close approximation to a root of the equation $f(x) = 0$, then Newton's method states that a better approximation is

$$\alpha_2 = \alpha_1 - \frac{f(\alpha_1)}{f'(\alpha_1)}.$$

If the calculated values of $f(\alpha_1)$ and $f'(\alpha_1)$ are too small and too large respectively, by equal amounts k, show that the calculated value of α_2 is too large by approximately an amount

$$\frac{k}{\{f'(\alpha_1)\}^2}\{f(\alpha_1) + f'(\alpha_1)\}.$$

4. Find dy/dx: (i) at the point $(-1, 3)$ on the curve

$$3x^4 - 2x^2 y + y^3 + xy^2 - 2x - 5y - 2 = 0;$$

(ii) at the point $(0, \pi/6)$ on the curve $\sin(xy) + \cos(2x + 3y) = 0$.

5. If $z = z(u, v)$ where $u = 2\sqrt{(xy)}$ and $v = 2/\sqrt{(xy)}$ show that

$$x\frac{\partial z}{\partial x} - y\frac{\partial z}{\partial y} = 0,$$

and

$$4xy\frac{\partial^2 z}{\partial x\, \partial y} = u^2\frac{\partial^2 z}{\partial u^2} - 2uv\frac{\partial^2 z}{\partial u\, \partial v} + v^2\frac{\partial^2 z}{\partial v^2} + u\frac{\partial z}{\partial u} + v\frac{\partial z}{\partial v}.$$

6. If $V = V(x, y)$ where $x = r\sec\theta$, $y = r\operatorname{cosec}\theta$, show that

$$r\frac{\partial V}{\partial r} = x\frac{\partial V}{\partial x} + y\frac{\partial V}{\partial y},$$

and

$$2r^2\operatorname{cosec}2\theta\frac{\partial V}{\partial \theta} = x^3\frac{\partial V}{\partial x} - y^3\frac{\partial V}{\partial y}.$$

7. Determine the nature of the stationary points for the function

$$xy^2 + 2xy + 3x^2 - 8x - 1.$$

8. Discuss the stationary values of the function

$$x^2 y - xy^2 + 3xy + \tfrac{1}{2}x^2 + x.$$

9. The equation of conduction of heat in a solid body is given by

$$\frac{\partial T}{\partial t} = k\left(\frac{\partial^2 T}{\partial x^2} + \frac{\partial^2 T}{\partial y^2} + \frac{\partial^2 T}{\partial z^2}\right),$$

where k is the thermal conductivity. Verify that a solution of this equation is

$$T = \frac{A}{t^{3/2}} \exp\left(\frac{-r^2}{4kt}\right)$$

where $r^2 = x^2 + y^2 + z^2$ and A is a constant.

10. For the flow of fluid through a heated tube of radius r, the rate of flow of heat Q, through a cylindrical surface in the fluid at a distance x from the wall of the tube, where the temperature is T, is given by

$$Q = -k(r - x)\frac{\partial T}{\partial x}$$

where k is a positive constant.

If T is also given by $T = ax^3 + bx^2 + cx$, and the temperature gradient at the axis of the tube is zero, show, by comparing the values of $\partial T/\partial x$, $\partial^2 T/\partial x^2$, respectively, that

$$a = -\frac{4T_0}{5r^3}, \qquad b = \frac{3T_0}{5r^2}, \qquad c = \frac{6T_0}{5r}$$

where T_0 is the temperature of the fluid at the axis of the tube.

11. A travelling microscope was used to determine the measurements of a crystal of triangular cross section. Denoting the vertices of the cross section by A, B, and C, it was found that the angle BAC was $60°$, and that the lengths of AB and AC were 3.5 mm and 4.5 mm respectively.

If errors of +0.100 mm and −0.100 mms were made in estimating AB and AC respectively, calculate the resultant error in BC.

12. The coefficient of viscosity η of a liquid is given by the formula

$$9\pi^3 \eta^2 \sigma^4 = 4mkT,$$

where m is the mass of a molecule, σ is its diameter, T the absolute temperature, and k is Boltzmann's constant.

If, in a subsequent calculation, the error in m is αm, the error in T is βT, and the error in σ is $\gamma\sigma$, show that the resultant error in η, is

$$\tfrac{1}{2}(\alpha + \beta - 4\gamma)\eta.$$

13. In the experimental formulation of the dependence of reaction rates on temperature, the Arrhenius equation is

$$\phi = v \exp\left(\frac{ST - H}{RT}\right),$$

where v is a frequency factor, S the entropy of activation, T the absolute temperature, H the enthalpy of activation, and R the gas constant.

If $T\partial S/\partial T = \partial H/\partial T$, show that

$$\frac{\partial \phi}{\partial T} = \phi\left(\frac{1}{v}\frac{\partial v}{\partial T} + \frac{H}{RT^2}\right).$$

14. The temperature T_i, of a substance was recorded at different intervals of time t_i, and the results were as shown:

t_i (units of 10 mins)	1	2	3	4	5	6	7	8
T_i (°C)	5	7	10	21	32	51	70	97

Assuming that the temperature varied according to the law

$$T = at^2 + bt + c,$$

find, by minimizing

$$\sum_{i=1}^{8} (T_i - at_i^2 - bt_i - c)^2,$$

the best values for a, b, c.
(Further theory may be found in Chapter 19.)

15. Show that the stationary points of the function $\phi(x, y, z) = x + y^2 + z^2$, subject to the condition $z^2 - x^2 = 2$, are $(-\frac{1}{2}, 0, \pm\frac{3}{2})$.

(i) by eliminating x between the two expressions and proceeding as in Section 6, and

(ii) by introducing a parameter λ (called a *Lagrange multiplier*), and considering the stationary points of the auxiliary function

$$\psi(x, y, z) = x + y^2 + z^2 + (z^2 - x^2 - 2).$$

(This technique is known as the method of *Lagrange multipliers*.)

16. Show that the maximum value of the function $\phi(x, y, z) = xyz$, subject to the condition $x^2 + 3y^2 + 9z^2 = 9a^2$, is a^3.

CHAPTER 10

Multiple Integrals

1. Double integrals

Let $f(x, y)$ be a single-valued function of x and y and let C be a closed curve lying in the plane $z = 0$. (See Fig. 10.1.)

If A, the area enclosed by C, is divided into small rectangular elements of area $\delta x \, \delta y$, then the volume δV of the cylinder on $\delta x \, \delta y$ as base and terminated by the surface $z = f(x, y)$ is

$$\delta V \simeq f(x, y) \, \delta x \, \delta y.$$

Adding the volumes of all such elemental cylinders gives approximately the volume V enclosed by the cylinder on C, the plane $z = 0$ and the surface $z = f(x, y)$.

Hence if $\delta x \to 0$, $\delta y \to 0$, so that the area enclosed by C is divided into an infinite number of rectangular elements, then

$$V = \lim_{\substack{\delta x \to 0 \\ \delta y \to 0}} \sum_A f(x, y) \, \delta x \, \delta y \tag{1}$$

the summation extending over the area enclosed by C.

Assuming that the function $f(x, y)$ is continuous throughout A, and conforming

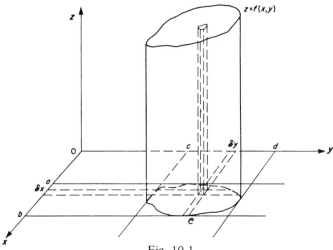

Fig. 10.1

242

to the notation in Section 6 of Chapter 5, the expression on the right-hand side of equation (1) is called the *double integral* of $f(x, y)$ over the region A, and is written

$$\lim_{\substack{\delta x \to 0 \\ \delta y \to 0}} \sum_A f(x, y)\, \delta x\, \delta y = \int_A f(x, y)\, dA = \int_A \int f(x, y)\, dx\, dy \tag{2}$$

where $\delta A = \delta x\, \delta y$, the area A being the *field of integration*.

If the function $f(x, y)$ is replaced by a function $f(x, y, z)$ of three variables then we introduce the concept of a *triple integral*

$$\iiint f(x, y, z)\, dx\, dy\, dz$$

the field of integration being a volume. By extending the process we obtain the concept of a *multiple integral*.

2. Repeated integrals

We now obtain an alternative expression for the volume discussed in Section 1, by deriving the summation of equation (1) in two stages. Let $x = a$ and $x = b$ be the extreme abscissae for the curve C (see Fig. 10.2) whose lower and upper portions XYX' and $XY'X'$ have equations $y = y_1(x)$ and $y = y_2(x)$ respectively.

As before, the volume of the cylinder on $\delta x\, \delta y$ as base is

$$\delta V \simeq f(x, y)\, \delta x\, \delta y.$$

The volume above the elemental section PQRS, for which x and δx are constant, is obtained by forming the sum of such elements δV over this section, as y varies from $y_1(x)$ to $y_2(x)$, and is

$$\sum_{y=y_1(x)}^{y=y_2(x)} f(x, y)\, \delta x\, \delta y = \delta x \sum_{y=y_1(x)}^{y=y_2(x)} f(x, y)\, \delta y.$$

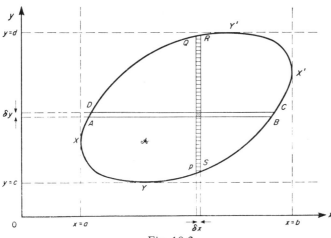

Fig. 10.2

Hence, in the limiting case $\delta y \to 0$, the volume above PQRS is

$$\lim_{\delta y \to 0} \delta x \sum_{y=y_1(x)}^{y=y_2(x)} f(x,y)\,\delta y = \delta x \int_{y=y_1(x)}^{y=y_2(x)} f(x,y)\,\mathrm{d}y = \delta x F(x), \quad \text{say.}$$

In order to include the whole of the area A enclosed by C we must form the sum of $F(x)\,\delta x$ as x varies from $x = a$ to $x = b$, i.e.

$$\sum_{x=a}^{x=b} F(x)\,\delta x.$$

Taking the limiting case $\delta x \to 0$, we obtain for the total volume V of Section 1, the expression

$$\lim_{\delta x \to 0} \sum_{x=a}^{x=b} F(x)\,\delta x = \int_a^b F(x)\,\mathrm{d}x$$

$$= \int_{x=a}^{x=b} \left(\int_{y=y_1(x)}^{y=y_2(x)} f(x,y)\,\mathrm{d}y \right) \mathrm{d}x \tag{3}$$

which is called a *repeated integral*. It is usually written in the form

$$\int_a^b \mathrm{d}x \int_{y_1(x)}^{y_2(x)} f(x,y)\,\mathrm{d}y, \quad \text{or} \quad \int_a^b \int_{y_1(x)}^{y_2(x)} f(x,y)\,\mathrm{d}y\,\mathrm{d}x$$

x being held constant whilst performing the integration with respect to y.

Since we have obtained an expression for the volume V in the form of a double integral and also as a repeated integral, we have

$$\iint_A f(x,y)\,\mathrm{d}x\,\mathrm{d}y = \int_a^b \int_{y_1(x)}^{y_2(x)} f(x,y)\,\mathrm{d}y\,\mathrm{d}x.$$

The following examples illustrate the method of evaluation.

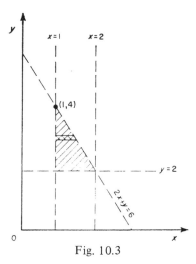

Fig. 10.3

Example 1

Evaluate

$$I = \iint_A \frac{dx\,dy}{(2x+y)^2}$$

where A is the area enclosed by the lines $x = 1$, $y = 2$ and $2x + y = 6$.

The field of integration is shown in Fig. 10.3.

Since the integral I is to be evaluated by first integrating with respect to x, we:

(i) fix y, and vary x from $x = 1$ to $x = \frac{1}{2}(6 - y)$; and

(ii) vary y from $y = 2$ to $y = 4$.

Hence

$$I = \int_2^4 \int_1^{(6-y)/2} \frac{dx\,dy}{(2x+y)^2}$$

$$= \int_2^4 \left(\frac{-\frac{1}{2}}{2x+y}\right)_{x=1}^{x=(6-y)/2} dy$$

$$= \int_2^4 \frac{1}{2}\left(\frac{1}{2+y} - \frac{1}{6}\right) dy = \left[\frac{1}{2}\log_e(2+y) - \frac{y}{12}\right]_{y=2}^{y=4}$$

$$= \frac{1}{2}\left\{\log_e\left(\frac{3}{2}\right) - \frac{1}{3}\right\}.$$

Example 2

Determine the volume enclosed by the paraboloid $z = 1 - x^2 - y^2/4$ and the plane $z = 0$.

The plane $z = 0$ intercepts the paraboloid $z = 1 - x^2 - y^2/4$ in the ellipse $x^2 + y^2/4 = 1$ (see Fig. 10.4(a)).

Since we have symmetry about the planes $x = 0$, $y = 0$ the volume required is equal to four times the volume enclosed by the paraboloid and the first quadrant of the ellipse.

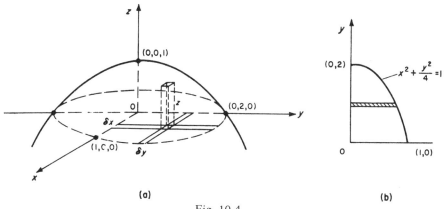

(a)

(b)

Fig. 10.4

Thus

$$V = 4 \iint_A z \, dx \, dy$$

where the field of integration A, is the area of the ellipse contained in the first quadrant (see Fig. 10.4(b)).

Hence:

(i) fix y, and vary x from $x = 0$ to $x = \sqrt{(1 - y^2/4)}$; and
(ii) vary y from $y = 0$ to $y = 2$;

so that

$$V = 4 \int_0^2 \int_0^{\sqrt{(1-y^2/4)}} \left(1 - x^2 - \frac{y^2}{4}\right) dx \, dy$$

$$= 4 \int_0^2 \left[x\left(1 - \frac{y^2}{4}\right) - \frac{x^3}{3} \right]_{x=0}^{x=\sqrt{(1-y^2/4)}} dy$$

$$= \frac{8}{3} \int_0^2 \left(1 - \frac{y^2}{4}\right)^{3/2} dy.$$

Let $y = 2 \sin \theta$ and $dy = 2 \cos \theta \, d\theta$, giving

$$V = \frac{16}{3} \int_0^{\pi/2} \cos^4 \theta \, d\theta = \pi \text{ cubic units.}$$

Example 3

In the extension of Stokes's law for ionic motion, the ion is assumed to be a sphere of radius a moving in a continuous medium.

If the rate of dissipation of energy induced by one ion is

$$\dot{E} = \mu \int_{-a}^{a} \int_{\sqrt{(a^2-x^2)}}^{\infty} \frac{x^2 y}{(x^2 + y^2)^4} \, dy \, dx + 2\mu \int_{a}^{\infty} \int_{0}^{\infty} \frac{x^2 y}{(x^2 + y^2)^4} \, dy \, dx$$

where μ is a parameter, show that $\dot{E} = 2\mu/9a^3$.

Let

$$\dot{E} = \mu(I_1 + 2I_2)$$

where

$$I_1 = \int_{-a}^{a} \int_{\sqrt{(a^2-x^2)}}^{\infty} \frac{x^2 y}{(x^2 + y^2)^4} \, dy \, dx$$

$$= \int_{-a}^{a} \left(-\frac{1}{6}\right) \frac{x^2}{(x^2 + y^2)^3} \Big|_{\sqrt{(a^2-x^2)}}^{\infty} dx = \frac{1}{6a^6} \int_{-a}^{a} x^2 \, dx = \frac{1}{9a^3},$$

and

$$I_2 = \int_{a}^{\infty} \int_{0}^{\infty} \frac{x^2 y}{(x^2 + y^2)^4} \, dy \, dx = \int_{a}^{\infty} \left(-\frac{1}{6}\right) \frac{x^2}{(x^2 + y^2)^3} \Big|_0^{\infty} dx$$

$$= \frac{1}{6} \int_a^\infty \frac{dx}{x^4} = \frac{1}{18a^3}$$

$$\therefore \ \dot{E} = \mu(I_1 + 2I_2) = \frac{2\mu}{9a^3}.$$

3. Change of order of integration

Referring again to Fig. 10.2, we see that if $y = d$ and $y = c$ give respectively the upper and lower extremities of the curve C, and $x = x_1(y)$ and $x = x_2(y)$ are the respective equations of the sections YXY', YX'Y' of the curve, then consideration of the elemental section ABCD leads to an alternative expression for V, namely

$$\int_{y=c}^{y=d} \left(\int_{x=x_1(y)}^{x=x_2(y)} f(x, y) \ dx \right) dy = \int_c^d dy \int_{x_1(y)}^{x_2(y)} f(x, y) \ dx$$

so that

$$\iint_A f(x, y) \ dx \ dy = \int_a^b \int_{y_1(x)}^{y_2(x)} f(x, y) \ dy \ dx = \int_c^d \int_{x_1(y)}^{x_2(y)} f(x, y) \ dx \ dy.$$

In other words we may first integrate either with respect to x or with respect to y, whichever is most convenient, provided that reference is made to the field of integration in order that the new limits may be determined.

One should however, first ensure that there are no discontinuities within the region A.

Example 4

Show that

$$\int_4^9 \int_x^9 \frac{\log_e y}{\sqrt{(y - x + 4)}} \ dy \ dx = \tfrac{4}{9}(24 \log_e 2 + 7).$$

The integral 'reads':

(i) fix x, and vary y from $y = x$ to $y = 9$; and
(ii) vary x from $x = 4$ to $x = 9$;

giving the field of integration shown in Fig. 10.5.

In order to reverse the order of integration:

(i) fix y, and vary x from $x = 4$ to $x = y$; and
(ii) vary y from $y = 4$ to $y = 9$.

Then

$$I = \int_4^9 \int_4^y \frac{\log_e y}{\sqrt{(y - x + 4)}} \ dx \ dy$$

$$= \int_4^9 \log_e y \left[-2\sqrt{(y - x + 4)} \right]_{x=4}^{x=y} \ dy$$

$$= 2 \int_4^9 (\sqrt{(y)} - 2) \log_e y \ dy.$$

l

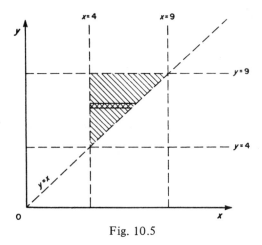

Fig. 10.5

Now

$$\int_4^9 \sqrt{(y)} \log_e y \; dy = \tfrac{2}{3} y^{3/2} \log_e y \; \Big|_4^9 - \tfrac{2}{3} \int_4^9 y^{1/2} \; dy$$

$$= 18 \log_e 9 - \frac{16}{3} \log_e 4 - \frac{76}{9}.$$

Similarly

$$\int_4^9 \log_e y \; dy = y \log_e y \; \Big|_4^9 - \int_4^9 dy$$

$$= 9 \log_e 9 - 4 \log_e 4 - 5$$

$$\therefore \; I = \tfrac{4}{9}(12 \log_e 4 + 7) = \tfrac{4}{9}(24 \log_e 2 + 7).$$

Example 5

Show that

$$\int_0^\infty \frac{1}{x} \left\{ \tan^{-1}\left(\frac{b}{x}\right) - \tan^{-1}\left(\frac{a}{x}\right) \right\} dx = \frac{\pi}{2} \log_e \left(\frac{b}{a}\right)$$

a and *b* being constants.

Since

$$\int_a^b \frac{du}{u^2 + x^2} = \frac{1}{x} \tan^{-1}\left(\frac{u}{x}\right) \Big|_a^b = \frac{1}{x} \left\{ \tan^{-1}\left(\frac{b}{x}\right) - \tan^{-1}\left(\frac{a}{x}\right) \right\}$$

we may express the given integral as a double integral:

$$I = \int_0^\infty \left(\int_a^b \frac{du}{u^2 + x^2} \right) dx = \int_0^\infty \int_a^b \frac{1}{u^2 + x^2} \; du \; dx.$$

The integral now 'reads':

  (i) fix x, and vary u from $u = a$ to $u = b$; and
  (ii) vary x from $x = 0$ to $x = \infty$.

In order to change the order of integration therefore (Fig. 10.6):

  (i) fix u, and vary x from $x = 0$ to $x = \infty$; and
  (ii) vary u from $u = a$ to $u = b$;

giving

$$I = \int_a^b \int_0^\infty \frac{1}{u^2 + x^2}\, dx\, du = \int_a^b \frac{1}{u}\left[\tan^{-1}\left(\frac{x}{u}\right)\right]_{x=0}^{x=\infty} du$$

$$= \int_a^b \frac{1}{u}\left(\frac{\pi}{2}\right) du = \frac{\pi}{2}\log_e\left(\frac{b}{a}\right).$$

Example 6

A molecule A moves over the surface of a sphere of radius a, whose centre is distance b from a fixed molecule $B(b > a)$.

If the mean mutual potential energy (PE) is given by

$$\text{PE} = \frac{\int_0^{2\pi} \int_0^\pi \phi(x) \sin x\, dx\, dy}{\int_0^{2\pi} \int_0^\pi \sin x\, dx\, dy}$$

show that

$$\text{PE} = \tfrac{1}{2} \int_0^\pi \phi(x) \sin x\, dx$$

and evaluate it for $\phi(x) = \sqrt{(a^2 + b^2 - 2\,ab\,\cos x)}$.

$$\int_0^{2\pi} \int_0^\pi \sin x\, dx\, dy = \int_0^{2\pi} (-\cos x)\Big|_0^\pi dy$$

$$= 2 \int_0^{2\pi} dy = 4\pi$$

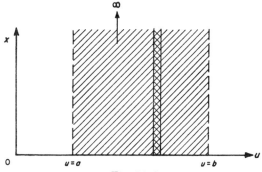

Fig. 10.6

and

$$\int_0^{2\pi} \int_0^{\pi} \phi(x) \sin x \, dx \, dy = \int_0^{\pi} \int_0^{2\pi} \phi(x) \sin x \, dy \, dx$$

$$= 2\pi \int_0^{\pi} \phi(x) \sin x \, dx.$$

$$\therefore \ \text{PE} = \tfrac{1}{2} \int_0^{\pi} \phi(x) \sin x \, dx.$$

In the special case given,

$$\text{PE} = \tfrac{1}{2} \int_0^{\pi} \sqrt{(a^2 + b^2 - 2ab \cos x)} \sin x \, dx$$

$$= \tfrac{1}{2}(a^2 + b^2 - 2ab \cos x)^{3/2} \left(\frac{1}{3ab}\right) \Big|_0^{\pi}$$

$$= \frac{1}{6ab} \{(a^2 + b^2 + 2ab)^{3/2} - (a^2 + b^2 - 2ab)^{3/2}\}$$

$$= \frac{1}{6ab} \{(b + a)^3 - (b - a)^3\}, \quad b > a$$

$$= \frac{1}{3b}(a^2 + 3b^2).$$

4. Transformation to polar coordinates

It is sometimes more convenient to evaluate a double integral by first transforming from *Cartesian* coordinates (x, y) to *polar* coordinates (r, θ), where

$$x = r \cos \theta, \quad y = r \sin \theta.$$

In Section 1, the expression for the volume V was obtained by dividing the area enclosed by C into rectangular elements of area $\delta x \, \delta y$, and then determining the volume of the cylinder on $\delta x \, \delta y$ as base.

If, however, polar coordinates are to be used, then the area is first divided into sectors of area $r \, \delta r \, \delta \theta$ (see Fig. 10.7). We then proceed as in Section 2.

The volume of the cylinder on $r \, \delta r \, \delta \theta$ as base, is

$$\delta V \simeq g(r, \theta) \, r \, \delta r \, \delta \theta$$

where $g(r, \theta) \equiv f(r \cos \theta, r \sin \theta)$.

The volume above the elemental section PQRS, for which θ and $\delta \theta$ are constant, is obtained by forming the sum of such elements δV over this section as r varies from $r = r_1(\theta)$ say, to $r = r_2(\theta)$, and is

$$\sum_{r=r_1(\theta)}^{r=r_2(\theta)} g(r, \theta) r \, \delta r \, \delta \theta = \delta \theta \sum_{r=r_1(\theta)}^{r=r_2(\theta)} g(r, \theta) r \, \delta r.$$

Hence, in the limit as $\delta r \to 0$, the volume above PQRS is

$$\lim_{\delta r \to 0} \delta \theta \sum_{r=r_1(\theta)}^{r=r_2(\theta)} g(r, \theta) r \, \delta r = \delta \theta \int_{r=r_1(\theta)}^{r=r_2(\theta)} g(r, \theta) r \, dr = \delta \theta \, G(\theta) \quad \text{say.}$$

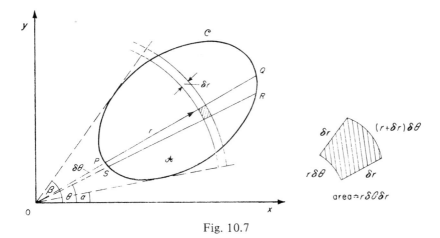

Fig. 10.7

If θ is now varied from $\theta = \alpha$ to $\theta = \beta$ in order to cover the whole of the field of integration A, we obtain the sum

$$\sum_{\theta=\alpha}^{\theta=\beta} G(\theta)\, \delta\theta.$$

Proceeding to the limiting case, as $\delta\theta \to 0$, gives the total volume V of Section 1 in the form

$$\lim_{\delta\theta \to 0} \sum_{\theta=\alpha}^{\theta=\beta} G(\theta)\, \delta\theta = \int_{\alpha}^{\beta} G(\theta)\, \mathrm{d}\theta = \int_{\alpha}^{\beta} \int_{r_1(\theta)}^{r_2(\theta)} g(r, \theta)\, r\, \mathrm{d}r\, \mathrm{d}\theta.$$

Thus

$$\iint_A f(x, y)\, \mathrm{d}x\, \mathrm{d}y = \iint_A f(r \cos \theta, r \sin \theta)\, r\, \mathrm{d}r\, \mathrm{d}\theta.$$

Example 7

Evaluate

$$\iint_A \frac{x^2 y}{\sqrt{(x^2 + y^2)}} \exp(x^2 + y^2)\, \mathrm{d}x\, \mathrm{d}y,$$

where A is the area in the first quadrant of the annulus formed by the two circles $x^2 + y^2 = 1$ and $x^2 + y^2 = 4$.

The field of integration is shown in Fig. 10.8, the polar form of the circles being $r = 1$ and $r = 2$ respectively.

In order to transform to polar coordinates:

(i) fix θ, and vary r from $r = 1$ to $r = 2$; and
(ii) vary θ from $\theta = 0$ to $\theta = \pi/2$.

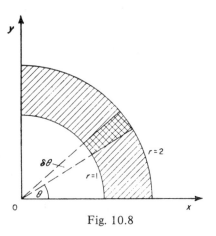

Fig. 10.8

Then

$$I = \int_0^{\pi/2} \int_1^2 \frac{r^2 \cos\theta \cdot r \sin\theta}{r} \cdot e^{r^2} \cdot r \, dr \, d\theta$$

$$= \int_0^{\pi/2} \int_1^2 \cos^2\theta \sin\theta \, r^3 e^{r^2} \, dr \, d\theta.$$

Now

$$\int_1^2 r^3 e^{r^2} \, dr = \int_1^2 r^2(re^{r^2}) \, dr = r^2(\tfrac{1}{2}e^{r^2}) \Big|_1^2 - \int_1^2 re^{r^2} \, dr$$

$$= 2e^4 - \tfrac{1}{2}e - \left(\frac{e^{r^2}}{2}\right)\Big|_1^2 = \frac{3e^4}{2}.$$

$$\therefore \ I = \frac{3e^4}{2} \int_0^{\pi/2} \cos^2\theta \sin\theta \, d\theta = -\frac{e^4}{2} \cos^3\theta \Big|_0^{\pi/2}$$

i.e.

$$I = \frac{e^4}{2}.$$

Example 8

Show that the volume common to the sphere $x^2 + y^2 + z^2 = 9$, and the cylinder $x^2 - 3x + y^2 = 0$ is $6(3\pi - 4)$ cubic units.

Referring to Fig. 10.9, we see that the volume required is equal to four times the volume obtained by considering that part of the cylinder in the first quadrant and in the sense of z increasing.

The volume, V say, is therefore given by

$$V = 4 \iint_A z \, dx \, dy = 4 \iint_A \sqrt{(9 - x^2 - y^2)} \, dx \, dy$$

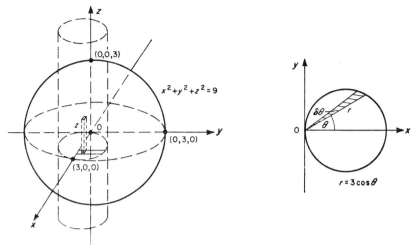

Fig. 10.9

where A is the area of that part of the circle $x^2 - 3x + y^2 = 0$ which lies in the first quadrant.

Since $x = r \cos \theta$ and $y = r \sin \theta$, this becomes $r = 3 \cos \theta$ in terms of polar coordinates.

In order to determine the appropriate limits of integration:

(i) fix θ, and vary r from $r = 0$ to $r = 3 \cos \theta$; and
(ii) vary θ from $\theta = 0$ to $\theta = \pi/2$.

Then

$$V = 4 \int_0^{\pi/2} \int_0^{3\cos\theta} \sqrt{(9 - r^2)}\, r\, dr\, d\theta$$

$$= 4 \int_0^{\pi/2} (-\tfrac{1}{3})(9 - r^2)^{3/2} \Big|_0^{3\cos\theta} d\theta$$

$$= 36 \int_0^{\pi/2} (1 - \sin^3 \theta)\, d\theta$$

$$= 6(3\pi - 4).$$

PROBLEMS

1. Determine the area enclosed by:

(i) the circle $x^2 + y^2 = a^2$;
(ii) the curves $y = x^2 + 1$, $y = x + 3$;

by means of double integration.

2. Evaluate

$$\int_0^{1/2} \int_{y^2}^{(1-y)} \frac{2y}{2x + 1}\, dx\, dy.$$

254

3. By transforming to polar coordinates, evaluate

$$\int_0^1 \int_0^{\sqrt{(1-y^2)}} \frac{x^2}{x^2+y^2} \exp\{\sqrt{(x^2+y^2)}\} \, dx dy.$$

4. Show that

$$\int_0^{\pi/2} \int_0^\infty \frac{\cos(2x \cos y)}{e^x} \, dx \, dy = \frac{\pi}{2\sqrt{5}}.$$

5. Sketch the field of integration for the double integral

$$\int_0^{a/\sqrt{2}} dy \int_y^{\sqrt{(a^2-y^2)}} \log_e(x^2+y^2) \, dx.$$

Transform to polar coordinates and hence evaluate the integral.

6. Show that

$$\int_1^{e^2} \int_{1/y}^y \frac{\log_e(xy)}{x} \, dx \, dy = 4(e^2 - 1).$$

7. Evaluate

$$\int_0^1 \int_0^{\pi/6} \frac{xy}{\cos(x^2y)} \{\cos(1-x^2)y - \cos(1+x^2)y\} \, dy \, dx.$$

8. By expressing

$$\int_0^\infty \frac{1}{x} (\text{sech } x - \text{sech } \lambda x) \, dx$$

as a double integral, show that its value is $\log_e \lambda$.

9. By considering the double integral

$$\iint_A \exp\{-a(x^2+y^2)\} \, dx \, dy$$

where A is the area in the whole of the first quadrant, show that

$$\int_0^\infty e^{-ax^2} \, dx = \frac{1}{2} \sqrt{\left(\frac{\pi}{a}\right)}.$$

10. If the circle $r = a$ is intersected by the line $r = b \sec \theta (b < a)$, determine the areas of the two segments into which the circle is divided.

11. Determine the volume of the cylinder $x^2 + y^2 + x = 0$ intercepted between the planes $x + y + z + 2 = 0$, $z = 0$.

12. Derive an expression in the form of a double integral for the volume common to the cylinder $x^2 + y^2 - 3y = 0$ and the sphere $x^2 + y^2 + z^2 - 9 = 0$.
   By first transforming to polar coordinates determine its value.

13. Determine the second moment of area of the area enclosed by the cardioid $2r = 1 + \cos \theta$, about an axis perpendicular to its plane and passing through the origin.

14. In the production of neutrons from uranium 235, the probability that one neutron escapes from a spherical zone of radius a, without encountering any collision, is

$$p = 1 - \frac{3}{8a^3} \int_0^{2a} \int_x^{2a} y \exp\left(\frac{-x}{\lambda}\right) dy \, dx$$

where λ is the mean free path.
   Evaluate the expression for p.

15. In the theory of quantum mechanics, the expectation value of the interaction energy between electrons is

$$\left(\frac{\lambda^3 e}{4}\right)^2 \int_0^\infty \int_0^y x^2 y \exp\{-\lambda(x+y)\}\, dx\, dy$$

$$+ \left(\frac{\lambda^3 e}{4}\right)^2 \int_0^\infty \int_y^\infty xy^2 \exp\{-\lambda(x+y)\}\, dx\, dy,$$

where λ is a parameter and $-e$ is the charge on an electron.

Denoting the above expression by I, show that

$$16I = 5\lambda e^2.$$

16. The total energy of a monatomic gas of volume v is

$$E = nv\alpha \int_{-\infty}^\infty \int_{-\infty}^\infty \int_{-\infty}^\infty \epsilon \exp(-\beta\epsilon)\, dx\, dy\, dz$$

where x, y, and z represent the momenta of a particle of mass m, α and β are constants, n is the number of molecules, and ϵ, the kinetic energy per molecule is given by

$$\epsilon = \frac{1}{2m}(x^2 + y^2 + z^2).$$

Show that

$$E = \frac{3nv}{2\beta}\left(\frac{2\pi m}{\beta}\right)^{3/2}.$$

(You may assume the result given in Question 9, above.)

17. The following equation occurs in the theory of the diffusion of caesium along tungsten filaments·

$$C = \frac{C_0}{4\pi Dt} \int_{-d/2}^{d/2} \int_{-d/2}^{d/2} \exp\left(-\frac{(a-x)^2 + (b-y)^2}{4Dt}\right) dx\, dy$$

a, b, C_0, D, and t being constants for the integral.

Show that

$$C = \frac{C_0}{4\sqrt{(\pi Dt)}} \exp\left(-\frac{a^2 + b^2}{4Dt}\right).$$

18. For the accelerating slit system of a mass spectrometer, the number of ions contained within unit solid angle of the electron beam, at the plane of the first slit. is

$$n = \frac{2n_c l}{\pi} \int_0^{2d} \int_0^\infty \frac{x^2\left(1 - \dfrac{y^2 a^2}{c^2}\right)}{(1 + x^2)\left(x^2 + \dfrac{y^2 a^2}{c^2}\right)}\, dx\, dy$$

where n_c is the number of ions, with initial velocity c, which are created per second, l is the length of the slit, d is the width of the slit, and a is a constant.

Show that

$$n = 2n_c dl\left(1 - \frac{ad}{c}\right).$$

CHAPTER 11

Fourier Series

1. Definition

In Chapter 5 the Taylor and Maclaurin expansions were determined for various functions. It is sometimes preferable to express a function $f(x)$ as a series of sine or cosine terms.

Suppose we can find constants $a_0, a_1, a_2, \ldots, b_1, b_2, \ldots$ such that

$$f(x) = a_0 + a_1 \cos x + a_2 \cos 2x + \cdots + b_1 \sin x + b_2 \sin 2x + \cdots$$

$$= a_0 + \sum_{n=1}^{\infty} (a_n \cos nx + b_n \sin nx). \tag{1}$$

Then the series given by equation (1) is called a *Fourier series*, a_n and b_n being the *Fourier coefficients*.

Replacing x by $x + 2k\pi$ where k is integral, equation (1) becomes

$$f(x + 2k\pi) = a_0 + \sum_{n=1}^{\infty} \{a_n \cos n(x + 2k\pi) + b_n \sin n(x + 2k\pi)\}$$

$$= a_0 + \sum_{n=1}^{\infty} (a_n \cos nx + b_n \sin nx)$$

as before, showing the function $f(x)$ is periodic and of *period* 2π.

An alternative form for the series may be obtained from equation (1) by putting

$$a_n = \lambda_n \sin \alpha_n, \quad b_n = \lambda_n \cos \alpha_n$$

where λ_n and α_n are constants, in which case the equation becomes

$$f(x) = a_0 + \sum_{n=1}^{\infty} \lambda_n (\sin \alpha_n \cos nx + \cos \alpha_n \sin nx)$$

$$= a_0 + \sum_{n=1}^{\infty} \lambda_n \sin(nx + \alpha_n).$$

The terms $\lambda_1 \sin(x + \alpha_1)$, $\lambda_2 \sin(2x + \alpha_2)$, $\ldots$, $\lambda_n \sin(nx + \alpha_n)$ are called the *first harmonic, second harmonic*, $\ldots$, and nth *harmonic* respectively.

2. Determination of the Fourier coefficients

In order to determine the values of a_n and b_n we require the following results.
If m and n, $\neq 0$, are integers, then

$$\int_0^{2\pi} \cos nx \, dx = \int_0^{2\pi} \sin nx \, dx = 0. \tag{2}$$

$$\int_0^{2\pi} \cos mx \cos nx \, dx = \frac{1}{2} \int_0^{2\pi} \{\cos(m+n)x + \cos(m-n)x\} \, dx$$

$$= \begin{cases} 0, & \text{if } m \neq n \\ \pi, & \text{if } m = n \end{cases} \tag{3}$$

$$\int_0^{2\pi} \sin mx \sin nx \, dx = \frac{1}{2} \int_0^{2\pi} \{\cos(m-n)x - \cos(m+n)x\} \, dx$$

$$= \begin{cases} 0, & \text{if } m \neq n \\ \pi, & \text{if } m = n \end{cases} \tag{4}$$

and

$$\int_0^{2\pi} \sin mx \cos nx \, dx = \frac{1}{2} \int_0^{2\pi} \{\sin(m+n)x + \sin(m-n)x\} \, dx$$

$$= 0. \tag{5}$$

Calculation of a_0. Integrating both sides of equation (1) with respect to x between limits $x = 0$ and $x = 2\pi$, gives

$$\int_0^{2\pi} f(x) \, dx = a_0 \int_0^{2\pi} dx + \sum_{n=1}^{\infty} \int_0^{2\pi} (a_n \cos nx + b_n \sin nx) \, dx$$

$$= 2\pi a_0$$

from equation (2), assuming that the integration of the right-hand side, term by term, is permissible. Thus

$$a_0 = \frac{1}{2\pi} \int_0^{2\pi} f(x) \, dx.$$

Calculation of a_n. Multiplying both sides of equation (1) by $\cos mx$ (m integral) and integrating with respect to x from $x = 0$ to $x = 2\pi$ gives

$$\int_0^{2\pi} f(x)\cos mx \, dx = a_0 \int_0^{2\pi} \cos mx \, dx + \sum_{n=1}^{\infty} a_n \int_0^{2\pi} \cos nx \cos mx \, dx$$

$$+ \sum_{n=1}^{\infty} b_n \int_0^{2\pi} \sin nx \cos mx \, dx$$

$$= a_0(0) + a_m(\pi) + b_n(0)$$

from equations (2), (3) and (5). Therefore

$$a_m \equiv a_n = \frac{1}{\pi} \int_0^{2\pi} f(x)\cos mx \, dx, \quad (m = n).$$

Calculation of b_n. Multiplying both sides of equation (1) by $\sin mx$ and integrating with respect to x from $x = 0$ to $x = 2\pi$ gives

$$\int_0^{2\pi} f(x)\sin mx \, dx = a_0 \int_0^{2\pi} \sin mx \, dx + \sum_{n=1}^{\infty} a_n \int_0^{2\pi} \cos nx \sin mx \, dx$$

$$+ \sum_{n=1}^{\infty} b_n \int_0^{2\pi} \sin nx \sin mx \, dx$$

$$= a_0(0) + a_n(0) + b_n(\pi)$$

from equations (2), (5) and (4). Thus

$$b_m \equiv b_n = \frac{1}{\pi} \int_0^{2\pi} f(x)\sin mx \, dx, \quad (m = n).$$

Similar results would have been obtained by integrating over any range of 2π, τ to $\tau + 2\pi$ say, the range $0 \leqslant x \leqslant 2\pi$ being chosen for convenience only.

The result may be stated as follows.

If $f(x)$ is a periodic function of x, of period 2π, then the corresponding Fourier series, if it exists, is given by

$$f(x) = a_0 + \sum_{n=1}^{\infty} (a_n \cos nx + b_n \sin nx)$$

where

$$a_0 = \frac{1}{2\pi} \int_0^{2\pi} f(x) \, dx, \quad a_n = \frac{1}{\pi} \int_0^{2\pi} f(x)\cos nx \, dx,$$

$$b_n = \frac{1}{\pi} \int_0^{2\pi} f(x)\sin nx \, dx \tag{6}$$

or more generally

$$a_0 = \frac{1}{2\pi} \int_\tau^{\tau+2\pi} f(x) \, dx, \quad a_n = \frac{1}{\pi} \int_\tau^{\tau+2\pi} f(x)\cos nx \, dx,$$

$$b_n = \frac{1}{\pi} \int_\tau^{\tau+2\pi} f(x)\sin nx \, dx. \tag{7}$$

For a function which is discontinuous within the given range, the coefficients a_0, a_n and b_n are expressed as the sum of two or more integrals, covering the whole range once only, as the following example illustrates.

Example 1

Find the Fourier series for the function $f(x)$ defined by

$$f(x) = \begin{cases} 1 & \text{for } -\pi < x < 0 \\ x & \text{for } 0 < x < \pi \end{cases}$$

$f(x + 2\pi) = f(x)$.

(See Fig. 11.1.)

From the previous section

$$a_0 = \frac{1}{2\pi} \int_{\tau}^{\tau+2\pi} f(x) \, dx = \frac{1}{2\pi} \int_{-\pi}^{\pi} f(x) \, dx$$

$$= \frac{1}{2\pi} \int_{-\pi}^{0} dx + \frac{1}{2\pi} \int_{0}^{\pi} x \, dx = \tfrac{1}{4}(\pi + 2).$$

$$a_n = \frac{1}{\pi} \int_{-\pi}^{0} \cos nx \, dx + \frac{1}{\pi} \int_{0}^{\pi} x \cos nx \, dx$$

$$= \frac{1}{n\pi} \sin nx \Big|_{-\pi}^{0} + \frac{1}{\pi} \left(\frac{x}{n} \sin nx \Big|_{0}^{\pi} - \frac{1}{n} \int_{0}^{\pi} \sin nx \, dx \right)$$

$$= \frac{1}{n^2 \pi} \cos nx \Big|_{0}^{\pi}$$

$$= \frac{1}{n^2 \pi} (\cos n\pi - 1)$$

so that $a_{2n} = 0$, and $a_{2n-1} = -2/\{\pi(2n-1)^2\}$.

Similarly

$$b_n = \frac{1}{\pi} \int_{-\pi}^{0} \sin nx \, dx + \frac{1}{\pi} \int_{0}^{\pi} x \sin nx \, dx$$

$$= -\frac{1}{n\pi} \cos nx \Big|_{-\pi}^{0} + \frac{1}{\pi} \left(-\frac{x}{n} \cos nx \Big|_{0}^{\pi} + \frac{1}{n} \int_{0}^{\pi} \cos nx \, dx \right)$$

$$= \frac{1}{n\pi} (\cos n\pi - 1) - \frac{1}{n} \cos n\pi.$$

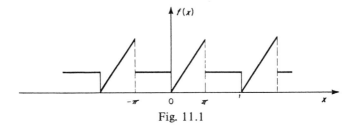

Fig. 11.1

Hence

$$f(x) = \frac{1}{4}(\pi + 2) - \frac{2}{\pi}\left(\cos x + \frac{1}{3^2}\cos 3x + \frac{1}{5^2}\cos 5x + \cdots\right)$$
$$+ \left(\frac{\pi - 2}{\pi}\sin x - \frac{1}{2}\sin 2x + \frac{\pi - 2}{3\pi}\sin 3x - \frac{1}{4}\sin 4x + \cdots\right).$$

3. Period 2T

In the general case, the function $f(x)$ is defined to be a periodic function of x of period $2T$, provided it satisfies the equation

$$f(x) = f(x + 2kT),$$

for all integral values of k. The corresponding Fourier series is then derived by using a simple transformation.

Let

$$u = \frac{\pi x}{T},$$

so that $0 \leqslant u \leqslant 2\pi$, as x varies from $x = 0$ to $x = 2T$, and

$$f(x) \equiv f\left(\frac{uT}{\pi}\right) \equiv \phi(u) \quad \text{say.}$$

Since the function $\phi(u)$ is of period 2π, we may express it as a Fourier series, as in Section 2, in the form

$$\phi(u) = a_0 + \sum_{n=1}^{\infty}(a_n \cos nu + b_n \sin nu) \qquad (8)$$

where

$$a_0 = \frac{1}{2\pi}\int_0^{2\pi}\phi(u)\,du, \quad a_n = \frac{1}{\pi}\int_0^{2\pi}\phi(u)\cos nu\,du,$$

$$b_n = \frac{1}{\pi}\int_0^{2\pi}\phi(u)\sin nu\,du.$$

Reverting now to the original variable x, with $du = (\pi/T)\,dx$, equation (8) gives the required form of the Fourier series, i.e.

$$f(x) = a_0 + \sum_{n=1}^{\infty}\left(a_n \cos\frac{n\pi x}{T} + b_n \sin\frac{n\pi x}{T}\right)$$

where

$$a_0 = \frac{1}{2T}\int_0^{2T}f(x)\,dx, \quad a_n = \frac{1}{T}\int_0^{2T}f(x)\cos\frac{n\pi x}{T}\,dx,$$

$$b_n = \frac{1}{T}\int_0^{2T}f(x)\sin\frac{n\pi x}{T}\,dx \qquad (9)$$

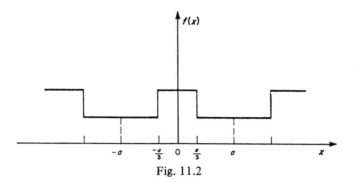

Fig. 11.2

or more generally

$$a_0 = \frac{1}{2T} \int_T^{T+2T} f(x)\, dx, \quad a_n = \frac{1}{T} \int_T^{T+2T} f(x)\cos\frac{n\pi x}{T}\, dx,$$

$$b_n = \frac{1}{T} \int_T^{T+2T} f(x)\sin\frac{n\pi x}{T}\, dx. \tag{10}$$

Example 2

Find the Fourier series for the function

$$f(x) = \begin{cases} 1, & \text{for } -a < x < -a/3 \\ 2 & \text{for } -a/3 < x < a/3 \\ 1 & \text{for } a/3 < x < a \end{cases}$$

$$f(x + 2a) = f(x).$$

(See Fig. 11.2.)

Equation (10) gives

$$a_0 = \frac{1}{2a} \int_T^{T+2a} f(x)\, dx = \frac{1}{2a} \int_{-a}^a f(x)\, dx$$

$$= \frac{1}{2a} \int_{-a}^{-a/3} dx + \frac{1}{2a} \int_{-a/3}^{a/3} 2\, dx + \frac{1}{2a} \int_{a/3}^a dx = \frac{4}{3}.$$

$$a_n = \frac{1}{a} \int_{-a}^{-a/3} \cos\frac{n\pi x}{a}\, dx + \frac{1}{a} \int_{-a/3}^{a/3} 2\cos\frac{n\pi x}{a}\, dx + \frac{1}{a} \int_{a/3}^a \cos\frac{n\pi x}{a}\, dx$$

$$= \frac{1}{n\pi} \left(-\sin\frac{n\pi}{3} + 2\sin\frac{n\pi}{3} + 2\sin\frac{n\pi}{3} - \sin\frac{n\pi}{3} \right)$$

$$= \frac{2}{n\pi} \sin\frac{n\pi}{3}.$$

$$b_n = \frac{1}{a} \int_{-a}^{-a/3} \sin \frac{n\pi x}{a} \, dx + \frac{1}{a} \int_{-a/3}^{a/3} 2 \sin \frac{n\pi x}{a} \, dx + \frac{1}{a} \int_{a/3}^{a} \sin \frac{n\pi x}{a} \, dx$$

$$= -\frac{1}{n\pi} \left(\cos \frac{n\pi}{3} - \cos n\pi + 2 \cos \frac{n\pi}{3} - 2 \cos \frac{n\pi}{3} + \cos n\pi - \cos \frac{n\pi}{3} \right)$$

$$= 0.$$

Hence

$$f(x) = \frac{4}{3} + \frac{2}{\pi} \sum_{n=1}^{\infty} \frac{1}{n} \sin \frac{n\pi}{3} \cos \frac{n\pi x}{a}.$$

4. Even and odd functions

The definitions of such functions have already been given in Section 4 of Chapter 1.

Whether a function is even or odd may also be determined geometrically, since the graph of an even function is symmetric about the y-axis, whilst that of an odd function is symmetric about the origin (see Fig. 11.3).

For a function $f(x)$ of period $2T$

$$\int_0^{2T} f(x) \, dx = \int_{-T}^{T} f(x) \, dx$$

$$= \int_{-T}^{0} f(x) \, dx + \int_0^{T} f(x) \, dx.$$

Replacing x by $-x$ in the first term on the right-hand side of the equation gives

$$\int_0^{2T} f(x) \, dx = \int_{T}^{0} f(-x)(-dx) + \int_0^{T} f(x) \, dx$$

$$= \int_0^{T} f(-x) \, dx + \int_0^{T} f(x) \, dx$$

$$= \int_0^{T} \{f(x) + f(-x)\} \, dx.$$

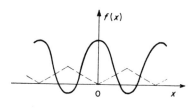

Even functions

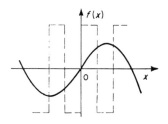

Odd functions

Fig. 11.3

Thus

$$\int_0^{2T} f(x)\, dx = \begin{cases} 2\int_0^T f(x)\, dx, & \text{if } f(x) \text{ is an even function} \\ 0, & \text{if } f(x) \text{ is an odd function.} \end{cases}$$

Since the product of two odd or two even functions is an even function, and the product of an even and odd function is an odd function, the following simplifications occur in the formulae for the Fourier coefficients.

(*i*) *For f(x) an even periodic function*

$$a_0 = \frac{1}{T} \int_0^T f(x)\, dx, \quad a_n = \frac{2}{T} \int_0^T f(x)\cos\frac{n\pi x}{T}\, dx, \quad b_n = 0 \tag{11}$$

so that the Fourier series contains cosine terms only, i.e.

$$f(x) = a_0 + \sum_{n=1}^{\infty} a_n \cos\frac{n\pi x}{T}.$$

(*ii*) *For f(x) an odd periodic function*

$$a_0 = 0, \quad a_n = 0, \quad b_n = \frac{2}{T} \int_0^T f(x)\sin\frac{n\pi x}{T}\, dx, \tag{12}$$

so that the Fourier series contains sine terms only, i.e.

$$f(x) = \sum_{n=1}^{\infty} b_n \sin\frac{n\pi x}{T}.$$

Example 3

Determine the Fourier series for the function $f(x)$ given by

$$f(x) = x^2, \quad -\pi < x < \pi$$

$$f(x + 2\pi) = f(x).$$

(See Fig. 11.4.)

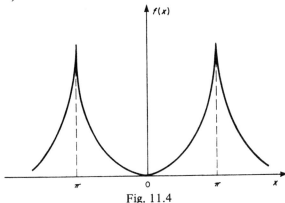

Fig. 11.4

Since $f(x)$ is an even function of x, the corresponding Fourier series will contain cosine terms only.

Equation (11) gives

$$a_0 = \frac{1}{\pi} \int_0^\pi x^2 \, dx = \frac{\pi^2}{3},$$

and

$$a_n = \frac{2}{\pi} \int_0^\pi x^2 \cos nx \, dx$$

$$= \frac{2}{\pi} \left(\frac{x^2}{n} \sin nx \bigg|_0^\pi - \frac{2}{n} \int_0^\pi x \sin nx \, dx \right)$$

$$= \frac{-4}{n\pi} \left(\frac{-x}{n} \cos nx \bigg|_0^\pi + \frac{1}{n} \int_0^\pi \cos nx \, dx \right)$$

$$= \frac{4}{n^2} \cos n\pi$$

$$= \frac{4(-1)^n}{n^2}.$$

Hence

$$f(x) = \frac{\pi^2}{3} + 4 \sum_{n=1}^\infty \frac{(-1)^n}{n^2} \cos nx.$$

5. The range $0 \leqslant x \leqslant T$

Suppose the function $f(x)$ is defined only within the range $x = 0$ to $x = T$.

Then we have a choice. Either we replace $f(x)$ by $f(-x)$ for the range $x = -T$ to $x = 0$, thereby creating an even function, or we replace $f(x)$ by $-f(-x)$ thereby creating an odd function.

The Fourier series is then given by either equation (11) or equation (12), and is equal to $f(x)$ only for the range $0 \leqslant x \leqslant T$. Such expansions are called *half-range* Fourier series since the integration is over only half the range.

Example 4

Determine a Fourier sine series for $f(x) = \cos x$ in the range $(0, \pi)$.

The Fourier sine series is obtained by choosing the function $f(x) = -\cos(-x) = -\cos x$ in the range $(-\pi, 0)$ so that the combination produces an odd function over the range $-\pi < x < \pi$:

$$f(x) = \begin{cases} \cos x, & \text{for } -\pi < x < 0 \\ -\cos x & \text{for } 0 < x < \pi, \end{cases}$$

$$f(x + 2\pi) = f(x).$$

(See Fig. 11.5.)

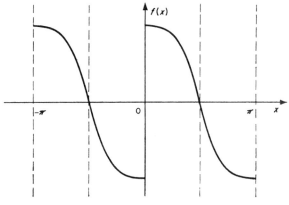

Fig. 11.5

Equation (12) gives

$$b_n = \frac{2}{\pi} \int_0^\pi \cos x \sin nx \, dx$$

$$= \frac{1}{\pi} \int_0^\pi \{\sin(n+1)x + \sin(n-1)x\} \, dx$$

$$= -\frac{1}{\pi} \left[\frac{\cos(n+1)x}{n+1} + \frac{\cos(n-1)x}{n-1} \right]_{x=0}^{x=\pi}$$

$$= -\frac{1}{\pi} \left(\frac{\cos(n+1)\pi}{n+1} - \frac{1}{n+1} + \frac{\cos(n-1)\pi}{n-1} - \frac{1}{n-1} \right)$$

$$\therefore \ b_{2n} = \frac{8n}{\pi(4n^2-1)} \quad \text{and} \quad b_{2n-1} = 0,$$

since $\cos(2n+1)\pi = \cos(2n-1)\pi = \cos\pi = -1$, and $\cos 2n\pi = \cos(2n-2)\pi = 1$.
Hence

$$f(x) = \frac{8}{\pi} \sum_{n=1}^{\infty} \frac{n}{4n^2-1} \sin 2nx.$$

6. Harmonic analysis

We now consider the case when the function $f(x)$ is not given explicitly, but is represented by a set of numerical values.

Suppose that the range $0 \leqslant x \leqslant 2T$ is divided into k equal parts by the abscissae $x_0, x_1, \ldots, x_k$ where $x_0 = 0$ and $x_k = 2T$, the corresponding values of the function $f(x)$ being known. Then the problem consists of calculating values for the coefficients a_0, a_n and b_n occurring in the Fourier series.

Since the distance between two consecutive points is $2T/k$, the corresponding formulae are obtained from equation (9) by replacing δx by $2T/k$ and the integral signs by summation signs extending over the given range.

Thus

$$
\left.
\begin{aligned}
a_0 &= \frac{1}{2T} \sum_{i=1}^{k} f(x_i) \cdot \frac{2T}{k} = \frac{1}{k} \sum_{i=1}^{k} f(x_i) \\
a_n &= \frac{1}{T} \sum_{i=1}^{k} f(x_i)\cos\left(\frac{n\pi x_i}{T}\right) \cdot \frac{2T}{k} = \frac{2}{k} \sum_{i=1}^{k} f(x_i)\cos\left(\frac{n\pi x_i}{T}\right)
\end{aligned}
\right\}
\tag{13}
$$

and

$$
b_n = \frac{1}{T} \sum_{i=1}^{k} f(x_i)\sin\left(\frac{n\pi x_i}{T}\right) \cdot \frac{2T}{k} = \frac{2}{k} \sum_{i=1}^{k} f(x_i)\sin\left(\frac{n\pi x_i}{T}\right)
$$

The following example will illustrate the method.

Example 5

The temperature θ_i °C of a substance at time t_i minutes after the start of an experiment is given by the following table:

t_i	5	10	15	20	25	30	35	40
θ_i	9.00	19.00	31.00	25.00	26.00	30.00	17.00	10.00

Determine a Fourier series which will represent the graph obtained by plotting t_i against θ_i.

With $k = 8$ and $2T = 40$, equation (13) gives

$$
a_0 = \frac{1}{8} \sum_{i=1}^{8} \theta_i, \quad a_n = \frac{1}{4} \sum_{i=1}^{8} \theta_i, \left(\frac{n\pi t_i}{20}\right) \cos \quad b_n = \frac{1}{4} \sum_{i=1}^{8} \theta_i \sin\left(\frac{n\pi t_i}{20}\right).
$$

Thus

	5	10	15	20	25	30	35	40
t_i	5	10	15	20	25	30	35	40
θ_i	9.00	19.00	31.00	25.00	26.00	30.00	17.00	10.00
$\cos\left(\dfrac{\pi t_i}{20}\right)$	0.7071	0	−0.7071	−1	−0.7071	0	0.7071	1
$\theta_i \cos\left(\dfrac{\pi t_i}{20}\right)$	6.36	0	−21.92	−25	−18.38	0	12.02	10.00
$\sin\left(\dfrac{\pi t_i}{20}\right)$	0.7071	1	0.7071	0	−0.7071	−1	−0.7071	0
$\theta_i \sin\left(\dfrac{\pi t_i}{20}\right)$	6.36	19.00	21.92	0	−18.38	−30.00	−12.02	0

whence

$$a_0 = \tfrac{1}{8}(167.00) = 20.87$$

$$a_1 = \frac{1}{4} \sum_{i=1}^{8} \theta_i \cos\left(\frac{\pi t_i}{20}\right) = \tfrac{1}{4}(-36.92) = -9.23$$

and

$$b_1 = \frac{1}{4} \sum_{i=1}^{8} \theta_i \sin\left(\frac{\pi t_i}{20}\right) = \tfrac{1}{4}(-13.12) = -3.28.$$

Proceeding in the same way leads to the following results:

$$a_2 = \frac{1}{4} \sum_{i=1}^{8} \theta_i \cos\left(\frac{\pi t_i}{10}\right) = \tfrac{1}{4}(-14.00) = -3.50$$

$$b_2 = \frac{1}{4} \sum_{i=1}^{8} \theta_i \sin\left(\frac{\pi t_i}{10}\right) = \tfrac{1}{4}(-13.00) = -3.25$$

$$a_3 = \frac{1}{4} \sum_{i=1}^{8} \theta_i \cos\left(\frac{3\pi t_i}{20}\right) = \tfrac{1}{4}(6.92) = 1.73$$

$$b_3 = \frac{1}{4} \sum_{i=1}^{8} \theta_i \sin\left(\frac{3\pi t_i}{20}\right) = \tfrac{1}{4}(8.88) = 2.22$$

and so on.

The corresponding Fourier series is thus

$$\theta = 20.87 - 9.23 \cos\left(\frac{\pi t}{20}\right) - 3.28 \sin\left(\frac{\pi t}{20}\right) - 3.50 \cos\left(\frac{\pi t}{10}\right)$$

$$-3.25 \sin\left(\frac{\pi t}{10}\right) + 1.73 \cos\left(\frac{3\pi t}{20}\right) + 2.22 \sin\left(\frac{3\pi t}{20}\right) + \cdots.$$

The more terms obtained, the more accurately the Fourier series corresponds to the graph of t_i against θ_i.

7. Partial differential equations

It is the aim of this section to introduce the method of solution of certain types of partial differential equations by means of Fourier series, rather than dealing with these equations in general.

Consider the one-dimensional wave equation

$$c^2 \frac{\partial^2 \phi}{\partial x^2} = \frac{\partial^2 \phi}{\partial t^2} \tag{14}$$

where ϕ is a function of the independent variables x and t, and c is a constant.

It is easily verified that the general solution of this equation is

$$\phi(x, t) = u(x + ct) + v(x - ct)$$

u and v being arbitrary functions of $x + ct$ and $x - ct$ respectively.

Suppose that a more particular solution is required which is to satisfy certain boundary and initial conditions. Let such a solution be

$$\phi(x, t) = X(x)T(t) \tag{15}$$

where $X = X(x)$ is a function of x only, $T = T(t)$ is a function of t only, and $\phi(0, t) = \phi(\lambda, t) = 0$ for all t.

Then

$$\frac{\partial^2 \phi}{\partial x^2} = T\frac{d^2 X}{dx^2}, \quad \frac{\partial^2 \phi}{\partial t^2} = X\frac{d^2 T}{dt^2}$$

and equation (14) becomes

$$c^2 T\frac{d^2 X}{dx^2} = X\frac{d^2 T}{dt^2}$$

i.e.

$$\frac{1}{X}\frac{d^2 X}{dx^2} = \frac{1}{c^2 T}\frac{d^2 T}{dt^2}. \tag{16}$$

Since the left-hand side of equation (16) is a function of x only and the right-hand side is a function of t only, each side must be equal to a constant, $-\omega^2$ say. Therefore

$$\frac{d^2 X}{dx^2} + \omega^2 X = 0, \quad \frac{d^2 T}{dt^2} + \omega^2 c^2 T = 0$$

solutions of which are, from Chapter 8,

$$X = A\cos\omega x + B\sin\omega x, \quad T = A'\cos\omega ct + B'\sin\omega ct$$

$$= C\cos(\omega x + \alpha) \qquad\qquad = C'\cos(\omega ct + \beta)$$

where C, C', α and β are arbitrary constants.

Equation (15) now gives

$$\phi(x, t) = K\cos(\omega x + \alpha)\cos(\omega ct + \beta) \tag{17}$$

where K is a constant.

The constants α and ω may now be determined by considering the given boundary conditions. Taking $\alpha = \pi/2$ and $\sin\omega\lambda = 0$, which satisfy the boundary conditions, leads to the particular solution

$$\phi(x, t) = K\sin\omega x \cos(\omega ct + \beta)$$

where $\omega = n\pi/\lambda$ $(n = 1, 2, \ldots)$.

The above method, known as *separation of variables*, thus gives the solution of equation (14) in the form

$$\phi(x, t) = \sum_{n=1}^{\infty} K_n \sin\left(\frac{n\pi x}{\lambda}\right) \cos\left(\frac{n\pi ct}{\lambda} + \beta_n\right) \tag{18}$$

the case $n = 0$ being omitted since it yields only the trivial solution $\phi(x, t) = 0$.

The values of the arbitrary constants $K_n (n = 1, 2, \ldots)$ are now determined as for a Fourier series, and are chosen such that the solution given in equation (18) satisfies the initial conditions.

The procedure is illustrated by the following examples.

Example 6

Find the solution of the wave equation (14) which satisfies the conditions

(i) $\phi(0, t) = \phi(\lambda, t) = 0,$ for $t \geq 0$

(ii) $\dfrac{\partial \phi}{\partial t} = 0$ for $t = 0$

(iii)
$$\phi(x, 0) = \begin{cases} x, & \text{for } 0 \leq x \leq \dfrac{\lambda}{2} \\ \lambda - x & \text{for } \dfrac{\lambda}{2} \leq x \leq \lambda. \end{cases}$$

Condition (i).   At this stage the solution is given by equation (18) above.

Condition (ii).   From equation (18)

$$\frac{\partial \phi}{\partial t} = -\frac{\pi c}{\lambda} \sum_{n=1}^{\infty} n K_n \sin\left(\frac{n\pi x}{\lambda}\right) \sin\left(\frac{n\pi ct}{\lambda} + \beta_n\right)$$

which is zero at $t = 0$ provided we choose $\beta_n = 0$ ($n = 1, 2, \ldots$).

The solution at this stage is therefore

$$\phi(x, t) = \sum_{n=1}^{\infty} K_n \sin\left(\frac{n\pi x}{\lambda}\right) \cos\left(\frac{n\pi ct}{\lambda}\right). \tag{19}$$

Condition (iii).   With $t = 0$ equation (19) becomes

$$\phi(x, 0) = \sum_{n=1}^{\infty} K_n \sin\left(\frac{n\pi x}{\lambda}\right).$$

We now choose K_n such that the value of $\phi(x, 0)$ given by this equation is consistent with that resulting from (iii). This is achieved by choosing K_n to be the coefficient of $\sin(n\pi x/\lambda)$ in the Fourier sine series for

$$\phi(x, 0) = \begin{cases} x, & 0 \leq x \leq \lambda/2 \\ \lambda - x, & \lambda/2 \leq x \leq \lambda. \end{cases}$$

Thus, referring to equation (12) of Section 4, we choose

$$K_n = \frac{2}{\lambda} \int_0^\lambda \phi(x, 0) \sin\left(\frac{n\pi x}{\lambda}\right) dx$$

$$= \frac{2}{\lambda} \int_0^{\lambda/2} x \sin\left(\frac{n\pi x}{\lambda}\right) dx + \frac{2}{\lambda} \int_{\lambda/2}^\lambda (\lambda - x) \sin\left(\frac{n\pi x}{\lambda}\right) dx$$

$$= \frac{4\lambda}{n^2 \pi^2} \sin\left(\frac{n\pi}{2}\right).$$

$$\therefore K_{2n} = 0 \quad \text{and} \quad K_{2n-1} = \frac{(-1)^{n-1} 4\lambda}{(2n-1)^2 \pi^2}.$$

Substituting into equation (19) now gives the required solution in the form

$$\phi(x, t) = \frac{4\lambda}{\pi^2} \sum_{n=1}^\infty \frac{(-1)^{n-1}}{(2n-1)^2} \sin\frac{(2n-1)\pi x}{\lambda} \cos\frac{(2n-1)\pi c t}{\lambda}.$$

Example 7

Use the method of separation of variables to solve the equation of linear heat flow

$$\frac{\partial \theta}{\partial t} = \kappa \frac{\partial^2 \theta}{\partial x^2}$$

given that

(i) $\dfrac{\partial \theta(0, t)}{\partial x} = \dfrac{\partial \theta(\pi, t)}{\partial x} = 0 \qquad$ for $t \geqslant 0$

(ii) $\theta(x, 0) = x(\pi - x) \qquad\qquad$ for $0 < x < \pi.$

If we take

$$\theta(x, t) = X(x)T(t)$$

as in Section 7, the equation corresponding to equation (16) is

$$\frac{1}{X}\frac{d^2 X}{dx^2} = \frac{1}{\kappa T}\frac{dT}{dt} = -\omega^2 \text{ say.}$$

$$\therefore \frac{d^2 X}{dx^2} + \omega^2 X = 0 \quad \text{and} \quad \frac{dT}{dt} + \omega^2 \kappa T = 0$$

giving

$$X = A \cos \omega x + B \sin \omega x, \quad T = C \exp(-\omega^2 \kappa t)$$

so that

$$\phi(x, t) = (A' \cos \omega x + B' \sin \omega x) \exp(-\omega^2 \kappa t).$$

Condition (i)

$$\frac{\partial\theta(x, t)}{\partial x} = \omega(B' \cos \omega x - A' \sin \omega x) \exp(-\omega^2 \kappa t).$$

Hence $\partial\theta(0, t)/\partial x$ is zero provided we choose $B' = 0$. Similarly $\partial\vartheta(\pi, t)/\partial x$ is zero provided $\sin \omega\pi = 0$, i.e. provided $\omega = 0, 1, 2, \ldots$.

The solution at this stage is thus

$$\theta(x, t) = \sum_{\omega=0}^{\infty} A_\omega \exp(-\omega^2 \kappa t) \cos \omega x. \tag{20}$$

Condition (ii). The initial configuration ($t = 0$) is

$$\theta(x, 0) = \sum_{\omega=0}^{\infty} A_\omega \cos \omega x.$$

We now choose A_ω to be the coefficient of $\cos \omega x$ in the Fourier cosine series for

$$\theta(x, 0) = x(\pi - x), \quad 0 < x < \pi,$$

i.e.

$$A_\omega = \frac{2}{\pi} \int_0^\pi x(\pi - x)\cos \omega x \, dx, \quad \omega = 1, 2, \ldots$$

$$= -\frac{2}{\omega^2}(1 + \cos \omega\pi)$$

so that

$$A_{2\omega} = -\frac{1}{\omega^2} \quad \text{and} \quad A_{2\omega-1} = 0.$$

Also

$$A_0 = \frac{1}{\pi} \int_0^\pi x(\pi - x) \, dx = \frac{\pi^2}{6}.$$

Equation (20) now gives

$$\theta(x, t) = \frac{\pi^2}{6} - \sum_{\omega=1}^{\infty} \frac{1}{\omega^2} \exp(-4\omega^2 \kappa t)\cos 2\omega x.$$

PROBLEMS

1. Determine the Fourier series for the function $f(x)$ defined by

$$f(x) = \begin{cases} \cos x, & \text{for } 0 < x < \pi/2 \\ \sin x, & \text{for } \pi/2 < x < \pi \end{cases}$$

$$f(x + \pi) = f(x).$$

2. If $f(x)$ is defined by

$$f(x) = \begin{cases} 1 + x, & \text{for } -1 < x < 0 \\ 1 - x, & \text{for } 0 < x < 1, \end{cases}$$

determine the Fourier series for $f(x)$ in the interval $-1 < x < 1$.

3. Determine the Fourier series for the periodic function $f(x)$, of period $2T$, defined by

$$f(x) = \begin{cases} -e^{-x}, & \text{for } -T < x < 0 \\ e^{-x}, & \text{for } 0 < x < T. \end{cases}$$

4. Determine a Fourier cosine series for $f(x) = \sin x$, in the range $(0, \pi)$.

5. Obtain the Fourier series for the function $f(x)$ defined by

$$f(x) = \begin{cases} 0, & \text{for } -\dfrac{\pi}{4} < x < -\dfrac{\pi}{8} \\[2mm] \dfrac{8}{\pi}, & \text{for } -\dfrac{\pi}{8} < x < \dfrac{\pi}{8} \\[2mm] 0, & \text{for } \dfrac{\pi}{8} < x < \dfrac{\pi}{4} \end{cases}$$

$$f\left(x + \frac{\pi}{2}\right) = f(x).$$

6. The function $f(x)$ is such that

$$f(x) = \pi^2 + 2\pi x - 2x^2 \quad \text{for } 0 < x < \pi.$$

Given that $f(x)$ is an even function of period 2π, obtain the corresponding Fourier series.

7. The temperature θ_i °C of a substance at time t_i minutes after the start of an experiment is given by the following table:

t_i	1	2	3	4	5	6	7	8	9	10	11	12
θ_i	5.2	6.0	7.1	8.0	9.4	15.7	16.0	14.9	10.0	7.5	5.8	5.0

Determine a Fourier series which will represent the graph obtained by plotting t_i against θ_i.

8. Solve the equation

$$\frac{\partial^2 \phi}{\partial x^2} + \frac{\partial^2 \phi}{\partial y^2} = 0,$$

given that

    (i) $\phi(0, y) = \phi(\pi, y) = \phi(x, \infty) = 0.$
    (ii) $\phi(x, 0) = 1.$

9. A solute diffuses from a solution bounded between the planes $x = 0$ and $x = a$, into a solvent bounded between the planes $x = a$ and $x = b$.

If

(i) $D \dfrac{\partial^2 C(x, t)}{\partial x^2} = \dfrac{\partial C(x, t)}{\partial t}$,

(ii) $C(0, t) = C(b, t) = 0$, and

(iii) $C(x, 0) = \begin{cases} C_0, & \text{for } 0 \leqslant x < a \\ 0, & \text{for } a < x \leqslant b \end{cases}$

where $b > a$, show that

$$C = \frac{C_0 a}{b} + \frac{2C_0}{\pi} \sum_{n=1}^{\infty} \frac{1}{n} \exp\left\{-\left(\frac{n\pi}{b}\right)^2 Dt\right\} \cos\left(\frac{n\pi x}{b}\right) \sin\left(\frac{n\pi a}{b}\right).$$

10. Show that $\theta = K \sin(ax + \alpha)\cosh(by + \beta)$ satisfies the heat equation

$$\frac{\partial^2 \theta}{\partial x^2} + \frac{\partial^2 \theta}{\partial y^2} = 0$$

provided $a^2 = b^2$.

A square plate is bounded by the lines $x = 0$, $x = \pi$, $y = 0$, and $y = \pi$, the edges $x = 0$ and $x = \pi$ being kept at zero temperature.

If the faces of the plate and the edge $y = 0$ are perfectly insulated, whilst the edge $y = \pi$ is maintained at θ_0, show that

$$\theta = \frac{4\theta_0}{\pi} \sum_{n=1}^{\infty} \frac{\sin(2n-1)x \cosh(2n-1)y}{(2n-1)\cosh(2n-1)\pi}.$$

CHAPTER 12

Group Theory

Group theory involves abstract mathematical ideas but has practical applications in physics and chemistry, particularly in molecular theory.

1. Definition of a group

A group G consists of a set of *elements* with a law of composition for any pair, the following four conditions having to be satisfied.

(The law of composition, that is the way two elements, say A and B, are combined, is represented by $A * B$ or more usually AB.)

Condition 1: To every pair of elements, A and B in G, there is another element C in G such that $C = AB$ (including AA, written as A^2).

Condition 2: If A, B, C are any three elements in G, not necessarily different then $(AB)C = A(BC)$ which is the associative law of algebra.

Condition 3: G contains a *unit element* E such that $AE = EA = A$ for every element A in G (E is sometimes called the *identity* or *unit element*).

Condition 4: For any element A in G, there is an *inverse element* also in G, written A^{-1}, such that

$$AA^{-1} = A^{-1}A = E.$$

AB is usually called the *product* of A and B although we cannot always think of this as a product in the usual sense of multiplying two numbers together. Multiplication may not be involved at all, as is shown in Example 2 below.

If a group has n elements, it is said to be of *order n*.

Example 1

Show that the set of four elements $\{1, i, -1, -i\}$ where $i^2 = -1$, form a group under multiplication, i.e. AB is the algebraic multiplication of A by B *in this case*.

First we must verify that the four conditions are satisfied

Now $1 \times i = i, i \times (-i) = 1, i^2 = -1, (-i)^2 = -1$, and so on for all 16 possible combinations, so that each product is another element of the set. Condition 1 is thus satisfied.

The associative law (condition 2) is always true for the multiplication of numbers.

The unit element E required for condition 3 is the integer 1.

Condition 4 requires each inverse to be a number in the set, and in this case they are the reciprocals

$$1/1 = 1, \quad 1/i = -i, \quad 1/-1 = -1, \quad 1/-i = i,$$

so that each element multiplied by its inverse is equal to one.

Thus all four conditions for a group are satisfied, and the four elements form a group of order 4 under multiplication.

Example 2

Show that the numbers $\{0, 1, 2\}$ form a group under addition (modulo 3).

(The constraint of modulo 3 means that if a number greater than 2 is obtained, then the number 3 is subtracted from it so that, for example, $5 = 2(\text{modulo } 3)$.)

In this example the product AB is found by adding the two numbers A and B, and then subtracting 3 if the number is greater than 2.

Thus, $0 + 2 = 2$, $1 + 2 = 0$, $1 + 1 = 2$, $2 + 2 = 1$, $1 + 0 = 1$, and so on, showing that the addition of any two numbers yields only one of the same three, and condition 1 is satisfied.

The associative law is always true for addition, so that condition 2 also holds. For example, $2 + (2 + 1) = 2 + 0 = 2$ and $(2 + 2) + 1 = 1 + 1 = 2$.

Condition 3 is satisfied, since E, the unit element, is the number 0, so that for example the product of 2 and 0 is $2 + 0 = 2$ (i.e. $AE = A$).

The inverses may be thought of as those numbers which reverse any change, so that $AA^{-1} = E$.

The inverse of 0 is 0 since $0 + 0 = 0$, the inverse of 1 is 2 since $1 + 2 = 0$, and the inverse of 2 is 1 since $2 + 1 = 0$.

Condition 4 is therefore, covered too, and the three numbers form a group of order 3 under addition (modulo 3).

Example 3

The elements E, A, B, C represent the rotation of a square about its centre O, through angles of $0, 90°, 180°, 270°$ respectively, the rotation being in a clockwise direction in the plane of the square.

If the product AB is understood to mean that rotation B is carried out first, followed by rotation A, show that E, A, B, C form a group under rotation.

Let the square be labelled PQRS and consider the four conditions necessary for a group.

Condition 1   Since B rotates the square through $180°$ and A through another $90°$, giving $270°$ in all, $AB = C$. This is shown in Fig. 12.1.

C^2 will be a rotation through $540°$ which is the same as one through $180°$, so that $C^2 = B$. Proceeding in this way we find that any product of E, A, B or C results in one of the same four elements.

Condition 2 $(AB)C = A(BC)$ since in both cases rotation C is carried out first, followed by B, then A. Similarly for any three elements.

Condition 3 E is the unit element since it leaves the square unchanged.

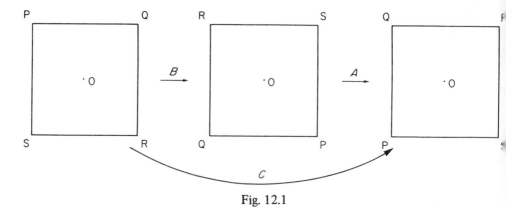

Fig. 12.1

Condition 4   The inverses will be the rotations which return the square to its original position.

Since A is a rotation through $90°$, A^{-1} is therefore a rotation through another $270°$, given by C, and hence $A^{-1} = C$. Similarly $B^{-1} = B$, $C^{-1} = A$, $E^{-1} = E$. Thus the four rotations form a group of order 4.

The convention used in this example, viz. operations carried out in reverse order, is most often used for operations on structures. Some authors, however, carry out the operations in the order as written.

Note that in the theory of groups it is not necessary for $AB = BA$, although *if* AB is equal to BA then the group is said to be *abelian*. The groups in the three examples so far covered are all abelian.

2. Group multiplication tables

The products of the various elements of the group can be conveniently represented in a table, called the group multiplication table.

For a group of order n, there will be $n \times n$ values in the table. Consider, for example, the group in Example 1, namely $\{1, i, -1, -i\}$. Under multiplication, the group table is

	1	i	(−1)	−i
1	1	i	−1	−i
(i)	i	−1	(−i)	1
−1	−1	−i	1	i
−i	−i	1	i	−1

the element in the table being the product AB, where A is the row heading and B is the column heading.

Thus $i \times (-1) = -i$ shown ringed in the table.

Later, we shall prove that in any group multiplication table, each element occurs once only in each row and once only in each column. If, as in the example above, the table is symmetric about its leading diagonal, that is, the table is unaltered if the rows are written as columns, then $AB = BA$ and the group is abelian.

Again, it must be stressed that the word 'multiplication' in group multiplication table does not imply that arithmetic multiplication occurs. Examples 2 and 3 are illustrations in which other operations are performed.

Example 4

Let A, B, C and D be the following matrices:

$$A = \begin{bmatrix} 1 & 0 \\ 0 & 1 \end{bmatrix}, \quad B = \begin{bmatrix} 0 & 1 \\ -1 & 0 \end{bmatrix}, \quad C = \begin{bmatrix} -1 & 0 \\ 0 & -1 \end{bmatrix} \text{ and } D = \begin{bmatrix} 0 & -1 \\ 1 & 0 \end{bmatrix}$$

By constructing the group multiplication table show that they form a group under matrix multiplication.

Now

$$BC = \begin{bmatrix} 0 & 1 \\ -1 & 0 \end{bmatrix} \times \begin{bmatrix} -1 & 0 \\ 0 & -1 \end{bmatrix} = \begin{bmatrix} 0 & -1 \\ 1 & 0 \end{bmatrix} = D$$

and

$$B^2 = \begin{bmatrix} 0 & 1 \\ -1 & 0 \end{bmatrix} \times \begin{bmatrix} 0 & 1 \\ -1 & 0 \end{bmatrix} = \begin{bmatrix} -1 & 0 \\ 0 & -1 \end{bmatrix} = C.$$

Carrying out similar calculations for all products, the group multiplication table is,

	A	B	C	D
A	A	B	C	D
B	B	C	D	A
C	C	D	A	B
D	D	A	B	C

Since all products are elements of the set $\{A, B, C, D\}$, condition 1 holds.

The associative law, required for condition 2, is true for matrix multiplication, and A is the unit matrix, which is, therefore, the unit element required by condition 3.

Since $BD = DB = A$, therefore $D = B^{-1}$ and $B = D^{-1}$. Also, $C^2 = A$, so that $C^{-1} = C$ and $A^{-1} = A$ (inverse of a unit matrix is the unit matrix). Thus condition 4 is also satisfied.

Example 5

Consider an equilateral triangle PQR with centre O and axes Ox, Oy, Oz as in Fig. 12.2. Suppose that the following operations can occur on the triangle:

    E the identity operation, leaving the triangle as it is;

    A a clockwise rotation of $120°$ about O in the plane of the triangle;

    B an anti-clockwise rotation of $120°$ about O in the plane of the triangle;

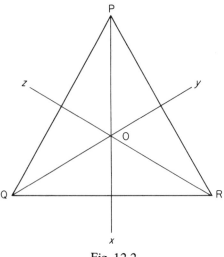

Fig. 12.2

    C a rotation of 180° about the axis O*x*;
    D a rotation of 180° about the axis O*y*;
    F a rotation of 180° about the axis O*z*.

For each of these operations the resultant orientation of the triangle is similar to that at the start, although of course, P, Q and R may be interchanged.

By forming a group multiplication table, show that these operations form a group of order 6.

As in Example 3, the product *AC* is understood to mean that operation *C* is first carried out followed by operation *A*, as shown in Fig. 12.3.

Since the final triangle is that which would have been obtained by operation *D* on the original triangle, we conclude that *AC* = *D*.

Similar calculations for the other products lead to the following group multiplication table:

	E	*A*	*B*	*C*	*D*	*F*
E	*E*	*A*	*B*	*C*	*D*	*F*
A	*A*	*B*	*E*	*D*	*F*	*C*
B	*B*	*E*	*A*	*F*	*C*	*D*
C	*C*	*F*	*D*	*E*	*B*	*A*
D	*D*	*C*	*F*	*A*	*E*	*B*
F	*F*	*D*	*C*	*B*	*A*	*E*

The product of any two elements of the six is seen to be also an element of the same six.

The associative law is true since both (*AB*)*C* and *A*(*BC*) indicate the performance of operation *C*, followed by *B*, and then *A*. Similarly for any three elements.

E is the unit element, and the inverses can easily be derived from the group

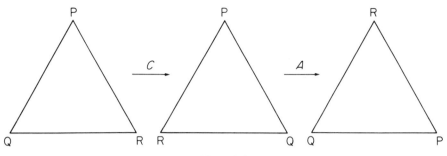

Fig. 12.3

multiplication table. For example, $AB = E$, and because $AA^{-1} = E$, then $A^{-1} = B$. Similarly $B^{-1} = A$, $C^{-1} = C$, $D^{-1} = D$ and $F^{-1} = F$.

All four conditions are satisfied, and the six elements form a group of order 6. It is non-abelian, since, for example, $CA = F$ but $AC = D$.

3. Subgroups

In Example 5, if operations E, A and B only are considered and a group multiplication table formed for these (derived from the previous table), it is found to be

	E	A	B
E	E	A	B
A	A	B	E
B	B	E	A

and all four conditions for a group are satisfied by these three elements. The group $\{E, A, B\}$ is then said to be a *subgroup* of the group $\{E, A, B, C, D, F\}$.

We see, therefore, that subgroups of group G consist of some of the elements of G which themselves form a group. Two trivial cases are the unit element E alone, which is a subgroup, and the set of all elements of G. These are called *improper subgroups*, others being called *proper subgroups*.

To verify that a number of elements of a group form a subgroup, it is only necessary to check that condition 1 is satisfied. If this is so, the subgroup will automatically contain the unit element and the inverses of the elements. The associative law will hold, since it held for G.

4. Properties

We now examine a few properties of groups, some of which the reader may have noticed already.

The rules for indices for a single element follow those of ordinary algebra. For example,

$$AAA = A^3, \quad AAAA = A^4, \quad A^{-1}A^{-1} = A^{-2}, \quad A^m A^n = A^{m+n}, \text{ etc.,}$$

with $A^0 = E$, the unit element.

For two elements, however, we must remember that in general $AB \neq BA$, so

K

that

$$(AB)^n = (AB)(AB)(AB) \ldots (AB) \neq A^n B^n.$$

However, if $AB = BA$, then $(AB)^n = A^n B^n$. More results are proved below.

(i) *There can be only one unit element in a group.*

Suppose there are two, say E_1 and E_2. Since the product of a unit element with any other element leaves that other element unchanged, we must have $E_1 E_2 = E_2$ and also $E_1 E_2 = E_1$.

Therefore, $E_1 = E_2$ and they are the same element.

(ii) $E^n = E$.

Since $E^m E = E^m$, i.e. $E^{m+1} = E^m$,

putting $m = 1, 2, \ldots, n$ respectively, gives

$$E = E^2 = E^3 = \ldots = E^{n-1} = E^n.$$

(iii) *Each row of the group multiplication table contains all the elements once only.*

Suppose that one element, say X_1 occurred twice, so that

$$AY_1 = AY_2 = X_1.$$

Then $A^{-1} A Y_1 = A^{-1} A Y_2$ giving

$$EY_1 = EY_2 \quad \text{and} \quad Y_1 = Y_2.$$

Thus two of the original elements of the group would be equal, which is untrue, and it follows that all products in a row are different. Since there are the same number of products in that row as there are elements in the group, all elements must occur once only.

A similar result holds for columns.

(iv) *If A, B are two elements of a group, $A^{-1} B$ and $B^{-1} A$ are also elements of the group.*

From condition 2, A^{-1} is an element of the group, and from condition 1 $A^{-1} B$ is also an element of the group.

Similarly for $B^{-1} A$, which may or may not be equal to the same element.

(v) $(AB)^{-1} = B^{-1} A^{-1}$.

Now $(AB)(B^{-1} A^{-1}) = ABB^{-1} A^{-1} = AEA^{-1} = AA^{-1} = E$ so that, pre-multiplying each side by $(AB)^{-1}$ gives

$$(AB)^{-1}(AB)(B^{-1} A^{-1}) = (AB)^{-1} E.$$

But $(AB)^{-1}(AB) = E$. Therefore

$$E(B^{-1} A^{-1}) = (AB)^{-1} E \quad \text{and} \quad B^{-1} A^{-1} = (AB)^{-1}.$$

5. Cyclic groups

If a group G consists of elements all of which can be expressed as powers of a single element, then G is called a *cyclic group*.

A cyclic group of order n will contain elements

$$E, A, A^2, A^3, \ldots, A^{n-1}, \quad \text{with } A^n = A^0 = E.$$

The group is said to be *generated* by the element A.

If a set $E, A, A^2, \ldots, A^{n-1}$ exists, with $A^n = E$, then conditions 1, 2, and 3 for groups clearly apply. Also, the inverse of A^m must be A^{n-m}, since $A^m A^{n-m} = A^n = E$.

Since $n - m \leqslant n$, A^{n-m} will also be an element of the set, which now forms a group. Moreover, cyclic groups must be abelian, since $A^l A^m = A^m A^l$.

An example of a cyclic group occurs in Example 3 where a square is rotated through $0°, 90°, 180°, 270°$, the operations being labelled E, A, B and C respectively.

Clearly $A^2 = B$, $A^3 = C$ and $A^4 = E$, so that A generates a cyclic group of order 4 under rotation, with elements E, A, A^2, A^3.

Not so obvious, is the group in Example 4, which was shown to have a group multiplication table

	A	B	C	D
A	A	B	C	D
B	B	C	D	A
C	C	D	A	B
D	D	A	B	C

We see from the table that

$$B^2 = C, \quad B^3 = BB^2 = BC = D \quad \text{and} \quad B^4 = B^2 B^2 = C^2 = A.$$

Thus the group is generated by element B, and writing $B^0 = A$ as E, the group multiplication table becomes

	E	B	B^2	B^3
E	E	B	B^2	B^3
B	B	B^2	B^3	E
B^2	B^2	B^3	E	B
B^3	B^3	E	B	B^2

It is now clear too, that the associative law holds since $B^i(B^j B^k) = B^{i+j+k} = (B^i B^j)B^k$.

Example 6

The molecular structure of benzene (C_6H_6) is symmetric about a centre point, as shown in Fig. 12.4, so that a rotation of $60°$ in the plane about the centre point brings the molecule into a position similar to the initial position.

If C_6^n represents the rotation through an angle of $n\pi/3$ for $n = 1, 2, \ldots, 6$, show, using the group multiplication table, that the six operations form a cyclic group.

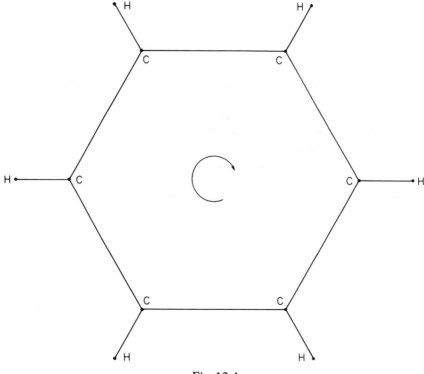

Fig. 12.4

Which of the following form a subgroup:

(i) $\{E, C_6^2, C_6^4\}$,   (ii) $\{C_6^2, C_6^3\}$,   (iii) $\{E, C_6^1, C_6^2, C_6^3\}$?

Find a subgroup of order 2.

As explained in Examples 3 and 5, the operations of rotation obey the associative law, have inverses, and the identity element $E = C_6^0 = C_6^6$.

The group multiplication table is

	E	C_6^1	C_6^2	C_6^3	C_6^4	C_6^5
E	E	C_6^1	C_6^2	C_6^3	C_6^4	C_6^5
C_6^1	C_6^1	C_6^2	C_6^3	C_6^4	C_6^5	E
C_6^2	C_6^2	C_6^3	C_6^4	C_6^5	E	C_6^1
C_6^3	C_6^3	C_6^4	C_6^5	E	C_6^1	C_6^2
C_6^4	C_6^4	C_6^5	E	C_6^1	C_6^2	C_6^3
C_6^5	C_6^5	E	C_6^1	C_6^2	C_6^3	C_6^4

All products form another element of the group, so that condition 1 is satisfied. Because $C_6^m = (C_6^1)^m$, the group can be generated by C_6^1 and so is cyclic.

(i) $\{E, C_6^2, C_6^4\}$. The multiplication table for this set is

	E	C_6^2	C_6^4
E	E	C_6^2	C_6^4
C_6^2	C_6^2	C_6^4	E
C_6^4	C_6^4	E	C_6^2

The four conditions are fulfilled and so the three elements form a subgroup.

(ii) $\{C_6^2, C_6^3\}$. The product C_6^2, C_6^3 is equal to C_6^5 which is an element outside the set. Further, there is no unit element, so that neither condition 1 nor condition 3 is satisfied, and C_6^2, C_6^3 does not form a subgroup.

(iii) $\{E, C_6^1, C_6^2, C_6^3\}$. The product $C_6^1 C_6^3 = C_6^4$ is not contained in the set, so these elements do not form a subgroup.

For a subgroup of order 2, one element must be E, and the other element must be its own inverse, which is only true of C_6^3. Therefore the subgroup is $\{E, C_6^3\}$, with multiplication table

	E	C_6^3
E	E	C_6^3
C_6^3	C_6^3	E

6. Isomorphic groups

If two groups have similarly structured group multiplication tables, then they are called *isomorphic*. Quite possibly, the elements of the two groups stand for very different things, they may even be combined in completely different ways. For example, in Example 1, we saw that $1, i, -1, -i$ formed a group under multiplication. Later we derived the group multiplication table which was

	1	i	-1	$-i$
1	1	i	-1	$-i$
i	i	-1	$-i$	1
-1	-1	$-i$	1	i
$-i$	$-i$	1	i	-1

Now consider the group table for the group in Example 3 concerned with the rotation of a square. The table is given below, and alongside, the above table for $1, i, -1, -i$ is reproduced, but with $a = 1, b = i, c = -1, d = -i$:

	A	B	C	D			a	b	c	d
A	A	B	C	D		a	a	b	c	d
B	B	C	D	A		b	b	c	d	a
C	C	D	A	B		c	c	d	a	b
D	D	A	B	C		d	d	a	b	c

It is now obvious that the two group tables have the same structure, so that the two groups are isomorphic. Comparison with the group table in Example 4, verifies that the group of matrices under multiplication in that example is isomorphic with these.

The advantage of isomorphism is that isomorphic groups must have the same group properties, even though the groups may arise in what appear to be completely different situations. Known results for one group will apply equally to the other group. For instance, we saw in Section 5 that the group formed by the rotations of a square is a cyclic group of order 4. It follows that, since the group $\{1, i, -1, -i\}$ is isomorphic to it, that this also is a cyclic group of order 4, and can be generated by i, a fact which the reader can verify. In this case, both groups can also be generated by the fourth element, that is $-i$ in the one case, and $D = $ rotation of $270°$ in the other case.

7. The order of an element

Consider a finite group G with unit element E. If A is an element of G, then so is $A^2, A^3, A^4, \ldots$.

But, since G is finite (i.e. it has a finite number of elements) they cannot all be different. Therefore for some n and m, $A^m = A^n$, and $A^{m-n} = E$, assuming $m > n$.

Thus, for every element in the group, some positive power of that element is equal to E.

If, for element A, α is the least positive integer for which $A^\alpha = E$, then α is called the *order* of A. Two results follow.

(i) *E is the only element of order 1*

This is proved by supposing that A has order 1. Then $A^1 = E$. That is, $A = E$ and so is the unit element.

(ii) *If A is of order α, then so is A^{-1}*

Now $A^\alpha = E$, and $(A^{-1})^\alpha = A^{-\alpha} = (A^\alpha)^{-1} = E^{-1} = E$, so that A^{-1} is of order α or less.

Suppose that A^{-1} is of order m where $m < \alpha$. Then $(A^{-1})^m = E$, and $A^\alpha (A^{-1})^m = A^{\alpha-m} = E$, indicating that A is of order $\alpha - m$, which is impossible.

Therefore A^{-1} is of order α.

8. Cosets

Consider a finite group G of order n, with elements $E, A, B, C, \ldots$, and suppose that G_1 is a subgroup of G with m elements, $E, a, b, \ldots, l$, say. (Remember that it must contain E to be a subgroup.)

Now take an element which is in G but not in G_1, let us suppose it is A. The product of A with each of the elements of G_1, gives

$$EA, aA, bA, \ldots, lA.$$

Then these elements are said to form the *right-coset* of G_1 with respect to A and this coset is denoted by $G_1 A$.

Similarly, the *left-coset* of G_1 with respect to A is $AE, Aa, Ab, \ldots, Al$ and is denoted by AG_1.

The following results lead to an important theorem.

All elements of the coset are different.

Suppose that two elements are the same, say $aA = bA$, then $aA\, A^{-1} = bA\, A^{-1}$ and $a = b$, which is untrue. Therefore all elements must be different.

The coset does not contain any element which is in G_1.

Suppose that an element of the coset is in G_1, say $aA = b$. Then

$$a^{-1} aA = a^{-1} b$$

that is,

$$A = a^{-1}b$$

which is an element of the subgroup G_1 (see result 4 of Section 4).

But this is not true, because A was chosen as an element of G but not G_1. Therefore no element of the coset can be in G_1.

Since E cannot be a member of the coset then $G_1 A_1$ cannot form a subgroup.

Suppose that another right-coset of G_1 is derived, say $G_1 B$ where B is neither in G_1 nor in $G_1 A$.

We now have three sets of elements

$$G_1 \ : \ E, a, b, c, \ldots, l$$
$$G_1 A : \ EA, aA, bA, cA, \ldots, lA$$
$$G_1 B : \ EB, aB, bB, cB, \ldots, lB$$

each containing m elements of group G.

$G_1 A$ and $G_1 B$ have no element in common.

Suppose that they do have an element in common, say $bB = aA$. Then

$$b^{-1} bB = b^{-1} aA \quad \text{and} \quad B = (b^{-1}a)A.$$

As stated in the previous result, $b^{-1}a$ will be an element of G_1, say c.

Thus $B = cA$ which is an element of the right-coset $G_1 A$, which contradicts the assumption that B is not a member of this coset.

The original statement that $G_1 A$ and $G_1 B$ do have an element in common must therefore be false, and we conclude that they have no common element.

It follows from these results that $G_1, G_1 A$ and $G_1 B$ will each have m different elements. Unless the total number of elements (n) in G is $3m$, yet another coset can be formed with a completely different set of elements. This procedure may be continued until all elements of G have been covered, which can happen only if m is a divisor of n.

This result is known as *Lagrange's theorem*: *The order of a subgroup of a finite group is a factor of the order of the group.*

In Section 3, we determined the subgroup $\{E, A, B\}$ of the group $\{E, A, B, C, D, F\}$ considered in Example 5. Since $n = 6$ for this group, Lagrange's theorem tells us that proper subgroups must be of the order 2 or 3. There cannot be subgroups of order 4 or 5. Other subgroups of this particular example are $\{E, C\}$, $\{E, D\}$ and $\{E, F\}$.

If the order (n) of a group G is a prime number, then the only subgroups that exist are the improper subgroups E (of order 1) and G itself (of order n). Now, in any finite group, there is always at least one cyclic subgroup, because A, A^2, $A^3, \ldots, A^m = E$ satisfy all the conditions of a group, where A is in G. For n prime, therefore, the group must be cyclic.

Example 7

The following group multiplication table occurs in the study of the structure of the $XeOF_4$ molecule:

	E	C_4^1	C_4^3	C_2^1	σ_v	σ_v'	σ_d	σ_d'
E	E	C_4^1	C_4^3	C_2^1	σ_v	σ_v'	σ_d	σ_d'
C_4^1	C_4^1	C_2^1	E	C_4^3	σ_d	σ_d'	σ_v'	σ_v
C_4^3	C_4^3	E	C_2^1	C_4^1	σ_d'	σ_d	σ_v	σ_v'
C_2^1	C_2^1	C_4^3	C_4^1	E	σ_v'	σ_v	σ_d'	σ_d
σ_v	σ_v	σ_d'	σ_d	σ_v'	E	C_2^1	C_4^3	C_4^1
σ_v'	σ_v'	σ_d	σ_d'	σ_v	C_2^1	E	C_4^1	C_4^3
σ_d	σ_d	σ_v	σ_v'	σ_d'	C_4^1	C_4^3	E	C_2^1
σ_d'	σ_d'	σ_v'	σ_v	σ_d	C_4^3	C_4^1	C_2^1	E

If $\{E, C_2^1\}$ is a subgroup G_1, find the right-cosets $G_1 C_4^1$, $G_1 \sigma_d$ and one other. Now

$$G_1 C_4^1 = \{E C_4^1, C_2^1 C_4^1\} = \{C_4^1, C_4^3\}$$

and

$$G_1 \sigma_d = \{E \sigma_d, C_2^1 \sigma_d\} = \{\sigma_d, \sigma_d'\}.$$

σ_v and σ_v' have not yet been covered.

$$G_1 \sigma_v = \{E \sigma_v, C_2^1 \sigma_v\} = \{\sigma_v, \sigma_v'\}$$

so that all eight elements have been derived.

9. Groups and molecular symmetry

Groups play an important part in the theory of molecular structure, particularly where it is applied to molecular vibration theory, crystallography and spectroscopy.

A central part of the theory is the idea of symmetrical operations in which atoms are interchanged in such a way that they leave the molecular system apparently unchanged. That is, because the molecule has some basic symmetry, atoms may be interchanged with similar atoms, without affecting the chemical or physical properties of the molecular system.

In this section, we examine the important symmetry operations and explain the notation used.

First, note that E is the operation which leaves the initial and final positions identical, that is, the 'do-nothing' operation. This has to be included to conform with the theory of groups.

Rotation about an axis of symmetry (C_n):

This is the operation in which a molecule is rotated about a line passing through the molecule and about which there is some symmetry, so that after a rotation through an angle θ, the molecule will be in a position similar to its original position. After a complete revolution ($\theta = 360°$), it must return to its original position. The line about which it is rotated is called a *symmetry axis*.

If n is the number of times that the operation must be carried out for the complete revolution, so that $n\theta = 2\pi$, then n is said to be the *order* of the symmetry axis. It is usual to denote the symmetry operation of rotation by C_n and also to label the axis C_n.

Imagine a molecule with like atoms at A, B, C and D where ABCD forms a square as in Fig. 12.5. The line passing through the centre of the square O and perpendicular to the plane containing the square is a symmetry axis, for which $\theta = 90°$. The order of the axis is, therefore, 4 and the axis labelled C_4.

Another symbol, C_4^m, is used to denote that the operation has been carried out m times, so that, for example C_4^3 indicates a revolution of 270°. Clearly $C_4^1 = C_4$.

C_n^m will therefore denote a rotation through an angle of $2\pi m/n$, so that $C_n^n = E$. Several equivalent notations exist. For example, since C_4^2 denotes a rotation through 180°, it is equivalent to C_2, i.e. $C_4^2 = C_2$ and similarly $C_9^6 = C_3^2$.

It is usual to write the symbol in its simplest form, so that, if, for example, $\theta = 45°$, then the rotation operations

$$C_8^1, C_8^2, C_8^3, C_8^4, C_8^5, C_8^6, C_8^7, C_8^8$$

are written

$$C_8, C_4, C_8^3, C_2, C_8^5, C_4^3, C_8^7, E.$$

It is also possible to consider rotations in the negative direction, indicated by negative indices.

Thus,

$$C_4^{-1} = C_4^3, \quad C_4^{-2} = C_4^2, \quad C_4^{-3} = C_4, \quad \text{etc.,}$$

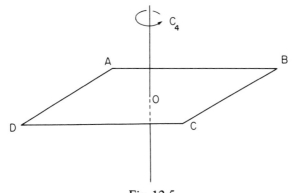

Fig. 12.5

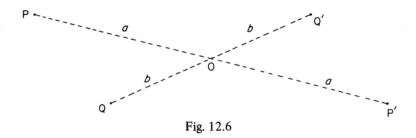

Fig. 12.6

so that, in general

$$C_n^{-k} = C_n^{n-k}.$$

We have shown in earlier parts of this chapter that such rotations form a group. In fact $\{E, C_n^2, C_n^3, \ldots, C_n^{n-1}\}$ form a cyclic group of order n with $C_n^n = E$.

Inversion through the centre of symmetry (i):

This is the operation in which each atom is interchanged with its image point with respect to a *centre of symmetry*. The resulting molecule is similar to the original.

A point P has an image point P' if POP' is a straight line and OP = OP', O being the centre of symmetry, as in Fig. 12.6. Similarly P is the image point of P'. If a molecule consists of like atoms at P and P', together with like atoms at Q and Q', then O is a centre of symmetry for this molecule. Thus a symmetry operation is possible for a molecule if every atom has a similar atom at its image point. Not all molecules have centres of symmetry. Notice that, in three dimensions, inversion with respect to the origin transforms a point (x, y, z) to the point $(-x, -y, -z)$.

The operation of inversion is represented by i, where a repeat of the operation returns every point to its original position. Thus $i^2 = E$, and since $i\,i^{-1} = E$, therefore $i = i^{-1}$.

E, i will clearly form a group under the operation of inversion.

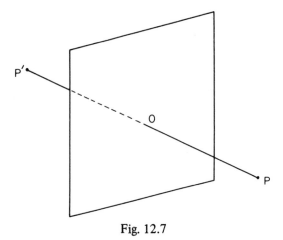

Fig. 12.7

Reflection in a symmetry plane (σ):

This is the operation in which the molecule is replaced by its mirror image in a *plane of symmetry*, the resulting molecule being similar to the original. The plane of symmetry is sometimes called a *mirror plane*.

In Fig. 12.7, the point P becomes P′ under this operation where POP′ is perpendicular to the plane and OP = OP′. If, therefore, there is an atom at P, there must be a similar atom at P′ for the plane to be a plane of symmetry.

The operation is represented by σ and the plane is also labelled σ. Since a repeat of the operation restores points to their original positions, $\sigma^2 = E$, $\sigma^{-1} = \sigma$ and E, σ form a group under the operation of reflection.

As an example of the three operations introduced so far, consider the ethylene molecule shown in Fig. 12.8.

There are three mutually perpendicular C_2 axes, C_2', C_2'' and C_2''', three σ planes of symmetry, labelled σ', σ'' and σ''', and a centre of symmetry i. It is usual to regard the C_n-axis with the largest value of n as a vertical axis and planes of symmetry containing this axis would also be vertical, and are labelled σ_v (see Fig. 12.14). If the axis is perpendicular to a plane of symmetry, the plane is labelled σ_h. In the ethylene molecule, the symmetry axes are all of the same order so that any of the three planes can be regarded as vertical.

Rotary reflection (S_n):

This operation, denoted by S_n, is the combination of a rotation about an axis through an angle $\theta = 2\pi/n$, followed or preceded by reflection in a plane perpendicular to the axis of rotation. Again, for some molecules, this operation transforms the molecule in such a way that only like atoms are interchanged. It is not necessary for the axis or the plane, individually, to be symmetrical with respect to the structure under consideration.

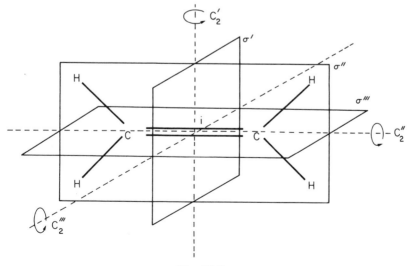

Fig. 12.8

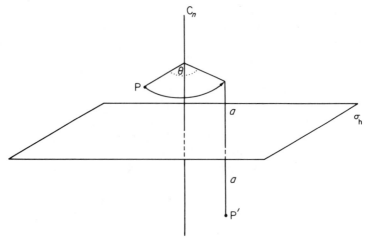

Fig. 12.9

Fig. 12.9 shows the effect of this operation on a point P, rotated through an angle θ about the axis C_n and then reflected in the plane σ_h, so that P is transformed to P'. If P is at height a above the plane, then P' is at depth a below the plane.

For this operation, $S_n = \sigma_h C_n$ where C_n denotes a rotation about the axis C_n and σ_h is reflection in the horizontal plane σ_h. Again n is the *order* of the axis.

Similarly, the symbol $S_n^m = \sigma_h^m C_n^m$ signifies m rotations through an angle θ (i.e. a rotation through the angle $m\theta$) and m reflections in the plane σ_h.

Since $\sigma_h^2 = E$, it follows that

$$S_n^m = \begin{cases} \sigma_h C_n^m & \text{if } m \text{ is odd,} \\ C_n^m & \text{if } m \text{ is even.} \end{cases}$$

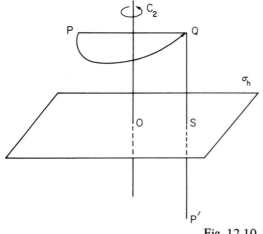

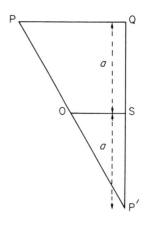

Fig. 12.10

Example 8

Show that $S_2 = i$ where the centre of symmetry is the point where the axis of rotation cuts the reflecting plane. $S_2 = \sigma_h C_2$, so that any point P is rotated through $180°$ about the axis C_2, to the point Q say, and then reflected in the plane σ_h to the point P'.

With O, S as in Fig. 12.10, PQOS and P' all lie in the same plane and O will be the mid-point of PP'. Inversion through O, therefore, converts P to P'. Thus $S_2 = i$.

Example 9

Find S_6^m for all values of m in terms of C_n^l and σ in the simplest form.

$$S_6 = \sigma C_6$$
$$S_6^2 = \sigma^2 C_6^2 = C_3$$
$$S_6^3 = \sigma^3 C_6^3 = \sigma C_2$$
$$S_6^4 = \sigma^4 C_6^4 = C_3^2$$
$$S_6^5 = \sigma^5 C_6^5 = \sigma C_6^5$$
$$S_6^6 = \sigma^6 C_6^6 = C_6^6 = E$$

with similar expressions for further revolutions.

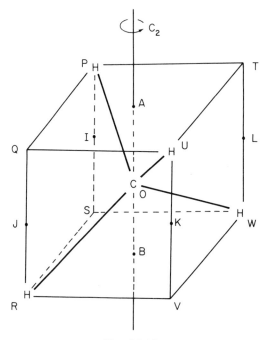

Fig. 12.11

292

Example 10

A methane molecule may be represented by four hydrogen atoms at the corners P, U, R and W of the cube shown in Fig. 12.11 with a carbon atom at the centre O of the cube. Find examples of the symmetry operations when they exist.

If A, B are centres of the upper and lower planes of the cube, the line through AB is a symmetry axis C_2, since operation C_2 interchanges the atoms at P and U, in addition to those at R and W, but leaves O unchanged. From symmetry, there are two more C_2-axes passing through the centre of the cube. There are also four C_3-axes. One passes through U, O and S, and in a similar way, the others pass through a hydrogen atom and the carbon atom.

An example of a symmetry plane is the plane containing points Q, T, W and R, since reflection in this plane interchanges the hydrogen atoms at P and U, but leaves the others unchanged. Altogether, there are six such planes, each plane containing two hydrogen atoms plus the carbon atom.

If I, J, K and L are mid-points of the lines on which they are situated, the plane σ_h through them also contains the point O, but it is not a symmetry plane of the molecule. However, $S_4 = \sigma_h C_4$, denoting a rotation of $90°$ about the axis AB and then reflection in the plane σ_h, is a symmetry operation. It transforms P to R, R to U, U to W, and W to P, leaving O unchanged.

There are three S_4 operations, each with $S_4^2 = C_2$.

Any inverse operation requires a centre of symmetry at O. Clearly this would not transform the molecule into itself, and so this is not a symmetry operation for this molecule.

10. Groups of symmetry operations

For any given molecule it is possible to list the symmetry operations which it possesses. It is also possible to find the product of any two operations as described

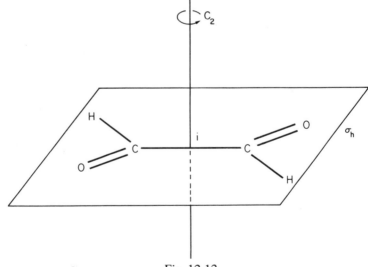

Fig. 12.12

earlier in this chapter. In fact, the symmetry operations for a molecule form a group, which is important because it enables certain physical properties of the molecule to be determined. The following examples show how the group multiplication table is derived.

Example 11

The molecule, glyoxal has a C_2-axis, a σ_h-plane and a centre of symmetry (i) as shown in Fig. 12.12. Find the group multiplication table for these operations.

To determine the outcome of multiplying two symmetry operators together consider their effect on some arbitrary point (x, y, z) in space. It is possible to write down the new coordinates of this point which is obtained by each of the four operations and hence by the product of any two operations. It is usual to take the centre of mass of the molecule as the origin, and for the z-axis to coincide with the symmetry axis of highest order. This leads to the placing of the coordinate axes as shown in Fig. 12.13, with the atoms of the molecule lying in the xy-plane.

The group is formed by the operators E, C_2, σ_h and i where

E leaves (x, y, z) unchanged,

C_2 changes it to $(-x, -y, z)$,

σ_h changes it to $(x, y, -z)$, and

i changes to to $(-x, -y, -z)$.

We note that $C_2^2 = E$, $\sigma_h^2 = E$ and $i^2 = E$.

$C_2\sigma_h$ means first consider the operation σ_h and then C_2, so that (x, y, z) becomes $(x, y, -z)$ and then $(-x, -y, -z)$. Since this is also the change made by operation i, we have $C_2\sigma_h = i$.

The effect on (x, y, z) of $C_2 i$ is to change it to $(x, y, -z)$, as for σ_h,

the effect on (x, y, z) of $\sigma_h C_2$ is to change it to $(-x, -y, -z)$, as for i,

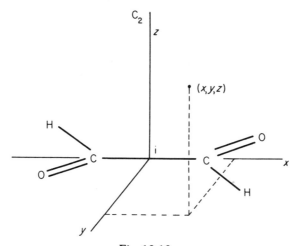

Fig. 12.13

the effect on (x, y, z) of $\sigma_h i$ is to change it to $(-x, -y, z)$, as for C_2,

the effect on (x, y, z) of $i\, C_2$ is to change it to $(x, y, -z)$, as for σ_h, and

the effect on (x, y, z) of $i\, \sigma_h$ is to change it to $(-x, -y, z)$, as for C_2.

The complete table is, therefore

	E	C_2	σ_h	i
E	E	C_2	σ_h	i
C_2	C_2	E	i	σ_h
σ_h	σ_h	i	E	C_2
i	i	σ_h	C_2	E

Example 12

The ammonia molecule, NH_3, is shown in Fig. 12.14, together with the three symmetry planes that it possesses and the one axis of symmetry.

Find the group multiplication table for the symmetry operations.

The three hydrogen atoms lie at the vertices of an equilateral triangle, with the nitrogen atom lying directly above the centre of the triangle.

The symmetry operations are

σ_v : reflection in plane σ_v, for σ'_v, σ''_v and σ'''_v,

C_3 : rotation of $120°$ about the axis C_3,

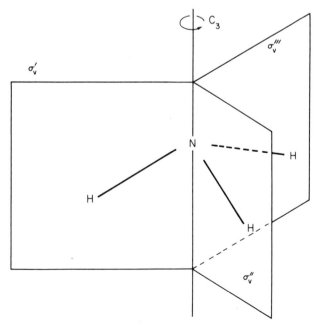

Fig. 12.14

C_3^2 : rotation of 240° about the axis C_3,

E : the identity operation.

Note that all operations leave N unchanged.

C_3 is drawn as the z-axis, but in this case to see the effect of the operations on a general point P, it is easier to look at the geometry of the transformation rather than use coordinates.

For example, Fig. 12.15 shows the transformation on P for the product $C_3 \sigma_v'$. Reflection in plane σ_v' transforms P to P', and the rotation C_3 transforms P' to P''. Since this is equivalent to the reflection of P in the plane σ_v''', it follows that $C_3 \sigma_v' = \sigma_v'''$.

Similar arguments hold for all the other products, where the transformation can be represented in two dimensions, because all movements take place in a plane.

The group multiplication table is given below:

	E	C_3	C_3^2	σ_v'	σ_v''	σ_v'''
E	E	C_3	C_3^2	σ_v'	σ_v''	σ_v'''
C_3	C_3	C_3^2	E	σ_v'''	σ_v'	σ_v''
C_3^2	C_3^2	E	C_3	σ_v''	σ_v'''	σ_v'
σ_v'	σ_v'	σ_v''	σ_v'''	E	C_3	C_3^2
σ_v''	σ_v''	σ_v'''	σ_v'	C_3^2	E	C_3
σ_v'''	σ_v'''	σ_v'	σ_v''	C_3	C_3^2	E

In this case, unlike the glyoxal molecule, the group is not abelian.

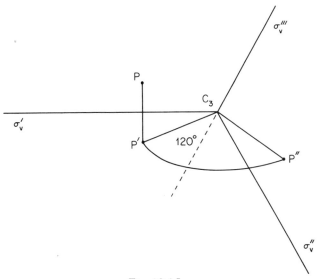

Fig. 12.15

PROBLEMS

1. For groups with the following group multiplication tables:

   (a) find the unit element, and all the inverses;
   (b) state whether or not the group is abelian;
   (c) find the element equal to BFD^2C.

	A	B	C	D	F
A	A	B	C	D	F
B	B	D	A	F	C
(i) C	C	A	F	B	D
D	D	F	B	C	A
F	F	C	D	A	B

	A	B	C	D	F	G
A	B	D	G	A	C	F
B	D	A	F	B	G	C
(ii) C	F	G	D	C	A	B
D	A	B	C	D	F	G
F	G	C	B	F	D	A
G	C	F	A	G	B	D

2. Investigate the following systems for group properties. For those which are groups, set out the group multiplication table.

   (a) 0, 2, 4, 6 (modulo 8) under addition (see Example 2),
   (b) 1/4, 1, 4 under multiplication,
   (c) −1, 1 under multiplication,

$$\text{(d) the matrices } \mathbf{A} = \begin{bmatrix} 1 & 0 \\ 0 & 1 \end{bmatrix}, \quad \mathbf{B} = \begin{bmatrix} 1 & 0 \\ 0 & -1 \end{bmatrix}, \quad \mathbf{C} = \begin{bmatrix} -1 & 0 \\ 0 & -1 \end{bmatrix},$$

$$\mathbf{D} = \begin{bmatrix} -1 & 0 \\ 0 & 1 \end{bmatrix}$$

under matrix multiplication.

3. Use Lagrange's theorem and group multiplication tables to find as many sub-groups as possible of

   (a) the group 1, 2, 3, 4 (modulo 5) under multiplication,
   (b) the group A, B, C, D, F, G with group table as in (ii) of Problem 1.

4. Construct a group multiplication table for the group $\{E, A, B\}$ of order 3, and show that there is only one possible table for such a group. Prove that the group must be cyclic.

5. Show that the groups with the following group multiplication tables are isomorphic:

	A	B	C	D
A	D	A	B	C
B	A	B	C	D
C	B	C	D	A
D	C	D	A	B

	a	b	c	d
a	a	b	c	d
b	b	a	d	c
c	c	d	b	a
d	d	c	a	b

6. Show that the two groups in Questions 2(a) and 2(d) are not isomorphic.

7. Show that the group $\{E, A, B, C, D, F\}$ is cyclic when the group multiplication table is

	E	A	B	C	D	F
E	E	A	B	C	D	F
A	A	C	F	B	E	D
B	B	F	E	D	C	A
C	C	B	D	F	A	E
D	D	E	C	A	F	B
F	F	D	A	E	B	C

8. The table of products for the set $\{E, A, B, C, D\}$ is

	E	A	B	C	D
E	E	A	B	C	D
A	A	E	C	D	B
B	B	D	A	E	C
C	C	B	D	A	E
D	D	C	E	B	A

Use Lagrange's theorem to show that under these combinations, E, A, B, C, D do not form a group.

9. For the atomic structure of benzene, shown in Fig. 12.4, consider the following operations:

i : inversion about the centre of the molecule;

σ_1 : reflection in a plane perpendicular to the plane of the molecule through two opposite carbon atoms;

σ_m : reflection in a plane perpendicular to the other two, and through the centre, so dividing the molecule into two equal parts.

These three operations together with E form a group for transformation of the atoms of the benzene molecule. Find its group multiplication table.

10. The water molecule is represented by O, H, H lying in the plane yz and symmetric about the z-axis as in the diagram. The symmetry operations which exist for the molecule are

E : the identity operator,

C_2 : rotation of $180°$ about the z-axis,

σ_v' : reflection in the plane xz,

σ_v'' : reflection in the plane yz.

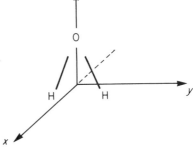

Consider an arbitary point (x, y, z) in space, and write down the coordinates of the point which is obtained by each of the four operations.

Find the group multiplication table for the group $E, C_2, \sigma_v', \sigma_v''$.

Show that the group is isomorphic with that in Question 9.

11. The ethylene molecule C_2H_4 may be considered to lie in a plane yz, symmetrical about the y- and z-axes as shown in the diagram. The symmetry operations for this molecule are

C_2, C_2', C_2'' : rotation of $180°$ about axis x, y, z respectively,

$\sigma_1, \sigma_2, \sigma_3$: reflection in planes xy, xz, yz respectively,

i : inversion about the origin,

E : the identity operator.

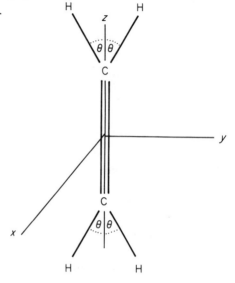

Consider a general point (x, y, z) and find the new coordinates after each of the operations.

Hence find the group multiplication table for the eight operations.

CHAPTER 13

Probability

In dealing with scientific experiments one is aware that the results obtained are very rarely exact. There is usually more than one possible result, and often a range of results. For instance, with an atmospheric pressure of 760 mm of mercury, water will boil at 100 °C, but if an experiment is carried out to determine the boiling point the recorded result may not be exactly 100 °C. There may be small errors in temperature or pressure measurements or there may be impurities in the water. Repeating the experiment a number of times, we would expect the results to be grouped about 100 °C. It could be said that the measured boiling temperature has more 'chance' of being near 100 °C than any other temperature.

In other types of experiment there will be a number of entirely different possibilities. A simple example of this is the tossing of a coin. Assuming the coin is unbiased it is as equally likely to fall with the head uppermost as with the tail. We would say then, that the 'chance' of a head is equal to the 'chance' of a tail.

To fully understand the methods of statistics, it is necessary first to consider the term *chance* or *probability* as it is called.

1. Probability

Imagine that a coin is tossed a number of times and that the numbers of heads and tails are recorded. The proportion of heads in a sequence of tosses of the coin is given by the number of heads divided by the total number of tosses of the coin. If this proportion is calculated at regular intervals, say, after every 5 tosses of the coin, then it will be found to be converging to some value. The variation in the calculated proportions will become smaller and smaller. The value to which the proportions are converging is the proportion of heads which would be obtained in an infinite sequence of tosses. Suppose that this value is 0.65, which may occur if the coin is a bent one, being fairly heavily biased in favour of heads.

This *proportion* in an infinite run is also said to be the *probability* of a head in a single toss of the coin, and for this particular coin, the probability of a head is 0.65. The probability in one test is, then, equivalent to the proportion in a large number of identical tests. For probability problems on coins, we often assume that the coin is 'unbiased' or 'ideal', that is, the coin is just as likely to come down as a head or a tail. In practice, no coin can be so perfect, but the bias is often so small that for practical purposes it can be ignored. Unless stated otherwise, in this chapter we assume that coins are unbiased.

Although it is often convenient to think of probabilities as proportions, it is not always feasible to do so, since it may not be possible to repeat a test, or even imagine that the tests are repeatable under identical conditions. The probability of a political party winning an election or a football team winning a cup final are such cases. Sometimes it is more convenient to think of probability as a degree of confidence in an event occurring, or the degree of belief, or even as the least odds one is willing to place for a bet on the event occurring. The numerical values of these probabilities may vary from one person to another and so are called subjective probabilities. For many years there have been arguments about the true nature of probability, but mathematicians now accept that probability can be thought of as different things in different situations. They therefore, do not try to define probability, but only give the rules (or axioms) which are obeyed.

2. Laws of probability

In many cases, when a probability is required, the situation in which it is to be applied is one in which several results are possible, but all are equally likely. The following rule is very useful in such cases.

The probability of an event occurring is given by the ratio of the number of ways that event can occur to the total number of possible ways, provided all ways are equally likely.

That is, if the number of ways the event can occur is m, and the total number of possible ways is n, then the probability (p) of the event occurring is given by

$$p = \frac{m}{n}. \tag{1}$$

The occurrence of the event is called a *success*.

For example, in the case of tossing a coin, to calculate the probability of obtaining a head, we note that $m = 1$ (head), $n = 2$ (head or tail), and therefore $p = \frac{1}{2}$.

To calculate the probability of throwing a number greater than 4 with a die (unbiased), we have $m = 2$ (numbers 5 or 6), $n = 6$ (numbers 1 to 6) and therefore $p = \frac{1}{3}$.

As m cannot be greater than n, i.e. $0 \leqslant m \leqslant n$, the following inequality holds:

$$0 \leqslant p \leqslant 1. \tag{2}$$

When $p = 0$, the event is impossible. When $p = 1$, the event is certain. The probability of a piece of granite turning overnight into a diamond, is zero. But the probability that water will eventually become ice if continually cooled is one. Values of p nearly one indicate that a success is almost certain, and p nearly zero indicates that a success is almost impossible.

3. Addition of probabilities

In the following sections, the theory holds whether or not possible results are equally likely.

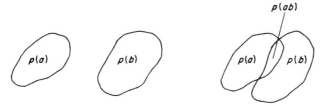

(i) *a* and *b* cannot occur together   (ii) *a* and *b* can occur together

Fig. 13.1. Addition of probabilities.

If the probability of event a occurring = p(a), and the probability of event b occurring = p(b), and the probability that both a and b occur = p(ab), then the probability that either a or b or both occur is given by

$$p(a + b) = p(a) + p(b) - p(ab). \tag{3}$$

If *a* and *b* cannot occur together then $p(ab) = 0$, and we have

$$p(a + b) = p(a) + p(b) \tag{4}$$

where $p(a + b)$ now implies the occurrence of either *a* or *b* since the two cannot occur together.

The law can be illustrated by representing the probabilities by areas as in Fig. 13.1.

The probability that at least one of the two events will occur is given by the total area enclosed in each case.

In (i) this is obviously $p(a) + p(b)$.

In (ii) $p(ab)$ is included with both $p(a)$ and $p(b)$ and must therefore be subtracted once, i.e. total area $= p(a) + p(b) - p(ab)$.

By representing the probabilities by areas we can extend the law to cover any number of possible occurrences. For example, the probability of at least one of events *a*, *b* or *c* occurring is given by the area in Fig. 13.2, i.e. $p(a + b + c) = p(a) + p(b) + p(c) - p(ab) - p(bc) - p(ac) + p(abc)$ assuming the usual notation.

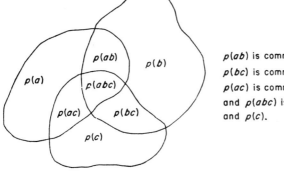

$p(ab)$ is common to $p(a)$ and $p(b)$, $p(bc)$ is common to $p(b)$ and $p(c)$, $p(ac)$ is common to $p(a)$ and $p(c)$, and $p(abc)$ is common to $p(a)$, $p(b)$ and $p(c)$.

Fig. 13.2. Addition of probabilities for 3 events.

Earlier, we calculated the probability of throwing a number greater than 4 with a die, as $\frac{1}{3}$. This could also have been calculated by the addition of probabilities as follows.

The probability of obtaining a $5 = \frac{1}{6}$.

The probability of obtaining a $6 = \frac{1}{6}$.

As a 5 and 6 cannot both occur the probability of obtaining a 5 or a 6 is $\frac{1}{6} + \frac{1}{6} = \frac{1}{3}$.

If, in a single trial or experiment a success is not achieved, the result is said to be a *failure*.

If the probability of a success is given by p, and the probability of a failure given by q, then as the result must be either a success or a failure

$$p + q = 1. \tag{5}$$

For example, the probability of throwing a 5 with a die is $\frac{1}{6}$. The probability of throwing any number except a 5 is therefore $1 - \frac{1}{6} = \frac{5}{6}$.

4. Multiplication of probabilities

If the probability of event a occurring is p(a), the probability of event b occurring is p(b), and p(b | a) denotes the probability of b occurring given that a has already happened, then the probability of a and b both occurring is given by

$$p(ab) = p(a) \cdot p(b|a). \tag{6}$$

If the occurrence of b is independent of a then $p(b \mid a)$ becomes $p(b)$, and

$$p(ab) = p(a) \cdot p(b). \tag{7}$$

For example, the probability of obtaining a 2 with the first throw of a die, followed by a 3 with the second throw is calculated as follows.

The probability of obtaining a 2 with the first throw is $\frac{1}{6}$.

The probability of obtaining a 3 with the second throw is $\frac{1}{6}$.

Therefore the probability of obtaining a 2 followed by a 3 is $\frac{1}{6} \cdot \frac{1}{6} = \frac{1}{36}$.

Example 1

If two coins are tossed, what is the probability of obtaining at least one head?

If a represents a head on the first coin and b represents a head on the second coin, the probability of either the first or second coins showing heads is given by equation (3), viz.

$$p(a + b) = p(a) + p(b) - p(ab)$$

where $p(a) = p(b) = \frac{1}{2}$.

Since the fall of the coins are independent

$$p(ab) = p(a) \cdot p(b)$$

$$\therefore \ p(a + b) = \frac{1}{2} + \frac{1}{2} - \frac{1}{2} \cdot \frac{1}{2}$$

$$= \frac{3}{4}.$$

Example 2

The letters of the word PREPARE are placed in a bag. What is the probability of drawing out the word RARE in the correct order?

The probability of drawing out R first is $\frac{2}{7}$.

The letters PPEEAR remain.

The probability of drawing out A from these six is $\frac{1}{6}$.

The letters PPEER remain.

The probability of drawing out R from these five is $\frac{1}{5}$.

Similarly the probability of drawing out E from the remaining four is $\frac{2}{4}$.

$\therefore$ the total probability $= \frac{2}{7} \cdot \frac{1}{6} \cdot \frac{1}{5} \cdot \frac{2}{4} = \frac{1}{210}$.

Example 3

The probability of a test tube breaking in a single experiment is 0.1. What is the probability of carrying out ten such experiments without breaking a test tube?

The probability of a test tube not breaking in a single experiment is $1 - 0.1 = 0.9$.

By the multiplication of probabilities, the probability of a test tube not breaking in two experiments is $0.9 \cdot 0.9 = (0.9)^2$.

The probability of a test tube not breaking in ten experiments is $(0.9)^{10} = 0.349$ to 3 decimal places.

Example 4

An experiment is equally likely to result as a success or a failure. A carries out the first experiment. If it fails B carries out the second experiment. If that fails A tries again, and so on. What is the probability that A succeeds first?

The probability of the first experiment being a success is $\frac{1}{2}$.

The probability of the first experiment being a failure and the second a success is $\frac{1}{2} \cdot \frac{1}{2}$.

The probability of the first two experiments being failures and the third a success is $\frac{1}{2} \cdot \frac{1}{2} \cdot \frac{1}{2}$ and so on.

The probability of the first n experiments being failures and the $(n + 1)$th a success is $1/2^{n+1}$.

The probability of A having the first success is given by the sum of the alternate probabilities, i.e.

$$\frac{1}{2} + \frac{1}{2^3} + \frac{1}{2^5} + \frac{1}{2^7} + \cdots \quad \text{(from equation 4).}$$

i.e.

$$p(A) = \frac{1}{2}\left(1 + \frac{1}{4} + \frac{1}{4^2} + \frac{1}{4^3} + \cdots\right).$$

This is a geometrical progression, and as $1 + x + x^2 + x^3 + \cdots = 1/(1 - x)$, we have

$$p(A) = \frac{1}{2}\left(\frac{1}{1 - \frac{1}{4}}\right) = \frac{1}{2} \cdot \frac{4}{3} = \frac{2}{3}.$$

The probability of B succeeding first is given by $1 - p(A) = \frac{1}{3}$.

5. Probabilities of different arrangements

In Example 2 the probability of obtaining the word RARE in the correct order was calculated when letters were taken at random from the word PREPARE. If, in the same way, we find the probability of obtaining the same four letters in a different order, say A, E, R, R, this is

$$\frac{1}{7} \cdot \frac{2}{6} \cdot \frac{2}{5} \cdot \frac{1}{4} = \frac{1}{210},$$

the same probability as that for RARE.

In fact, the probability of obtaining these four letters in any given order will be 1/210. The reason is that any arrangement of the four letters is as likely to occur as any other arrangement. Since this is true of any similar situation, it proves useful when probabilities are required for any possible order, as in the following example.

Example 5

Three balls are taken at random from a box containing two red and four blue balls. What is the probability that two blue balls and one red ball is obtained?

It is immaterial whether the balls are taken out together or one at a time, the answer will be the same. For convenience, we think of them as being taken out one at a time.

Now the probability of drawing a red, followed by two blue balls is $\frac{2}{6} \cdot \frac{4}{5} \cdot \frac{3}{4} = \frac{1}{5}$.

We have imposed an order on the selection. But two blue balls and one red ball can be obtained in three different ways, namely

    (i) red, blue, blue,
    (ii) blue, red, blue,
    (iii) blue, blue, red.

Since all will have the same probability of $\frac{1}{5}$, the final answer is $3 \cdot \frac{1}{5} = \frac{3}{5}$.

6. Choice of ways

If there are u ways of performing one operation and v ways of performing a second operation, then the number of ways of performing the two operations is uv.

For example, if there are three routes from X to Y, and four routes from Y to Z, the number of different routes from X to Z is $3 \cdot 4 = 12$.

7. Permutations

A *permutation* is an arrangement obtained by taking some or all of a number of items.

Therefore the number of permutations which can be formed with a group of unlike items is given by the number of ways of arranging them.

For example, consider the letters a, b, c, d. Suppose we require the number of permutations taking 3 letters at a time. There are 24 of these as shown:

abc	*abd*	*acd*	*bcd*
acb	*adb*	*adc*	*bdc*
bac	*bad*	*cad*	*cbd*
bca	*bda*	*cda*	*cdb*
cab	*dab*	*dac*	*dbc*
cba	*dba*	*dca*	*dcb*

It follows that the number of permutations of 4 unlike items taken 3 at a time is 24.

This is written

$$_4P_3 = 24.$$

To calculate $_4P_3$ without writing down all the different permutations and counting them, the following argument may be used.

There are 4 different letters from which we are choosing permutations of 3 letters.

Therefore the first letter of the three can be chosen in any of 4 ways, leaving 3 letters.

The second letter of the three can be chosen in any of 3 ways, leaving 2 letters.

The third letter of the three can be chosen in any of 2 ways.

Hence the total number of different ways of choosing the three letters is $4 \cdot 3 \cdot 2 = 24 = {}_4P_3$.

In the general case, if we require permutations of r letters from n: the first letter can be chosen in n ways, the second letter can be chosen in $(n - 1)$ ways, and so on, giving

$$_nP_r = n(n - 1)(n - 2) \ldots (n - r + 1)$$
$$= \frac{n(n - 1) \ldots 3 \cdot 2 \cdot 1}{(n - r) \cdot (n - r - 1) \ldots 3 \cdot 2 \cdot 1}.$$

i.e.

$$_nP_r = \frac{n!}{(n - r)!}. \tag{8}$$

It follows that

$$_nP_n = \frac{n!}{0!} = n! \quad (\text{as } 0! = 1)$$

where $_nP_n$ is the number of ways of arranging n unlike items.

The solutions of problems can often be tackled in different ways. This is illustrated by Examples 6 and 8 in which two methods are given.

Example 6

A class of six students is to be given six different experiments to carry out. In how many different ways may the experiments be allotted to the students?

First method. The number of ways is given by the number of permutations of 6 from 6, i.e.

$$_6P_6 = 6! = 720.$$

Second method. Working from first principles.
The first student may be given any one of 6 experiments.
This leaves 5 experiments to be allocated.
The second student may be given any one of these 5 experiments.
The third student may be given any one of 4 experiments, and so on.
The total number of ways is given by

$$6 \cdot 5 \cdot 4 \cdot 3 \cdot 2 \cdot 1 = 720.$$

Example 7

In a class of eight students, first, second and third prizes are to be awarded for Physics, and first and second prizes for Chemistry. In how many ways may the prizes be awarded?
The number of ways of awarding the Physics prizes is

$$_8P_3 = \frac{8!}{5!} = 8 \cdot 7 \cdot 6 = 336.$$

The number of ways of awarding the Chemistry prizes is

$$_8P_2 = \frac{8!}{6!} = 8 \cdot 7 = 56.$$

The total number of ways is $336 \cdot 56 = 18{,}816$.

8. Combinations

A *combination* is a selection obtained by taking some or all of a number of items.
Therefore the number of combinations which can be formed with a group of unlike items is given by the number of ways of selecting them.
For example consider again the 4 letters a, b, c, d. The number of combinations taking 3 letters at a time is 4, namely

$$abc, \quad abd, \quad acd, \quad bcd.$$

It follows that the number of combinations of 4 unlike items taken 3 at a time is 4.
This is written

$$_4C_3 = 4.$$

Note that each of the 4 combinations has 6 permutations, as shown in the four columns on page 305. Each combination of 3 items has therefore 3! permutations.

Each combination of r items will similarly have $r!$ permutations, so that there will be $r!$ times as many permutations as combinations.

Therefore

$$_nC_r = \frac{_nP_r}{r!}$$

i.e.

$$_nC_r = \frac{n!}{(n-r)!\,r!}\,.$$
(9)

It follows that

$$_nC_n = \frac{n!}{0!\,n!} = 1,$$

$$_nC_0 = \frac{n!}{n!\,0!} = 1,$$

and

$$_nC_{n-r} = \frac{n!}{(n-(n-r))!(n-r)!} = \frac{n!}{r!\,(n-r)!} = {_nC_r}.$$

(Sometimes written $\binom{n}{r}$.)

To avoid confusion as to whether permutations or combinations are required in the solution of problems, it is worth remembering that for permutations, order is important, whereas for combinations order is of no importance. Thus ab and ba are different permutations but the same combination.

Example 8

Three microscopes are to be chosen from five different types for use in a laboratory. In how many different ways can the microscopes be chosen?

First method. The number of ways of choosing the microscopes is given by the number of combinations of 3 from 5, i.e.

$$_5C_3 = \frac{5!}{3!\,2!} = 10.$$

Second method. Working from first principles: the first microscope can be chosen in 5 ways, the second microscope can be chosen in 4 ways, and the third microscope can be chosen in 3 ways.

The total number of ways is $5 \cdot 4 \cdot 3 = 60$.

But this assumes that they are chosen in order. As order is not important, divide by the number of ways in which the three could be re-arranged, i.e. 3!

Thus the number of different ways is $60/3! = 10$.

Example 9

Nine jars containing nine different liquids are to be split up into three groups, with two groups containing four jars and the remaining group one jar. In how many ways can this be done?

One group of 4 may be selected in $_9C_4$ ways:

$$_9C_4 = \frac{9!}{5!\,4!} = 126 \text{ ways.}$$

This leaves 5 jars.
Another group of 4 may be selected in $_5C_4$ ways:

$$_5C_4 = \frac{5!}{4!\,1!} = 5 \text{ ways.}$$

The third group of 1 jar is automatically selected.

The total number of ways is given by the number of ways of selecting the first group multiplied by the number of ways of selecting the second group multiplied by the number of ways of selecting the third group. That is,

$$126 \cdot 5 \cdot 1 = 630.$$

9. Arrangements containing like quantities

We have shown that the number of ways of arranging n unlike items is $_nP_n$ or $n!$. If r of n items are alike, any re-arrangement of the r like items gives the same arrangement. Now there are $r!$ ways of arranging r items. Therefore the number of different ways of arranging n items, with r alike is $n!/r!$.

The extension of this argument gives the following theorem.

If there are n items, with r alike of one kind, s alike of another kind, t alike of yet another kind, and so on, then the total number of different ways (N) of arranging them is given by

$$N = \frac{n!}{r!\,s!\,t!\ldots} \tag{10}$$

where $r + s + t + \ldots = n$.

Example 10

In how many ways can 2 red balls, 4 blue balls and 5 yellow balls be arranged in a line?

Substitution in formula (10) gives

$$N = \frac{11!}{2!\,4!\,5!} = 6{,}930.$$

Other Examples

Example 11

What is the probability of having four aces in a hand of 13 cards?

Using equation (1), $p = m/n$, with m as the number of hands containing 4 aces. As the other 9 cards in the hand can have any value, they can be chosen from the

48 remaining cards in the pack in $_{48}C_9$ ways. n is the total number of different hands. This is given by the number of ways of choosing 13 cards from 52, i.e. $_{52}C_{13}$. Therefore

$$p = \frac{m}{n} = \frac{_{48}C_9}{_{52}C_{13}} = \frac{48! \, 39! \, 13!}{39! \, 9! \, 52!} = 0.00264.$$

Example 12

In Example 2, what is the probability of drawing out the word RARE in any order?

The probability of obtaining the four letters in the order R, A, R, E is 1/210, found in Example 2.

From (10), the number of rearrangements of the four letters is $4!/2! \, 1! \, 1! = 12$.

The probability of obtaining the four letters in any order is equal to the probability in a particular order multiplied by the number of re-arrangements, i.e.

$$\frac{1}{210} \cdot 12 = \frac{2}{35}.$$

Example 13

An experiment to purify an amount of liquid is a success on average three times out of five. Each time the experiment succeeds, the profit made is £2, but £1 is lost each time it fails. Find the expectation of profit after two experiments.

After one experiment:

the probability of £2 profit is $\frac{3}{5}$.

the probability of £1 loss is $\frac{2}{5}$.

$\therefore$ expectation is $£\frac{3}{5} \times 2 - £\frac{2}{5} \times 1 = £\frac{4}{5}$.

In two experiments, the expectation is $£\frac{4}{5} \times 2 = £\frac{8}{5} = £1 \ 60p$.

PROBLEMS

1. What is the probability of obtaining:
(i) an odd number with one throw of a die?
(ii) a total score of 10 with two dice?
(iii) a total score greater than 9 with two dice?

2. On average, one in four sheets of plastic made by a certain process are almost perfect, whilst one in three have only one small flaw. The remainder have two or more flaws. What is the probability that a sheet has less than two flaws?

3. A firm owns two machines, one older than the other. The new one has a probability of 0.8 of working throughout a week without the need for repair. The probability for the other is 0.6.
(i) What is the probability of both machines working satisfactorily throughout the week?
(ii) What is the probability of at least one of the machines requiring repair?

4. Two cards are taken at random from a pack. What is the probability that they are both spades:
(i) if the first card is replaced before the second is taken?
(ii) if the first card is not replaced?

5. The letters of the word STATISTICS are placed in a box. One letter is taken from the box at a time. If four letters are taken out altogether, what is the probability that they spell out the word SITS in the correct order?

6. Out of ten flasks, three are defective. Two are chosen at random. What is the probability that both are:
(i) defective?
(ii) not defective?

7. If the probability that any one person requires medical attention in one particular year is $\frac{1}{5}$, what is the probability that a family of six will live throughout that year without the need of medical attention?

8. Four players each take a card from a pack, each player replacing his card before the next player takes one. One player has an 8. What is the probability that his card is the highest of the four, if aces are assumed to be high cards?

9. Four types of substances are to be tested using three types of acid at four different temperatures. How many tests are required to cover every combination of substance, acid and temperature?

10. Find the value of (i) $_8P_3$, (ii) $_7C_3$, (iii) $_{3n}P_{3n-2}$, (iv) $_nC_{n-1}$.

11. Ten different experiments are to be carried out in succession. In how many different ways may they be arranged?
If only seven of the experiments were to be performed, in how many ways could these be arranged?

12. If eight precious stones are to be listed by experts in order of value, in how many different ways could the three most valuable be chosen?

13. In how many different ways can five people be seated at a round table?

14. An industrial firm intends to buy twenty machines, all different, from another firm. Before doing so, three of the machines may be chosen and tested for a week. In how many ways can the machines be chosen?
If one of the twenty machines is faulty, what is the probability that one of the selected ones will be the faulty one?

15. A mixture is to be made up, consisting of four acids and three powders, to be chosen from eight acids and five powders. How many different mixtures can be made?

16. How many different arrangements can be made from the letters of the word THERMOMETER?
How many have O and H as the two end letters?
How many have the two M's separated?

17. A drum contains a large number of catalyst pellets, of which $\frac{1}{3}$ are iron. If five pellets are taken at random from the drum, what is the probability that all five are different from iron?
What is the probability of at least four of them being different from iron?

18. Four cards are drawn at random from a pack of cards. What are the probabilities of the following?
(i) Ace, King, Queen, and Jack of the same suit.
(ii) Four cards in a single suit.
(iii) Four cards of equal value.
(iv) Four cards in sequence in a single suit.

19. A quantity of oil is required with sulphur content less than 1%. Analyses show that quantities of oil from firm A would meet the requirements in $\frac{1}{2}$ of the cases, while oil from firm B would meet the requirements in $\frac{3}{4}$ of the cases. A quantity is taken from each firm and tested to see if it is acceptable. If not, another quantity is taken, and so on, until an acceptable amount is found. What is the probability that an acceptable amount is first found from firm B?

20. Five different dyes are recommended for five different types of cloth. If the instructions are lost, in how many different ways can all five dyes be used simultaneously such that a wrong dye is used on each cloth?

CHAPTER 14

Frequency Distributions

Statistical methods are concerned with the extraction and concise presentation of all the useful information from a group of data, which exists initially in the form of a table of numbers. For example, if the manager of a factory producing a certain type of measuring instrument requires information about the errors in the measurements given by the instruments he will not get a clear picture from a list of the errors for a number of such instruments.

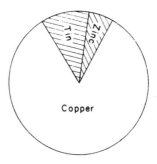

Fig. 14.1

The best method of obtaining a good overall picture is to represent the data in diagrammatic form. One example of diagrammatic representation is shown in Fig. 14.1, which illustrates the percentages of copper, tin and zinc found in certain samples of bronze. The different areas represent the amounts of the respective metals, in this case 85% copper, 10% tin, and 5% zinc.

1. Frequency diagrams

Histogram.   One type of diagram that is widely used is the *histogram,* an example illustrated by Fig. 14.2.

The value plotted along the x-axis is the *observation,* and the number of times the given observation occurs, i.e. the *frequency* (f) of that observation , is plotted along the vertical axis. The areas of the rectangles are proportional to the frequencies, but since, in general, the rectangles are all of equal width, their heights can be used as an exact measure of the frequency.

Recorded data arise in two forms. They may be *continuous,* that is the

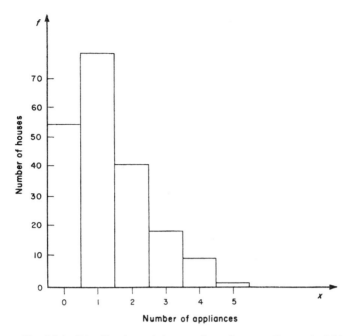

Fig. 14.2. Distribution of the number of gas appliances in 200 houses.

observations can theoretically take any value, or they may be *discrete*, in which case the observations are confined to certain numbers. In the distribution shown in Fig. 14.2, the abscissae were integers. An example of continuous data is that of the distribution shown in Fig. 14.3 where the percentage of copper could theoretically

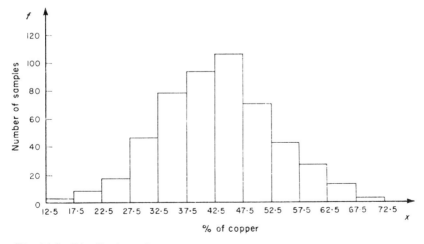

Fig. 14.3. Distribution of the percentage of copper in 500 samples of copper ore.

have any value between 0 and 100. In the diagram, observations have been grouped together so that the histogram could be drawn. Again, because the rectangles are all of the same width, the frequencies are given by their heights.

In practice all data are discrete since the accuracy of an observation is limited by that of the measuring instrument. In the distribution shown in Fig. 14.3, the initial observations were found to the nearest whole number, so that the values in this case were limited to integers. The first column, for instance, includes observations at $x = 13, 14, 15, 16,$ and 17%.

In general, to construct a histogram, we divide the range covered by the observations into a convenient number of groups and count the number of observations falling in each group. The number of groups must be small enough to give a compact representation of the data, but not so small that important information is concealed. A convenient practical rule is to use between 10 and 20 groups.

The group limits must be clearly defined so that there is no doubt into which group any observed value falls. If, for example, group limits were taken as 0–10, 10–20, 20–30, etc., then it is not clear into which group the value 10 is allocated. Limits could be taken as 0–9.9, 10–19.9, 20–29.9, etc. There is now no ambiguity but there is a discontinuous scale along the measurement axis. A method which will give a continuous scale is to use an additional decimal place. Suppose measurements were recorded to the nearest tenth. Any actual measurement between 0.05 and 0.15 would therefore be recorded as 0.1 and so on. Thus if the group interval was to be 5 units, group limits might be 5.05–10.05, 10.05–15.05, etc. It is now quite clear into which group each of the values will fall.

Having determined the range of the observations, decided upon the group interval and hence the number of groups, the lower limit of the lowest value is usually taken as the lower limit of the lowest group.

If, for example, the range is 15.3 to 35.7, i.e. 20.4, suitable group intervals would be 2.0, giving 11 groups.

The lowest value is 15.3, and any measurement between 15.25 and 15.35 will be recorded as 15.3.

Thus the lower limit of the lowest group is taken as 15.25, and the groups are 15.25–17.25, 17.25–19.25, 19.25–21.25, etc.

Example 1

Construct a histogram for the following list of 100 laboratory determinations of the specific gravity of sulphuric acid:

1.84	1.79	1.86	1.90	1.76	1.79	1.78	1.78	1.82	1.91
1.81	1.76	1.81	1.81	1.81	1.85	1.83	1.79	1.80	1.91
1.86	1.76	1.81	1.72	1.74	1.68	1.88	1.85	1.84	1.80
1.83	1.81	1.85	1.81	1.83	1.88	1.86	1.81	1.83	1.82
1.80	1.81	1.81	1.83	1.70	1.82	1.78	1.86	1.82	1.75
1.78	1.80	1.82	1.87	1.88	1.82	1.81	1.91	1.87	1.80
1.82	1.70	1.86	1.84	1.85	1.76	1.72	1.82	1.84	1.76
1.83	1.80	1.79	1.79	1.85	1.78	1.83	1.85	1.81	1.81

The range is 1.91 − 1.68 = 0.23

A group interval of 0.02 gives 12 groups.

The lowest recorded value is 1.68, so that the lowest group may be taken as 1.675–1.695. Writing down the groups as shown, the easiest method of finding the frequencies is to go through the list of figures once, putting a mark at the side of each group as a figure in that group is read. Every fifth mark is used to cross out the previous four. The resultant table is set out below.

Group limits		Frequency
1.675–1.695	1	1
1.695–1.715	11	2
1.715–1.735	11	2
1.735–1.755	11	2
1.755–1.775	̶1̶1̶1̶1̶1	5
1.775–1.795	̶1̶1̶1̶1̶1 ̶1̶1̶1̶1̶1	10
1.795–1.815	̶1̶1̶1̶1̶1 ̶1̶1̶1̶1̶1 ̶1̶1̶1̶1̶1 1111	19
1.815–1.835	̶1̶1̶1̶1̶1 ̶1̶1̶1̶1̶1 ̶1̶1̶1̶1̶1	15
1.835–1.855	̶1̶1̶1̶1̶1 ̶1̶1̶1̶1̶1	10
1.855–1.875	̶1̶1̶1̶1̶1 11	7
1.875–1.895	111	3
1.895–1.915	1111	4

The histogram is shown in Fig. 14.4.

Frequency curve. Consider a distribution similar to that in Fig. 14.3. Now suppose that more observations are taken and that they can be found more accurately. Suppose further that many more subdivisions are made. Then the histogram would have a shape similar to that in Fig. 14.3, but the rectangles would be narrower and the steps smaller. In the limiting case where an infinite number of observations are made and the width of the rectangles approach zero, the distribution can be represented by a smooth curve.

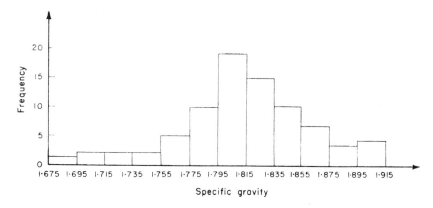

Fig. 14.4.

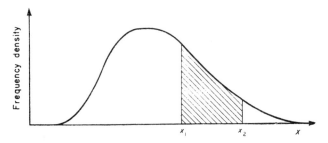

Fig. 14.5. A typical frequency curve.

Distributions represented by a curve in this way are purely theoretical. As shown in Fig. 14.5, the ordinate now becomes a *frequency density*, and proportional frequencies are measured by areas.

The proportion of observations of x lying between x_1 and x_2 is then given by the shaded area divided by the total area under the curve. If the frequency density scale is chosen so that the total area under the curve is equal to unity, then the frequency density becomes a *probability density* and the proportion is given by the shaded area. It follows therefore, that the shaded area represents the probability that any observation x, chosen at random, lies between x_1 and x_2.

This type of distribution has an important place in the theory of statistics and we shall refer to it again later.

2. Estimate of the 'centre' of the distribution

Although frequency diagrams give an idea of the overall picture of a distribution it is often necessary to describe the distribution in mathematical terms, especially if comparisons are to be made with other distributions.

The first important value to be estimated is the location of the 'centre' or 'middle' of the distribution. For instance, the statement may be made, 'Englishmen tend on average to be taller today than 20 years ago.' This average referred to, is an estimate of the middle of the distribution of the heights of Englishmen. Now before the centre can be calculated, it has to be defined.

There are, in statistics, three commonly used definitions as follows.

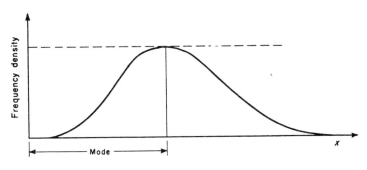

Fig. 14.6.

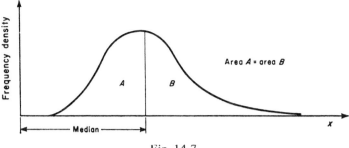

Fig. 14.7.

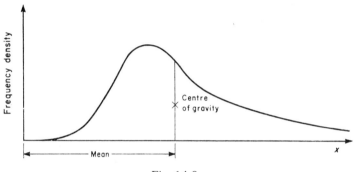

Fig. 14.8.

Mode.   *The mode is the observation having maximum frequency.* (See Fig. 14.6.)

Median.   *The median is the observation which divides the distribution into two equal parts* (see Fig. 14.7): That is, if the observations are arranged in order of magnitude, there are an equal number above and below the median. This definition is complete if there are an odd number of observations. With an even number the median is taken as the average of the two central values.

Mean.   *The mean is the arithmetic average.*

To see this value pictorially, imagine a solid body with the same shape as the distribution and of uniform thickness. Then the mean is the x coordinate of its centre of gravity (see Fig. 14.8).

3. Comparison of the mode, median and mean

Each of the three definitions for the centre of a distribution is of some importance and the decision to use any particular one depends on the problem at hand. The mode and median, for example, are easy to calculate, but for some distributions these estimates for the centre are misleading. Two examples in which this is the case are shown in Fig. 14.9.

Calculation of the mean involves more computation, but because it is dependent on each value of the distribution, it usually gives a more accurate assessment of the centre. However, there are certain cases when this is also misleading. For instance, a survey of the cost of equipment in a laboratory may show that all the units cost

318

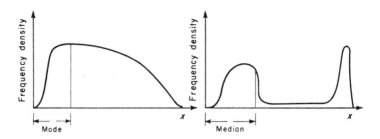

Fig. 14.9. Distributions in which the mode and median are
misleading as estimates of the centre.

between £10 and £100, except for a digital computer which cost £50,000. The
mean cost of a unit of equipment is rather meaningless in this case, and probably
the median would give a more sensible estimate.

The median is often used for comparing wages of different groups of workers,
when a few extreme values can distort the mean. One advantage of the median in
this case is that it seems fair from a logical point of view, because 50% of the
workers will have pay above the median, and 50% will be paid less than the median.
Another case where it is useful is when comparing lifetimes of people (or machines).
For example in a comparison of smokers' and non-smokers' lifetimes, once half of
the people of each type kept under observation have died, then the medians are
known and can be tested for differences. The means can only be calculated when
all the people have died.

The mean is most important throughout the theory of statistics, and we shall be
concerned mainly with this parameter. For symmetrical distributions the mode,
median and mean have the same value, and for near symmetrical distributions they
will be almost the same. In some cases, therefore, the mode or median, being easier
to calculate, would suffice as good estimates of the mean.

4. Formulae for the mean

The mean of the values $x_1, x_2, x_3, \ldots, x_n$ is denoted by $\bar{x}$ where

$$\bar{x} = \frac{x_1 + x_2 + x_3 + \ldots x_n}{n},$$

i.e.

$$\bar{x} = \frac{\sum_{i=1}^{n} x_i}{n}. \tag{1}$$

If values of x are repeated, the mean may be calculated as follows. Let f_i be the
frequency of x_i with i taking the values from 1 to n. Then the mean $\bar{x}$ is given by

$$\bar{x} = \frac{f_1 x_1 + f_2 x_2 + f_3 x_3 + \ldots + f_i x_i + \ldots + f_n x_n}{f_1 + f_2 + f_3 + \ldots + f_n}$$

i.e.

$$\bar{x} = \frac{\sum\limits_{i=1}^{n} f_i x_i}{\sum\limits_{i=1}^{n} f_i}. \tag{2}$$

Note that equation (1) is a special case of equation (2), namely when $f_i = 1$ for all i.

In subsequent theory, $\sum_{i=1}^{n} f_i x_i$ will be written as $\sum f_i x_i$. Other summations will be expressed using a similar notation, it being assumed that the summation extends over all possible values of i.

Example 2

The densities of five liquids are given as $0.71, 0.82, 0.63, 0.91, 0.58$ g cm^{-3}. Find the mean density.

$$\bar{x} = \frac{\sum x_i}{n} = \frac{3.65}{5} = 0.73 \text{ g cm}^{-3}.$$

Example 3

For 50 families living in one street, the number of persons per family was as follows:

No. per family (x)	1	2	3	4	5	6
No. of families (f)	2	20	12	10	5	1

Find the mean number per family for this distribution.

The solution is best set out as follows with the column fx calculated by multiplying each x by the value of f in the same row:

x	f	fx
1	2	2
2	20	40
3	12	36
4	10	40
5	5	25
6	1	6
	50	149

Then $\sum f_i = 50$

and $\sum f_i x_i = 149$.

$$\therefore \ \bar{x} = \frac{\sum f_i x_i}{\sum f_i} = \frac{149}{50} = 2.98.$$

The mean number per family = 2.98.

Example 4

Find the values of the mode and median for the data in Example 3.

The maximum frequency is 20, which occurs at $x = 2$. From the definition therefore, the mode is 2.

To find the median, form a column F, the *cumulative frequency*, with $F_1 = f_1$ and $F_i = F_{i-1} + f_i$ for $i > 1$, as follows.

x	f	F	
1	2	2	$F_1 = f_1 = 2.$
2	20	22	$F_2 = F_1 + f_2 = 2 + 20 = 22$
3	12	34	$F_3 = F_2 + f_3 = 22 + 12 = 34$
4	10	44	$F_4 = F_3 + f_4 = 34 + 10 = 44$
5	5		
6	1		

From the definition of the median, the value of x is required which divides the distribution into two equal halves of 25 observations. Thus x_i is required when $F = 25$. This occurs when 3 of the 12 observations at $x = 3$ are added to F_2. Thus the observation dividing the distribution into two equal halves is at $x = 3$, and we conclude that the median is 3.

5. Estimate of the 'spread' of the distribution

Having decided the centre of the distribution, the next important piece of information which may be obtained is an estimate of the 'spread' of the distribution. In some cases all the data tend to cluster round the mean, in others they are widely dispersed. As for the centre, there are several methods of estimating the spread. Five commonly used in Statistics are defined as follows.

Range. *The range is the difference between the maximum and minimum observations* (see Fig. 14.10).

Interquartile range. *The interquartile range is the difference between the upper and lower quartiles*, where the upper and lower quartiles are the observations x which cut off 25% of the distribution (see Fig. 14.11).

Mean deviation. *The mean deviation is the mean difference between the observed values and the mean.*

Referring to Fig. 14.12, the mean deviation of the values x_1, x_2, x_3, x_4 is the mean value of the positive lengths y_1, y_2, y_3, y_4, i.e. $(y_1 + y_2 + y_3 + y_4)/4$.

The mean deviation of $x_1, x_2, \ldots, x_i, \ldots, x_n$ is given by

$$\frac{\Sigma |x_i - \bar{x}|}{n} \tag{3}$$

or if x_i occurs f_i times, by

$$\frac{\Sigma f_i |x_i - \bar{x}|}{\Sigma f_i}. \tag{4}$$

Variance. *The variance is the mean square difference of the observed values from the mean.*

Referring to Fig. 14.12, this would be the mean of $y_1^2 + y_2^2 + y_3^2 + y_4^2$, i.e. $(y_1^2 + y_2^2 + y_3^2 + y_4^2)/4$.

The variance is denoted by σ^2. Hence the variance of $x_1, x_2, \ldots, x_i, \ldots x_n$ is

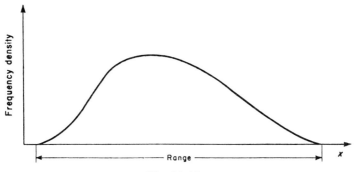

Fig. 14.10.

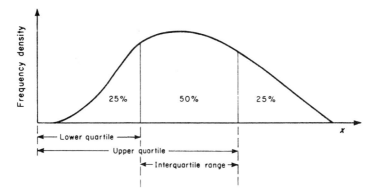

Fig. 14.11.

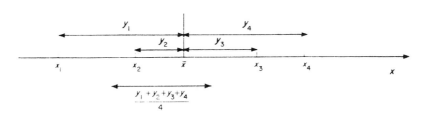

Fig. 14.12.

given by

$$\sigma^2 = \frac{\Sigma (x_i - \bar{x})^2}{n},\tag{5}$$

or if x_i occurs f_i times, by

$$\sigma^2 = \frac{\Sigma f_i(x_i - \bar{x})^2}{\Sigma f_i}.\tag{6}$$

Standard deviation. *The standard deviation is the positive square root of the variance,* i.e. σ.

6. Comparison of the different estimates for the spread

The parameter to be used for the estimation of the spread of the distribution must be decided when each problem is under consideration.

Both the range and interquartile range are easy to calculate. The range however depends only on two observations, the maximum and minimum. For this reason, it is unsatisfactory in most cases, as chance high or low readings could make a large difference to its value. The interquartile range is also inadequate for many distributions because it gives no indication of the manner of distribution within or outside the quartiles. The mean deviation takes every observation into account and is therefore usually a much better estimate. It is however unsatisfactory as a parameter in sampling theory and other parts of statistics. Because of their importance in the theory of statistics and because they give good estimates of the spread, the standard deviation and variance are the most frequently used parameters. The spread of a distribution is usually expressed in terms of the standard deviation rather than the variance so that it is in the same units as the measure of the centre.

Alternative formulae for σ. Equation (6) gave the formula for calculating the variance in the form

$$\sigma^2 = \frac{\Sigma f_i(x_i - \bar{x})^2}{\Sigma f_i}.$$

Therefore

$$\sigma^2 = \frac{\Sigma f_i(x_i^2 - 2x_i\bar{x} + \bar{x})^2}{\Sigma f_i}$$

$$= \frac{\Sigma f_i x_i^2 - 2\bar{x} \Sigma f_i x_i + \bar{x}^2 \Sigma f_i}{\Sigma f_i}$$

as $\bar{x}$ is a constant.

Now

$$\bar{x} = \frac{\Sigma f_i x_i}{\Sigma f_i}.$$

$$\therefore \ \sigma^2 = \frac{\Sigma f_i x_i^2}{\Sigma f_i} - 2\left(\frac{\Sigma f_i x_i}{\Sigma f_i}\right)^2 + \left(\frac{\Sigma f_i x_i}{\Sigma f_i}\right)^2.$$

That is

$$\sigma^2 = \frac{\Sigma f_i x_i^2}{\Sigma f_i} - \left(\frac{\Sigma f_i x_i}{\Sigma f_i}\right)^2. \tag{7}$$

Putting $f_i = 1$ for all i, the formula corresponding to equation (5) (i.e. for single values of x) becomes

$$\sigma^2 = \frac{\Sigma x_i^2}{n} - \left(\frac{\Sigma x_i}{n}\right)^2. \tag{8}$$

In practice, equations (7) and (8) are preferred to (5) and (6) as in almost all cases the calculations are easier.

Example 5

The following data give the number of injuries in a factory per day over a period of 80 days. Find the standard deviation of the distribution.

No. of injuries (x)	0	1	2	3	4	5	6
No. of days (f)	9	24	29	11	5	1	1

The solution may be set out as follows with each $f_i x_i^2$ calculated by multiplying $f_i x_i$ by x_i.

x	f	fx	fx^2
0	9	0	0
1	24	24	24
2	29	58	116
3	11	33	99
4	5	20	80
5	1	5	25
6	1	6	36
	80	146	380

$$\Sigma f_i = 80,$$
$$\Sigma f_i x_i = 146,$$
$$\Sigma f_i x_i^2 = 380.$$

$$\sigma^2 = \frac{\Sigma f_i x_i^2}{\Sigma f_i} - \left(\frac{\Sigma f_i x_i}{\Sigma f_i}\right)^2$$

$$= \frac{380}{80} - \left(\frac{146}{80}\right)^2 = 1.42.$$

The standard deviation, $\sigma = 1.19$ to two decimal places.

7. Change of origin and scale

In order to simplify the calculations of the mean and variance, it is often advisable to make a change of variable, say from x to y where

$$x = a + by \quad (a, b \text{ constants}).$$

This has the effect of changing both the origin and the scale of the original distribution. Consider the mean and standard deviation.

Mean. Let $\bar{x}$ be the mean of the distribution in terms of the x variable.
Let $\bar{y}$ be the mean of the distribution in terms of the y variable.
Now $x_i = a + by_i$ for each observation, so that

$$\bar{x} = \frac{\Sigma f_i x_i}{\Sigma f_i} = \frac{\Sigma f_i(a + by_i)}{\Sigma f_i} = \frac{a\,\Sigma f_i + b\,\Sigma f_i y_i}{\Sigma f_i} \quad (a, b \text{ constants})$$

$$= a + b\bar{y} \quad \text{as } \bar{y} = \frac{\Sigma f_i y_i}{\Sigma f_i}.$$

$$\therefore \ \bar{x} = a + b\bar{y}. \tag{9}$$

Standard deviation.   Let σ_x be the standard deviation of the distribution in terms of the x variable.

Let σ_y be the standard deviation of the distribution in terms of the y variable.

Now

$$\sigma_x^2 = \frac{\Sigma f_i(x_i - \bar{x})^2}{\Sigma f_i}.$$

But

$$x_i = a + by_i$$

and

$$\bar{x} = a + b\bar{y}.$$

$$\therefore \ x_i - \bar{x} = b(y_i - \bar{y})$$

and

$$\sigma_x^2 = \frac{\Sigma f_i b^2 (y_i - \bar{y})^2}{\Sigma f_i}$$

$$= b^2 \frac{\Sigma f_i (y_i - \bar{y})^2}{\Sigma f_i} \quad (b \text{ constant})$$

$$= b^2 \sigma_y^2.$$

Taking the square root, with the standard deviation positive, we have

$$\sigma_x = |b| \, \sigma_y. \tag{10}$$

Another method of simplifying the computational work is by grouping, as mentioned in the discussion of histograms. Again, the ideal number of groups lies between 10 and 20. With fewer than 10 groups, too much of the information is lost, whilst the table becomes too large to handle satisfactorily with greater than 20 groups.

Example 6

The proportions of useful mineral deposits in 500 samples of earth were determined. The following distribution was obtained, where readings were taken to the nearest hundredth, and then grouped as shown:

Proportion	0.045–0.095	0.095–0.145	0.145–0.195	0.195–0.245		
Frequency	1	5	21	54		
Proportion	0.245–0.295	0.295–0.345	0.345–0.395	0.395–0.445	0.445–0.495
Frequency	86	121	98	68	36
Proportion	0.495–0.545	0.545–0.595			
Frequency	8	2			

Find the mode, median, mean, range and standard deviation.

The calculations may be set out as shown below, where the observations are regarded as occurring at the centre of that group. These are the values x shown in the third column.

A transformation to the y system simplifies the calculations. The best way of doing this is to choose $y = 0$ at about the centre of the distribution and then to write down the numbers $-1, -2, -3, \ldots, 1, 2, 3, \ldots$, as shown in the fourth column. The constants a and b in the equation $x_i = a + by_i$ can then be calculated later.

Group limits	f	x	y	fy	fy^2	F
0.045–0.095	1	0.07	−5	−5	25	1
0.095–0.145	5	0.12	−4	−20	80	6
0.145–0.195	21	0.17	−3	−63	189	27
0.195–0.245	54	0.22	−2	−108	216	81
0.245–0.295	86	0.27	−1	−86	86	167
0.295–0.345	121	0.32	0	0	0	288
0.345–0.395	98	0.37	1	98	98	
0.395–0.445	68	0.42	2	136	272	
0.445–0.495	36	0.47	3	108	324	
0.495–0.545	8	0.52	4	32	128	
0.545–0.595	2	0.57	5	10	50	
	500			384	1468	
				−282		
				102		

Now

$$\bar{y} = \frac{\Sigma f_i y_i}{\Sigma f_i} = \frac{102}{500} = 0.204,$$

and

$$\sigma_y^2 = \frac{\Sigma f_i y_i^2}{\Sigma f_i} - \left(\frac{\Sigma f_i y_i}{\Sigma f_i}\right)^2 = \frac{1468}{500} - \left(\frac{102}{500}\right)^2 = 2.894.$$

$$\therefore \; \sigma_y = 1.701.$$

Since

$$x_i = a + by_i,$$

at

$$y_i = 0, x_i = 0.32 \quad \text{and} \quad \therefore a = 0.32,$$

and at

$$y_i = 1, x_i = 0.37 \quad \text{and} \quad 0.37 = 0.32 + b, \quad \text{i.e. } b = 0.05.$$

From equation (9),

$$\bar{x} = a + b\bar{y} = 0.32 + (0.05)(0.204) = 0.3302.$$

Therefore the mean of the distribution is 0.3302.
From equation (10),

$$\sigma_x = |b|\sigma_y = (0.05)(1.701) = 0.085 \quad \text{to three decimal places.}$$

Hence the standard deviation of the distribution is 0.085.
As the largest frequency ($f = 121$) is at $x = 0.32$, the mode is 0.32.
From column F, we see that the 250th observation occurs at $x = 0.32$, and therefore the median is 0.32.

The maximum and minimum values of x were respectively 0.57 and 0.07. Therefore the range is $0.57 - 0.07 = 0.5$.

8. Samples

In this chapter we have considered distributions in general, and various parameters have been calculated for these distributions. In practice however, the distribution which is obtained is often just part of a much larger distribution. For instance, information may be required on the error of thermometer readings at 60 °C. If 50 thermometers, say, were taken and their errors found by experiment, these 50 readings form a distribution. This is, however, just part of a larger distribution, consisting of readings of all thermometers of this make, or even all existing thermometers. The 50 thermometers are called a *sample* and the group of all thermometers a *population*.

The important point is that although the parameters are calculated from the sample distribution, in most cases the parameters required are those for the population distribution. The sample parameters however may be considered to be estimates of the population parameters. Thus the mean of the sample is an estimate of the mean of the population, and is calculated by the same formula, i.e. mean $= \Sigma x_i/n$. When estimating the variance of the population from a sample, the best estimate is given by the formula

$$\text{variance} = \frac{\Sigma(x_i - \bar{x})^2}{n - 1} \quad \text{or} \quad \frac{\Sigma x_i^2 - n\bar{x}^2}{n - 1},$$

that is, dividing by $n - 1$ instead of n. (This formula is derived on page 366.)

Although this estimate will sometimes be less than the true population variance and at other times will be greater than the true value, division by $n-1$ ensures that it is correct 'on the average'. For large samples, the difference is small as $n \simeq n - 1$, but the difference is important when dealing with small samples.

APPENDIX

Charlier's check

Charlier's check is sometimes used as a check on the working when calculating the variance of a distribution.

Based on the fact that

$$\Sigma f_i(y_i + 1)^2 = \Sigma f_i y_i^2 + 2\Sigma f_i y_i + \Sigma f_i ,$$

the check involves calculating the sum of an extra column, $f(y + 1)^2$.

Suppose that the check is to be used on the data in Example 6. The work could then be set out as follows.

Group limits	f	x	y	fy	fy^2	$f(y + 1)^2$
0.045–0.095	1	0.07	−5	−5	25	16
0.095–0.145	5	0.12	−4	−20	80	45
0.145–0.195	21	0.17	−3	−63	189	84
0.195–0.245	54	0.22	−2	− 108	216	54
0.245–0.295	86	0.27	−1	−86	86	0
0.295–0.345	121	0.32	0	0	0	121
0.345–0.395	98	0.37	1	98	98	392
0.395–0.445	68	0.42	2	136	272	612
0.445–0.495	36	0.47	3	108	324	576
0.495–0.545	8	0.52	4	32	128	200
0.545–0.595	2	0.57	5	10	50	72
	500			102	1468	2172

Then

$$\Sigma f_i y_i^2 + 2 \Sigma f_i y_i + \Sigma f_i = 1468 + 2(102) + 500 = 2{,}172$$
$$= \Sigma f_i(y_i + 1)^2 ,$$

showing that the values for Σf_i, $\Sigma f_i y_i$ and $\Sigma f_i y_i^2$ are correct.

Sheppard's correction

The method of grouping causes the calculated mean and variance to be approximations only, and they will almost certainly differ by small amounts from the true mean and variance.

If the distribution is a continuous one, that is the observations may take any value, then the true mean is given by

$$\frac{\int_{-\infty}^{+\infty} fx \, dx}{\int_{-\infty}^{+\infty} f \, dx} ,$$

whereas the calculated mean uses the formula $\Sigma f_i x_i / (\Sigma f_i)$. Similarly for the variance. The errors therefore may be said to arise due to using summation series as an approximation to an integral. W. F. Sheppard studied this problem and found that although the formula already used for the mean was the best possible, a more accurate value for the variance would be obtained by using the formula

$$\sigma^2 = \frac{\Sigma f_i(x_i - \bar{x})^2}{\Sigma f_i} - \frac{h^2}{12},$$

where h is the width of the class interval.

In Example 6, $h = 0.05$ and $h^2/12 \simeq 0.000208$

As $\sigma_x^2 = (0.085)^2 = 0.007225$, the corrected value is therefore

$$0.007225 - 0.000208 = 0.007017,$$

giving the standard deviation as 0.084 to three decimal places.

PROBLEMS

1. The number of breakages per day in a laboratory was recorded for 500 days as follows.

Number (x)	0	1	2	3	4	5	6	7	8
Frequency (f)	98	122	108	82	57	26	5	0	2

Represent the distribution by a histogram and calculate the mean and variance.

2. A chemical test was carried out on 12 pieces of steel to find traces of cobalt. The percentage of cobalt found in each were 0.072, 0.069, 0.070, 0.072, 0.076, 0.075, 0.065, 0.068, 0.071, 0.069, 0.073, 0.072.

Find the mean, range, and variance for these values.

3. A number of chemical determinations of the atomic weight of hydrogen were carried out, giving the following results for w, where the atomic weight = 1.007770 + $w \times 10^{-5}$. w: 1.1, 0.9, -0.5, -1.8, -0.1, 1.9, -0.4, 0.1, 0.2.

Find the median and the standard deviation for the results to seven decimal places.

4. Find the standard deviation of the numbers 1, 2, 3, ... , 17.

5. 60 weighing machines were tested for accuracy. The data show the distribution of errors when weights of 100 g were put on the scales.

Error in grams (x)	-0.25	-0.15	-0.05	0.05	0.15	0.25	0.35	0.45
No. of machines (f)	2	6	14	18	12	5	2	1

Find the standard deviation and the range.

6. The breaking loads of 600 samples of a plastic were measured to the nearest kg. The results were then grouped as follows.

Breaking load (kg)	No. of samples	Breaking load (kg)	No. of samples
1,020–39	2	1,160–79	95
1,040–59	9	1,180–99	77
1,060–79	22	1,200–19	56
1,080–99	47	1,220–39	20
1,100–19	70	1,240–59	12
1,120–39	86	1,260–79	3
1,140–59	101		

Draw a histogram of the distribution. Find
(i) the mode, median, mean.
(ii) the standard deviation and range.

7. Show that

(i) $\Sigma f_i(x_i - \bar{x}) = 0$.

(ii) $\dfrac{\Sigma f_i(x_i - \bar{x})^2}{\Sigma f_i - 1} = \dfrac{\Sigma f_i x_i^2 - \bar{x}^2 \Sigma f_i}{\Sigma f_i - 1}$.

8. The yields of a chemical process under constant conditions were

38, 40, 42, 46, 42, 44, 39, 47, 45%.

(i) Find the mean and standard deviation of this distribution.
(ii) If this is a sample from a large population, find the estimated mean and standard deviation of that population.

9. Show that the mean and variance of the distribution consisting of the numbers 1, 2, 3, . . . , $n - 1$, n are $\frac{1}{2}(n + 1)$ and $\frac{1}{12}(n^2 - 1)$ respectively.

10. A liquid is sold in flasks of nominal capacity 1 litre. 120 of the flasks were selected at random and the capacity of each one measured, giving the following data. Prepare a frequency table from the results, grouping them such that one group contains the values 995.0 cm^3 to 995.9 cm^3 inclusive. Calculate the mean capacity and the standard deviation.

Capacity (cm^3)

997.7	1,000.8	1,001.3	997.5	999.3	1,003.4	999.3	996.4
999.0	1,005.5	995.7	1,005.9	999.5	997.2	1,000.5	1,002.7
1,002.9	999.7	1,002.6	1,000.3	998.1	1,001.5	998.2	997.3
1,000.7	1,003.2	998.4	1,000.6	996.0	999.5	1,000.5	1,005.0
1,005.2	998.4	1,000.3	995.3	1,002.3	999.2	998.8	1,000.2
996.2	1,003.8	1,003.3	998.7	999.1	1,002.9	995.8	1,004.7
1,001.8	1,002.4	1,001.2	1,000.0	1,001.6	998.7	999.0	1,001.0
1,003.6	999.7	996.5	1,003.5	996.3	1,001.3	1,002.1	999.9
1,001.2	999.6	996.9	999.7	1,000.4	1,004.4	998.5	1,004.8
1,005.3	1,002.4	1,004.5	997.2	999.9	997.1	1,001.6	998.6
995.1	1,004.1	1,000.6	997.7	1,001.8	998.7	1,000.2	1,002.2
1,001.8	999.5	997.9	999.1	1,005.7	1,000.1	1,002.9	995.8
1,002.1	1,002.0	998.8	998.4	1,000.1	1,002.8	998.4	1,003.2
998.9	1,004.1	1,001.4	1,003.5	998.5	1,000.8	1.001.0	999.0
1,000.1	997.4	999.1	997.2	1,003.0	1,001.4	996.9	1,001.1

Binomial, Poisson and Normal Distributions

Certain types of distributions appear in many branches of the sciences, and although concerned with widely different physical relationships, they are defined by a single equation. We shall deal with three of the more important distributions in this chapter.

1. Binomial distribution

Suppose that a trial or experiment is carried out, the result of which must be either a success or a failure. Denoting the probability of success by p and the probability of failure by q, then $p + q = 1$. The experiment may simply consist of spinning a coin, where a head may be regarded as a success and a tail as a failure. In this case $p = q = \frac{1}{2}$. Alternatively the experiments may involve taking an item at random from a box in which 20% of the items are defective. In this case, if we regard choosing a defective as a success, $p = 0.2$ and $q = 0.8$.

Suppose now that the trial or experiment is repeated, with the same initial conditions. Again, it may result in a success or a failure. Taking the two experiments together, there are now four possibilities, listed below with their respective probabilities.

1st experiment	2nd experiment	probability
success	success	p^2
	failure	pq
failure	success	qp
	failure	q^2

Thus the probability of:

$$\text{two successes} = p^2$$
$$\text{one success and one failure} = 2pq$$
$$\text{two failures} = q^2.$$

After three experiments, the table giving each probability reads as follows:

1st experiment	2nd experiment	3rd experiment	probability
success	success	success	p^3
		failure	p^2q
	failure	success	pqp
		failure	pq^2
failure	success	success	qp^2
		failure	qpq
	failure	success	q^2p
		failure	q^3

Again, the probability of

$$\text{three successes} = p^3$$
$$\text{two successes and one failure} = 3p^2q$$
$$\text{one success and two failures} = 3pq^2$$
$$\text{three failures} = q^3.$$

Note that the three terms obtained after two experiments are given by the expansion of $(p + q)^2$, i.e.

$$(p + q)^2 = p^2 + 2pq + q^2,$$

and the four terms obtained after three experiments are given by the expansion of $(p + q)^3$, i.e.

$$(p + q)^3 = p^3 + 3p^2q + 3pq^2 + q^3.$$

Similarly

$$(p + q)^4 = p^4 + 4p^3q + 6p^2q^2 + 4pq^3 + q^4$$

will give the respective probabilities after four experiments, where, for example, $4pq^3$ is the probability of 1 success and 3 failures.

Extending the theory to the case of n experiments, the probabilities of $0, 1, 2, \ldots, n$ successes will be given by the successive terms in the expansion of $(p + q)^n$, where

$$(p + q)^n = {}_nC_0q^n + {}_nC_1pq^{n-1} + {}_nC_2p^2q^{n-2} + \cdots {}_nC_rp^rq^{n-r} + \cdots {}_nC_np^n. \quad (1)$$

The general term ${}_nC_rp^rq^{n-r}$ gives the probability of r successes and $n - r$ failures. This is stated in Bernoulli's theorem as follows.

If the probability of success and failure in a single trial are p and q respectively, then the probability of r successes in n trials is ${}_nC_rp^rq^{n-r}$, *i.e.*

$$P_r = {}_nC_rp^rq^{n-r}. \quad (2)$$

Example 1

A die is thrown three times. Draw a histogram of the probabilities of obtaining 0, 1, 2, 3 sixes.

In a single throw, the probability of obtaining a six is $\frac{1}{6}$, and of obtaining a number other than six is $\frac{5}{6}$.

Therefore, the probability of success is $p = \frac{1}{6}$, the probability of failure is $q = \frac{5}{6}$, and $n = 3$.

From equation (2), the probability of no success is

$$P_0 = {}_3C_0\left(\frac{1}{6}\right)^0\left(\frac{5}{6}\right)^3 = \frac{5^3}{6^3} = 0.579,$$

the probability of one success is

$$P_1 = {}_3C_1\left(\frac{1}{6}\right)^1\left(\frac{5}{6}\right)^2 = \frac{3.5^2}{6^3} = 0.374,$$

the probability of two successes is

$$P_2 = {}_3C_2\left(\frac{1}{6}\right)^2\left(\frac{5}{6}\right) = \frac{3.5}{6^3} = 0.069,$$

and the probability of three successes is

$$P_3 = {}_3C_3\left(\frac{1}{6}\right)^3\left(\frac{5}{6}\right)^0 = \left(\frac{1}{6}\right)^3 = 0.005.$$

The histogram is shown in Fig. 15.1.

Note that the sum of the probabilities is 1. This must be so since the three trials are certain to yield one of the four results. This is a useful check on the calculations.

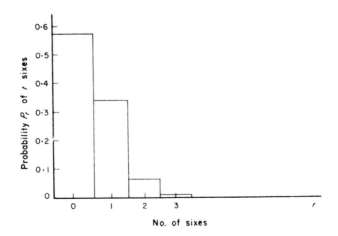

Fig. 15.1.

Example 2

10% of items resulting from a production process are defective. If samples of size 12 are taken, what are the probabilities of 0, 1, 2 defectives? If 100 samples are taken, in how many of them would at least 3 defective items be expected?

If the probability of a defective item is termed a success, then $p = 0.1$ and $q = 1 - p = 0.9$.

Using equation (2), $P_r = {}_nC_rp^rq^{n-r}$, the probability of no defectives is

$$P_0 = {}_{12}C_0(0.1)^0(0.9)^{12} = (0.9)^{12} = 0.282,$$

the probability of one defective is

$$P_1 = {}_{12}C_1(0.1)^1(0.9)^{11} = 12(0.1)(0.9)^{11} = 0.377,$$

and the probability of two defectives is

$$P_2 = {}_{12}C_2(0.1)^2(0.9)^{10} = 66(0.1)^2(0.9)^{10} = 0.230.$$

The probability of at least 3 defective items in a sample is given by

$$P_3 + P_4 + P_5 + \cdots + P_{11} + P_{12}.$$

As

$$P_0 + P_1 + P_2 + \cdots + P_{11} + P_{12} = 1,$$

we require the value of

$$1 - (P_0 + P_1 + P_2) = 1 - 0.889 = 0.111.$$

In 100 samples, the expected number is therefore

$$100 \times 0.111 = 11.1.$$

Mean of a binomial distribution. Given a distribution of the variate x, we have seen that the mean is given by the formula $\Sigma f_i x_i / \Sigma f_i$, where f_i is the frequency of x_i.

Now

$$\frac{\Sigma f_i x_i}{\Sigma f_i} = \Sigma \frac{f_i}{\Sigma f_i} x_i = \Sigma p_i x_i,$$

where p_i is the probability of obtaining x_i.

For the binomial distribution, P_r is the probability of obtaining r successes, where $P_r = {}_nC_rp^rq^{n-r}$.

Thus the mean is given by $\Sigma_{r=0}^n P_r r$.
From equation (1),

$$\Sigma P_r = \Sigma {}_nC_rp^rq^{n-r} = (p + q)^n = 1, \quad \text{as } p + q = 1.$$

Consider the expansion of $(pt + q)^n$ where t is a parameter:

$$(pt + q)^n = \Sigma_n C_r (pt)^r q^{n-r}.$$

Differentiating both sides of this equation with respect to t gives

$$np(pt + q)^{n-1} = \Sigma_n C_r p^r t^{r-1} q^{n-r} r \qquad (3)$$

Putting $t = 1$, gives

$$np = \Sigma_n C_r p^r q^{n-r} r$$

$$= \Sigma P_r r = \text{mean}.$$

$\therefore$ mean $= np$. $\qquad (4)$

Variance of a binomial distribution. The general formula for the variance

$$\frac{\Sigma f_i x_i^2}{\Sigma f_i} - \bar{x}^2,$$

becomes, for the binomial distribution

$$\Sigma P_r r^2 - (np)^2. \qquad (5)$$

To find $\Sigma P_r r^2$ in terms of n, p and q, multiply equation (3) by t, giving

$$npt(pt + q)^{n-1} = \Sigma_n C_r p^r t^r q^{n-r} r.$$

Differentiating with respect to t gives

$$np(pt + q)^{n-1} + n(n - 1)p^2 t(pt + q)^{n-2} = \Sigma_n C_r p^r t^{r-1} q^{n-r} r^2.$$

Putting $t = 1$, gives

$$np + n(n - 1)p^2 = \Sigma_n C_r p^r q^{n-r} r^2 = \Sigma P_r r^2.$$

Substituting for $\Sigma P_r r^2$ in equation (5) gives the variance as

$$np + n^2 p^2 - np^2 - (np)^2 = np(1 - p).$$

$\therefore$ variance $= npq$. $\qquad (6)$

Example 3

Find the mean and variance for the distribution in Example 1.

In Example 1, $n = 3$, $p = \frac{1}{6}$, $q = \frac{5}{6}$.

Therefore the mean, $np = 3 \cdot \frac{1}{6} = \frac{1}{2}$, and the variance, $npq = 3 \cdot \frac{1}{6} \cdot \frac{5}{6} = \frac{5}{12}$.

Example 4

From a large batch of articles, a sample of 10 is chosen at random. If these 10 are acceptable the batch is accepted. If two or more articles are defective the batch is rejected. If just one is defective, another sample of 10 is taken. The batch is then accepted if all 10 are acceptable, otherwise it is rejected. If 12% of the batch are defective, what is the probability that the batch is: (i) accepted; (ii) rejected after taking one sample.

Let the probability of choosing a defective when one article is taken be $p = 0.12$. Then $q = 1 - p = 0.88$.

Consider the first sample of 10 articles.

The probability of no defectives is $_{10}C_0(0.12)^0(0.88)^{10} = 0.279$, and the probability of one defective is $_{10}C_1(0.12)^1(0.88)^9 = 0.380$.

Then the probability of two or more defectives is

$$1 - (0.279 + 0.380) = 0.341.$$

Therefore the probability of acceptance after taking one sample is 0.279, and the probability that a second sample has to be taken is 0.380.

The probability that this second sample is acceptable is 0.279.

The probability that the batch is accepted is therefore

$$0.279 + 0.380 \times 0.279 = 0.385.$$

The probability that the batch is rejected after taking the first sample is 0.341.

2. Poisson distribution

The Poisson distribution arises from the binomial distribution when the number of trials, n, approaches infinity, the probability of success, p, approaches zero, but the mean np remains finite.

Consider the distribution of the number of times during the period of one day that an ambulance station has to send out an ambulance. One day there may be ten calls for ambulances, another day only three. Now the total *possible* number of ambulances required in any one day could be infinite, although the *actual* number required will be finite. It is, in fact, impossible to give values to n, p and q. The only information available is a count of the number of times the event occurs within a given interval. This is an example of the Poisson distribution.

To deduce the formula which describes the Poisson distribution, consider the probability P_r of r successes in n trials as given by the binomial distribution, i.e.

$$P_r = {_nC_r}p^r q^{n-r}$$

$$= \frac{n!}{(n-r)!r!}p^r q^{n-r}$$

$$= \frac{n(n-1)(n-2)\cdots(n-r+1)}{r!}p^r(1-p)^{n-r}$$

$$= \frac{1\left(1-\dfrac{1}{n}\right)\left(1-\dfrac{2}{n}\right)\cdots\left(1-\dfrac{r-1}{n}\right)}{r!}(np)^r(1-p)^n(1-p)^{-r}.$$

Now let $n \to \infty$, $p \to 0$, $\lim_{n\to\infty,\ p\to 0}np = \lambda$(finite).

In the limiting case,

$$1\left(1 - \frac{1}{n}\right)\left(1 - \frac{2}{n}\right)\cdots\left(1 - \frac{r-1}{n}\right) = 1,$$

$$(1 - p)^{-r} = 1, \quad \text{as } r \text{ is finite,}$$

$$(np)^r = \lambda^r,$$

and

$$(1 - p)^n = \left(1 - \frac{\lambda}{n}\right)^n = e^{-\lambda}$$

(See Chapter 1). Thus

$$P_r = \frac{\lambda^r e^{-\lambda}}{r!}. \tag{7}$$

The value of P_r calculated from equation (7) gives therefore the probability of the event occurring r times if the results follow a Poisson distribution.

Note that P_r is entirely dependent on λ. For example if $\lambda = 2$,

$$P_0 = e^{-2} = 0.135, \quad P_4 = \frac{2^4 e^{-2}}{4!} = 0.90, \text{ etc.}$$

The mean and variance of a binomial distribution were found to be np and npq respectively. As $n \to \infty, p \to 0$,

$$np \to \lambda \quad \text{and} \quad npq = np(1 - p) \to \lambda.$$

Thus, for a Poisson distribution,

$$\text{mean} = \text{variance} = \lambda. \tag{8}$$

The expansion of e^λ as a power series gives

$$e^\lambda = 1 + \lambda + \frac{\lambda^2}{2!} + \frac{\lambda^3}{3!} + \cdots + \frac{\lambda^r}{r!} + \cdots.$$

Multiplying throughout by $e^{-\lambda}$ gives

$$e^\lambda e^{-\lambda} = e^{-\lambda} + \frac{\lambda^2 e^{-\lambda}}{2!} + \frac{\lambda^3 e^{-\lambda}}{3!} + \cdots \frac{\lambda^r e^{-\lambda}}{r!} + \cdots$$

i.e.

$$1 = P_0 + P_1 + P_2 + P_3 + \cdots + P_r + \cdots.$$

This verifies that the sum of all the probabilities is equal to 1 as would be expected.

For the purpose of calculating successive probabilities, it is worth noting that

$$P_r = \frac{\lambda}{r} P_{r-1},$$
(9)

so that each probability can be calculated from the preceding one.

Example 5

The average sales of a certain product are 1.5 per day. Assuming that the distribution of the number of sales each day follows the Poisson law, find the first seven probabilities and represent them in a diagram.

With $\lambda = 1.5$, the probability of no sales is $P_0 = e^{-1 \cdot 5} = 0.223$.

Using equation (9), $P_1, P_2, P_3, P_4, P_5, P_6$ and P_7 are 0.335, 0.251, 0.126, 0.047, 0.014, 0.003, 0.001 respectively.

These are represented in the histogram shown in Fig. 15.2.

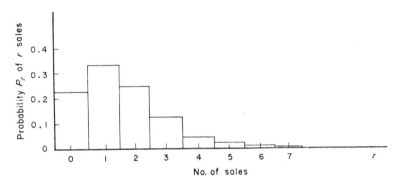

Fig. 15.2.

The Poisson distribution is obtained if events occur at random with the probability of an event occurring in a given interval of time remaining constant over the period under consideration. If, for example a product is as likely to be sold on any one day as another, then the distribution of sales per day will be a Poisson distribution with λ equal to the average number of sales per day. In practice, observed sales would consist of a sample from the basic population which is Poisson.

Example 6

On average, a pipeline develops 5.6 faults per week, and faults occur at random.

(i) What proportion of days may be expected to be trouble free?

(ii) In how many days of a year may three or more faults be expected?

The mean numbers of faults per day are

$$\frac{5.6}{7} = 0.8 = \lambda.$$

From equation (7),

$$P_r = \frac{\lambda^r e^{-\lambda}}{r!},$$

the probability of no faults in one day is $P_0 = e^{-0.8} = 0.449$, i.e. the percentage of trouble-free days is 44.9%.

From equation (9),

$$P_r = \frac{\lambda}{r} P_{r-1},$$

the probability of one fault in one day is

$$P_1 = \frac{0.8 \times 0.449}{1} = 0.359$$

and the probability of two faults in one day is

$$P_2 = \frac{0.8 \times 0.359}{2} = 0.144.$$

Then the probability of three or more faults is

$$P_3 + P_4 + P_5 + \cdots = 1 - (P_0 + P_1 + P_2)$$
$$= 0.048.$$

i.e. three or more faults may be expected on 4.8% of the days.

In one year, this is $0.048 \times 365 = 17.52$ days.

Example 7

Breakdowns of a machine occur at random with a mean number of three per week. What is probability of an interval of:

    (i) 2 days or more;

    (ii) between 2 and 3 weeks;

elapsing between 2 consecutive failures of the machine, assuming that it is in use for 5 days each week?

    (i) In order to tackle the problem, let us suppose that a break-down has just occurred. Then the question asks for the probability that no other break-downs occur within the next two days.

The mean number of break-downs for an interval of two days is $3 \cdot 2/5 = 1.2$.

The probability of no break-downs is $P_0 = e^{-1.2} = 0.301$.

    (ii) Again suppose that a break-down has just occurred. The question now asks for the probability that no break-downs occur for two weeks, but at least one occurs during the third week.

Now the mean λ for two weeks is 6, so that the probability of no break-downs for two weeks is $P_0 = e^{-6} = 0.0025$.

For one week $\lambda = 3$, and the probability of one or more break-downs during 1 week is $1 - P_0 = 1 - e^{-3} = 1 - 0.0498 = 0.9502$.

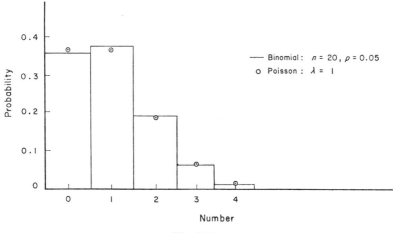

Fig. 15.3.

The required probability is therefore 0.0025 x 0.9502 = 0.00238.

The Poisson distribution as an approximation of the binomial distribution. The Poisson distribution arises from the binomial distribution as $n \to \infty$ and $p \to 0$. In fact, if p is small, whilst n is fairly large, the Poisson distribution may be used as an approximation of the binomial distribution. Fig. 15.3 compares the two for $p = 0.05$, and a distribution of size 20.

Comparison of data with a theoretical Poisson distribution. Sometimes considerations of the conditions under which data are obtained lead one to suspect that the data should follow a certain distribution. Or possibly, little may be known about the test conditions, but investigation of the data may suggest that it follows such a distribution. In either case, it may be advantageous to compare the data with a theoretical set of similar data that follow completely the law of the distribution. Naturally one does not expect the results to tally exactly, as this would be contrary to probability principles. Just as it is unlikely that in 100 spins of a coin exactly 50 heads are obtained, so it is unlikely that the two sets of figures would agree perfectly. However they may be alike, thus indicating that the data do follow that particular distribution.

Consider the following data, giving the number of faults in equal lengths of plastic piping. For 70 lengths of piping, the distribution was:

No. of faults (x)	0	1	2	3	4	5	6	
No. of pipes (f)	14	22	17	9	5	2	1	Total = 70

These figures appear to follow a Poisson distribution. The mean λ is estimated as

$$\frac{\Sigma f_i x_i}{\Sigma f_i} = \frac{119}{70} = 1.70.$$

The theoretical distribution with the same mean may be calculated from the equations

$$P_r = \frac{\lambda^r e^{-\lambda}}{r!} \quad \text{and} \quad P_r = \frac{\lambda}{r} P_{r-1}$$

giving P_0, P_1, P_2, etc. as 0.183, 0.311, 0.264, 0.150, 0.064, 0.022, 0.006, 0.001 respectively.

The theoretical results are then found by multiplying the respective probabilities by the total frequency 70.

Thus $P_0 \times 70 = 0.183 \times 70 = 12.8$ etc.

The complete results are:

x	0	1	2	3	4	5	6	7
observed	14	22	17	9	5	2	1	0
theoretical	12.8	21.7	18.5	10.5	4.5	1.5	0.4	0.1

The two sets of results are very similar, and it would seem reasonable to suppose that the data do arise from a Poisson distribution. A method of comparing the observed results against the theoretical ones is given in Chapter 17 (page 392).

Similar methods can also be used for other types of distributions.

3. Normal distribution

The binomial and Poisson distributions apply to discrete data. The normal distribution, or Gaussian distribution as it is sometimes called, applies to continuous data. It is symetrical about the mean μ, and always has a characteristic bell-shape, as shown in Fig. 15.4.

The line giving the shape of the distribution, the *normal curve*, has the equation

$$y = \frac{1}{\sigma\sqrt{(2\pi)}} \exp\left(-\frac{(x-\mu)^2}{2\sigma^2}\right), \tag{10}$$

with μ = mean, σ = standard deviation.

The distribution is therefore completely described by the mean and standard

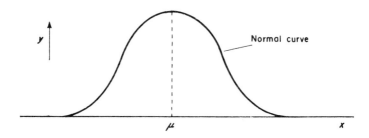

Fig. 15.4. The normal distribution.

deviation. Note that the normal curve theoretically touches the x-axis at $\pm\infty$, i.e. the variate x ranges from $\pm\infty$.

y now represents the *probability density* and is such that the total area under the curve is equal to 1, i.e.

$$\int_{-\infty}^{\infty} y \, dx = 1.$$

As explained in Chapter 14 (page 316), the probability of an observation x having a value between x_1 and x_2 is given by the area under the curve between x_1 and x_2, i.e.

$$P(x_1 \leqslant x \leqslant x_2) = \int_{x_1}^{x_2} y \, dx$$

$$= \frac{1}{\sigma\sqrt{(2\pi)}} \int_{x_1}^{x_2} \exp\left(-\frac{(x-\mu)^2}{2\sigma^2}\right) dx. \tag{11}$$

To calculate this probability, therefore, the integral in equation (11) has to be evaluated. If this is carried out for certain values of x_1 and x_2, the following properties of the distribution may be deduced:

approximately 68% of the distribution lies within $\mu \pm \sigma$,
approximately 95% of the distribution lies within $\mu \pm 2\sigma$, and
approximately 99.7% of the distribution lies within $\mu \pm 3\sigma$.

These results are shown in Fig. 15.5.

Thus, although the distribution theoretically ranges from $-\infty$ to $+\infty$, it is almost entirely confined within the limits $\mu \pm 3\sigma$. Only 0.3% lies outside these limits, and

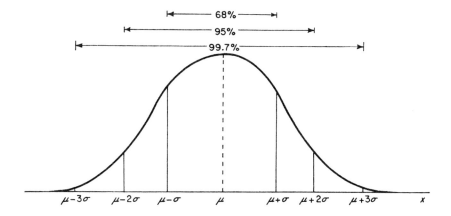

Fig. 15.5. The normal distribution related to μ and σ.

only 0.006% outside $\mu \pm 4\sigma$. For all practical purposes therefore, the distribution is confined to a fairly narrow range.

Standard normal distribution.   Normal distributions will of course differ for different means(μ) and standard deviations(σ). However, it is found convenient to make the substitution

$$u = \frac{x - \mu}{\sigma},$$

which has the effect of transforming all normal distributions into a single distribution, called the *standard normal distribution*.

Equation (11) then becomes

$$P(u_1 \leqslant u \leqslant u_2) = \frac{1}{\sqrt{(2\pi)}} \int_{u_1}^{u_2} \exp\left(-\frac{u^2}{2}\right) du, \quad \text{as } du = dx/\sigma. \tag{12}$$

This is the probability that u, chosen at random, has a value between u_1 and u_2.

It can be seen from equation (12) that the *standard normal curve* has the equation

$$y = \frac{1}{\sqrt{(2\pi)}} \exp\left(\frac{-u^2}{2}\right), \tag{13}$$

also that the mean of this distribution is zero and the standard deviation is unity. It follows from properties of the normal distribution that

approximately 68% of the distribution lies within $u = \pm 1$,

approximately 95% of the distribution lies within $u = \pm 2$, and

approximately 99.7% of the distribution lies within $u = \pm 3$.

These results are shown in Fig. 15.6.

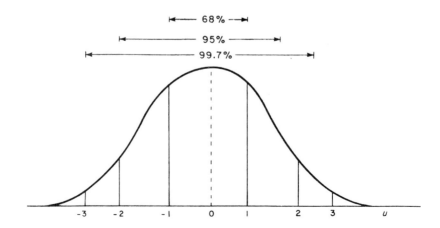

Fig. 15.6. The standard normal distribution.

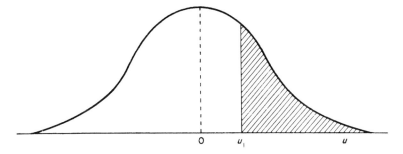

Fig. 15.7.

Evaluation of the integral

$$\frac{1}{\sqrt{(2\pi)}} \int_{u_1}^{\infty} \exp\left(\frac{-u^2}{2}\right) du$$

gives the probability of obtaining $u > u_1$. The results for u_1 from 0 to 5 are given in Table 1 on page 472. All required probabilities within the range $u = \pm 5$ can be found from this table. To see how this is done note that the tabulated value gives the proportion of the distribution which lies to the right of $u = u_1$ (shaded in Fig. 15.7).

Referring to Table 1, for $u_1 = 1.54$, the proportion to the right of u_1 is 0.0618, for $u_1 = 4.8$, the proportion is 7.93×10^{-7}, etc. Because the distribution is symmetrical, the proportion of the distribution lying to the left of $u_1 = -1.54$ is also 0.0618 or 6.18% (see Fig. 15.8.).

It is fairly easy, by simple mathematical calculations to find the proportion of the distribution between any two values of u. As the total area under the standard normal curve is equal to unity, this proportion gives the probability of obtaining a result, chosen at random, between the two values of u. The procedure, then, for dealing with most problems, is to transform into standard normal form and use Table 1 to find the required quantity.

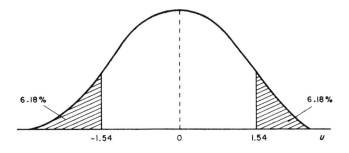

6.18%   6.18%

-1.54   0   1.54 u

Fig. 15.8.

Example 8

What proportion of the standard normal distribution lies between
     (i) $u = 1.32$ and 2.32.
     (ii) $u = -1.35$ and 0.85?
  (i)   The proportion required is shaded in Fig. 15.9.

From Table 1, the proportion to the right of $u = 2.32$ is 0.0102, and the proportion to the right of $u = 1.32$ is 0.0934.

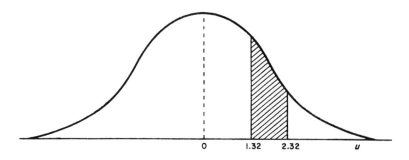

Fig. 15.9.

The required proportion is therefore

$$0.0934 - 0.0102 = 0.0832.$$

(ii) The proportion required is shaded in Fig. 15.10.

From Table 1, the proportion to the right of $u = 0.85$ is 0.1977, and the proportion to the left of $u = -1.35$ is 0.0885.

The required proportion is thus

$$1 - 0.1977 - 0.0885 = 0.7138.$$

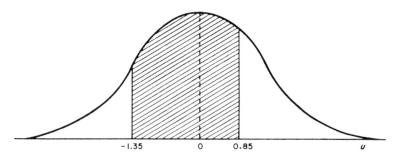

Fig. 15.10.

Example 9

The weights of cartons of soap have a mean value of 4 kg and a standard deviation of 0.2 kg. Assuming that the weights are normally distributed and a

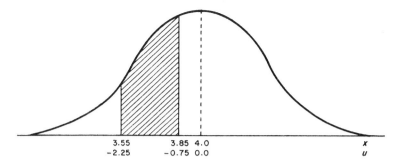

Fig. 15.11.

carton is chosen at random, what is the probability that it
    (i) has a weight between 3.55 and 3.85 kg,
    (ii) differs from the average by less than 0.35 kg?
    If 1000 cartons of soap are taken, how many may be expected to weigh less than 3.7 kg?
    We have $\mu = 4.0$ and $\sigma = 0.2$.
    (i) The required probability is given by the proportion of the normal distribution lying between 3.55 and 3.85 (shaded in Fig. 15.11).
    Transforming into standard normal form by the equation

$$u = \frac{x - \mu}{\sigma},$$

at $x = 3.85$,

$$u = \frac{3.85 - 4.0}{0.2} = -0.75,$$

and at $x = 3.55$,

$$u = \frac{3.55 - 4.0}{0.2} = -2.25$$

From Table 1,

    the proportion to the left of $u = -0.75$ is 0.2266, and

    the proportion to the left of $u = -2.25$ is 0.0122.

    The required proportion is therefore

    $0.2266 - 0.0122 = 0.2144.$

    (ii) The required probability is given by the proportion of the normal distribution lying between 3.65 and 4.35 (shaded in Fig. 15.12).
    Now at

$$x = 4.35, \quad u = \frac{4.35 - 4.0}{0.2} = 1.75,$$

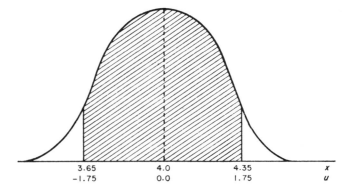

Fig. 15.12.

and at

$$x = 3.65, \quad u = \frac{3.65 - 4.0}{0.2} = -1.75.$$

From Table 1, the proportion to the right of $u = 1.75$ is 0.0401.
    By symmetry, this is also the proportion to the left of $u = -1.75$.
    The proportion required, therefore, is

$$1 - 2(0.0401) = 0.9198.$$

    To find the number of cartons expected to weigh less than 3.7 kg, consider the proportion of the normal distribution less than 3.7 (shaded in Fig. 15.13).

    At $x = 3.7$,

$$u = \frac{3.7 - 4.0}{0.2} = -1.5.$$

From Table 1, the proportion to the left of $u = -1.5$ is 0.0668.
In 1000 cartons, the expected number in this region is $1000 \times 0.0668 = 66.8$.

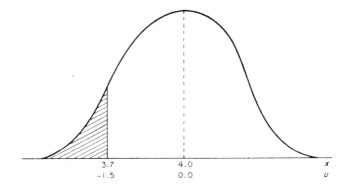

Fig. 15.13.

Example 10

A sulphuric acid plant makes acid with a mean concentration of 60%.
Assuming a normal distribution, what is the maximum value of the standard
deviation to ensure that 99% or more of the acid has a concentration between
56% and 64%?

Let x be the percentage of concentration.
By symmetry the proportion of the distribution greater than $x = 64$ is 0.005.
When $x = 64$,

$$u = \frac{x - \mu}{\sigma} = \frac{64 - 60}{\sigma} = 2.575$$

from Table 1, giving a standard deviation $\sigma = 0.64$ to two decimal places.

Uses of the normal distribution. Many distributions which arise in science and
engineering are normal or almost normal. Distributions ranging from heights or
weights of animals to dimensions of articles produced by an automatic process often
closely follow the law. So also, do human and other errors in physical and chemical
readings. Data from such distributions can therefore be tackled using normal distribution theory. Even when the distribution is not normal, sometimes a simple transformation, such as taking logs of the data, or square roots of the data, will produce
a normal distribution. Another important use is an approximation of the binomial
and Poisson distributions under certain conditions. This is discussed in the next
section.

The normal distribution has, therefore, immense importance in Statistics and as
we shall see later, this is especially so in the theory of sampling.

*The normal distribution as an approximation of the binomial and Poisson distri-
butions.* The normal distribution may be derived from the binomial distribution
under certain conditions, and so in some cases it may be used as an approximation
of the binomial distribution. In general the binomial distribution for a given p,
approaches the normal, with mean np and variance npq, as n increases. This is
especially so the nearer p and q are to 0.5, i.e. for fairly symmetrical distributions.

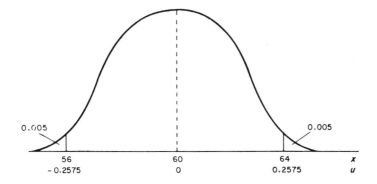

Fig. 15.14.

However, provided that n is fairly large, the approximation is good for other values of p and q.

The two distributions have the following probabilities for $n = 20, p = q = 0.5$:

Number	3	4	5	6	7	8	9	10
Binomial	0.001	0.005	0.015	0.037	0.074	0.120	0.160	0.176
Normal	0.001	0.005	0.015	0.037	0.073	0.119	0.160	0.177

Number	11	12	13	14	15	16	17
Binomial	0.160	0.120	0.074	0.037	0.015	0.005	0.001
Normal	0.160	0.119	0.073	0.037	0.015	0.005	0.001

The two sets of results are in very close agreement.

For $n = 20, p = 0.2, q = 0.8$, the probabilities are:

Number	0	1	2	3	4	5	6	7
Binomial	0.012	0.058	0.137	0.205	0.218	0.175	0.109	0.055
Normal	0.019	0.056	0.120	0.189	0.220	0.189	0.120	0.056

Number	8	9	10
Binomial	0.022	0.007	0.002
Normal	0.019	0.005	0.001

The results are still fairly close although n is quite small.

It follows that the Poisson distribution also approaches the normal as its mean λ increases, so that for relatively large values of the mean, the Normal distribution may also be used as an approximation of the Poisson.

Example 11

Requests for a component kept in a store total 1200 in a year. If the store stocks up at the beginning of each month, and requests are random, what number of articles should be stocked so that the chance of running out of the components in a month is less than $\frac{1}{10}$?

The probabilities follow a binomial distribution.

As requests occur at random, considering any one month, $p = \frac{1}{12}$, while $n = 1200$. The mean, $np = 100$, and $q = 1 - p = \frac{11}{12}$.

It will be necessary to use the normal as an approximation of the binomial, where the distribution has a mean $\mu = 100$, and standard deviation $\sigma = \sqrt{(npq)} = 9.58$.

Referring to Fig. 15.15, the value of x required is that which makes the shaded portion $\frac{1}{10}$ of the whole distribution.

From Table 1, the value of u with $\frac{1}{10}$ of the distribution to the right is 1.28.

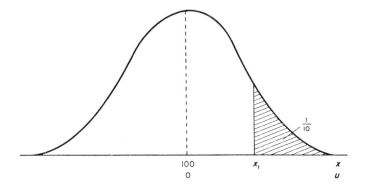

Fig. 15.15.

Now

$$u = \frac{x - \mu}{\sigma}.$$

$$\therefore 1.28 = \frac{x_1 - 100}{9.58},$$

giving

$$x_1 = 112.3$$

The number of articles to be stocked each month should therefore be 113.

PROBLEMS

1. A coin is tossed six times. What is the probability of obtaining
(i) exactly three heads,
(ii) more than one head?

2. In a large number of experiments a poison was found to kill 80% of rats on which it was tried. If it is used on a group of 5 rats, what are the probabilities of 0, 1, 2, 3, 4, 5 being killed? Draw a histogram of the resultant probabilities.

3. A process for making rubber valves results in 40% defectives. If a sample of 10 is taken at random, what are the expected number of faulty valves? What is the probability thst the number obtained will differ from this by only 1?

4. Instruments which are cleaned in batches of 10, are found, on average, to have 1.5 instruments which fail to be cleaned satisfactorily. In 100 batches, how many would be expected.
(i) to be completely clean,
(ii) to contain two or more unclean instruments?

5. Given a binomial distribution. What is the value of p for the probability of 4 successes in 10 trials to be twice the probability of 2 successes in 10 trials?

6. A survey of corrosion in water pumps revealed that 1% failed during a certain interval of time. A number of such pumps are to be bought, but with the stipulation that the probability that they should all work satisfactorily is greater than 0.9. What is the maximum number which can be bought?

7. The probability that an engine runs non-stop for 8 hours using fuel A is $\frac{2}{3}$, whilst with fuel B it is $\frac{3}{4}$. If 6 such engines are run for 8 hours, 3 on each fuel, what is the probability that more of those fail using fuel A than those using fuel B?

8. The average number of bacteria in a given area when viewed through a microscope is found to be 2.5. If a sample area is chosen at random, what is the probability that it contains 0, 1, 2, 3, 4, 5 bacteria? Draw a histogram of the probabilities.

9. The mean number of break-downs in a plant manufacturing sulphuric acid is 3.5 per week (7 days). If breakdowns occur at random, what is the probability of no breakdowns on a particular day?
In how many days of the year (365 days) may two or more breakdowns be expected?

10. A laboratory has 2 weighing machines. These can be used by any of the staff, but have to be booked for one hour. The average number of times one is required in an hour is 0.8. Assuming that needs occur at random, in how many hours per 40 hour week would they both be expected to be in use, and in how many would neither be used?

11. Apparatus to supply hydrogen sulphide, requires attention at random and on average 1.2 times per day. In a working week of 5 days, what is the probability that the apparatus first requires attention on the third day? What is the probability that the apparatus works satisfactorily for the whole week?

12. Determine the mean λ of a Poisson distribution with 10% of the frequencies having the value $x = 0$.

13. Show that the mean is approximately equal to the variance in the following distribution.
Assuming a Poisson distribution with the same mean, calculate the expected frequencies of plastic sheets with a given number of imperfections.

No. of imperfections per plastic sheet	0	1	2	3	4	5	6
No. of plastic sheets	19	27	32	12	6	3	1

14. The number of alpha particles emitted/unit time by a radio-active substance were counted for several periods of time. The following table records the frequencies (f) on which r alpha particles were emitted:

r	0	1	2	3	4	5	6	7	8	9
f	92	208	239	146	72	28	10	4	1	0

Calculate the mean number of alpha particles emitted/unit time and find the frequencies of the Poisson distribution having the same mean.

15. In a large batch of components 1% are defective. From this batch, 5 cases each containing 120 components are selected at random, and sent to a buyer. How many of these cases would be expected to have
(i) no defective components,
(ii) just 2 defective components?
What is the probability that at least one case contains 3 or more defectives?

16. A normal distribution has a mean of 80 and a standard deviation of 5. If a value is taken at random, what is the probability that it is
(i) greater than 90,
(ii) less than 74.8,
(iii) within 14.7 of 80,
(iv) between 73.6 and 83.6,
(v) between 64 and 74,
(vi) greater than 78.8?
Between what values does the middle 50% of the distribution lie?

17. The percentage of chlorine present in tablets produced and sold in boxes of 200 follows a normal distribution. The average amount of chlorine present is 30% and the standard deviation is 8%. In a box of 200, how many tablets may be expected to
(i) contain less than 35.52%,
(ii) contain between 24.8% and 47.2%,
(iii) contain between 22% and 28%,
(iv) differ from the mean by more than 8%?

18. Oxygen is prepared and stored in cylinders for later use. The average time before a cylinder is used is 10 days, with a standard deviation of $2\frac{1}{2}$ days. If the storage time follows a normal distribution, find the time
(i) to use up 20% of a batch of cylinders,
(ii) to use up all but 5% of a batch of cylinders.

19. Bottles containing a liquid are filled to the required level and then sealed by an automatic process. Inspection of a large number of bottles reveal that errors from the mean value are normally distributed with a standard deviation of 0.4 cm^3. If the mean value can be adjusted, what is its minimum value, if it is necessary that only 0.5% of the bottles contain less than 199.00 cm^3?

20. 800 results for the density of a liquid obtained by students were normally distributed, 12 giving the density less than 0.830 g cm^{-3}, and 20 giving the density greater than 0.870 g cm^{-3}. What was the mean density?

21. Records for several years show that the probability of a person in a chemical industry requiring attention for acid burns in any one year is 0.001. If the firm has 30,000 employees, what is the probability that 40 or more of these require attention in any given year?

CHAPTER 16

Quality Control

1. Distribution of sample means

If samples of size n are taken from an existing distribution, then the means of these samples will have a distribution which is dependent on the original one. The original distribution is called a *population*, its mean and standard deviation being denoted by μ and σ respectively. It can be shown (proof in appendix) that the distribution of means of samples also has a mean value of μ, but its standard deviation (usually called the *standard error of the mean*) is $\sigma/\sqrt{n}$, i.e.

$\text{S.E.}(\bar{x}) = \sigma/\sqrt{n}$. (assuming an infinite population)

Furthermore, if the original distribution is normal, then so is the distribution of means. In fact, whatever the form of the original distribution, for $n \geqslant 4$, the distribution of sample means can usually be assumed normal.

This information enables us to use samples to deduce certain facts about the distribution from which they are taken. Also as we see in this chapter, samples are sometimes used to control production processes.

Example 1

The lengths of items produced on a machine are distributed normally with mean 1.05 cm and a standard deviation of 0.008 cm. What is the probability that the mean of a sample of four items taken at random differs from the overall mean by more than 0.01 cm?

The means of samples follow a normal distribution with mean $\mu = 1.05$ cm and standard error,

$$\text{S.E.}(\bar{x}) = \frac{\sigma}{\sqrt{n}} = \frac{0.008}{\sqrt{4}} = 0.004 \text{ cm.}$$

The required probability is given by that part of the distribution less than 1.04 cm or greater than 1.06 cm, as shaded in Fig. 16.1.

Transforming into standard normal form, by the appropriate equation

$$u = \frac{\bar{x} - \mu}{\text{S.E.}(\bar{x})},$$

352

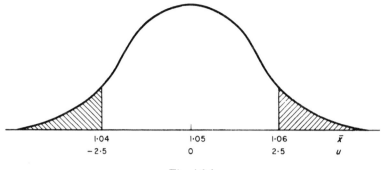

Fig. 16.1.

when $\bar{x} = 1.06$,

$$u = \frac{1.06 - 1.05}{0.004} = 2.5.$$

From symmetry, when $\bar{x} = 1.04$, $u = -2.5$, and from Table 1 on page 472, the shaded areas $= 2(0.00621) = 0.01242$.

Hence the required probability is 0.01242.

Example 2

A process produces items whose lengths are distributed normally with mean 4.50 units and a standard deviation of 0.05 units. Samples of 6 items are taken at random. Between what limits do 90% of the sample means lie? (See Fig. 16.2.)

The distribution of means of samples has a mean value $\mu = 4.50$ units and standard error,

$$\text{S.E.}(\bar{x}) = \frac{\sigma}{\sqrt{n}} = \frac{0.05}{\sqrt{6}} = 0.0204 \text{ units.}$$

From Table 1, the values of u cutting off 10% of the standard normal distribution are $u = \pm1.645$.

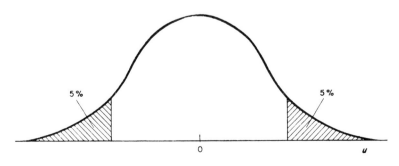

Fig. 16.2.

Now

$$u = \frac{\bar{x} - \mu}{\text{S.E.}(\bar{x})} \ ,$$

i.e.

$$\pm 1.645 = \frac{\bar{x} - 4.50}{0.0204} \ ,$$

giving $\bar{x} = 4.466$ and 4.534 units as the limiting values.

2. Measurement control charts

It is advisable and often necessary to keep some control on machines or production plants to ensure a satisfactory end product. For any production process, there will be variations due to many small factors, such as differences in materials, different operators of the machine, wear on parts of the machine, varying temperatures, and so on. Thus variations between products are inevitable.

It is, however, advisable to keep these differences to a minimum. Usually tolerances are placed on dimensions, and so long as the values stay within these tolerances, from a practical point of view, the product is satisfactory. Even if all the items being produced at any given time are well within the tolerances it is still worthwhile keeping a check so that any change can be noted almost immediately. In this way the machine may be corrected before items are made outside the tolerance limits.

The only way of ensuring that the machine behaves satisfactorily at all times is to measure every item, but in most cases this is impracticable as it would be both time consuming and costly. In most cases, however, the dimensions of the product are distributed normally. If, therefore, samples of several readings are taken at intervals, the means of these samples will also be distributed normally, and, providing that the mean (μ) and standard deviation (σ) of the original distribution are known, then the mean (μ) and standard deviation ($\sigma/\sqrt{n}$) of the distribution of means can be calculated. Fortunately, even if the original distribution is not normal, the means of the samples will be distributed nearly normally. It is possible, therefore, to calculate limits, within which, providing the process is steady, a certain proportion of the means will lie.

The limits which are generally adopted, having been found to work well in practice are

$$\mu \pm \frac{1.96\sigma}{\sqrt{n}} \qquad \text{called } warning \ limits$$

and

$$\mu \pm \frac{3.09\sigma}{\sqrt{n}} \qquad \text{called } action \ limits.$$

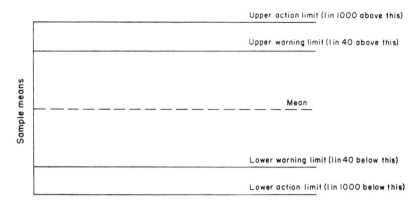

Fig. 16.3. The limits used on average control charts.

These contain 95% and 99.8% of the distribution respectively.

It follows that 2.5%, i.e. 1 in 40, of the sample means would be expected above the upper warning limit and 1 in 1000 above the upper action limit when the process is working normally. Similarly for the lower limits (see Fig. 16.3).

The names given to the limits serve to indicate the method in which a control chart is used. For example, if a sample mean falls outside the warning limits but inside the action limits, warning is given that the process may be going out of control. However if the next sample gives a good result it will no doubt be assumed that the previous reading was just one of the 20 which may be expected outside the limits. If, however, a value is obtained outside the action limits, then action will most likely be taken by obtaining another sample immediately and checking whether this shows the same trend. The process may even be stopped while this investigation is taking place. Obviously there are not strict rules to be applied, as consideration must be given to the position of preceding points.

By these methods, good control is maintained and any alteration in the setting of the machine can be corrected almost immediately.

Example 3

Pellets made by a certain process have a mean chlorine content of 30% with a standard deviation of 0.42%. An average control chart is to be used to control the chlorine content. Calculate control limits for samples of size 5 and prepare a control chart.

$$\text{The warning limits} = \mu \pm \frac{1.96\sigma}{\sqrt{n}}$$

$$= 30 \pm \frac{1.96 \cdot 0.42}{\sqrt{5}}$$

$$= 29.63\% \quad \text{and} \quad 30.37\%.$$

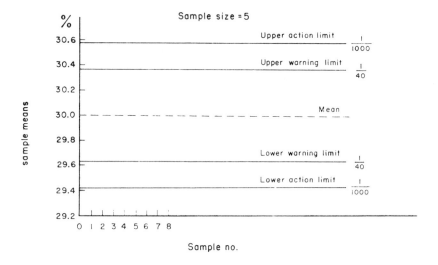

Fig. 16.4. The average control chart for chlorine content of pellets.

The action limits $= \mu \pm \dfrac{3.09\sigma}{\sqrt{n}}$

$$= 30 \pm \frac{3.09 \cdot 0.42}{\sqrt{5}}$$

$$= 29.42\% \quad \text{and} \quad 30.58\%.$$

The control chart is shown in Fig. 16.4.

Use of range. Before the limits for the chart can be evaluated, the mean and standard deviation of the measurements have to be calculated, and this would be necessary after each re-setting of a machine. A number of samples would be taken immediately and the limits calculated from these, at the same time making sure that the machine is within control, i.e. the results do appear to follow a normal distribution. It is found more convenient in practice to calculate the limits from the means and ranges of the samples, the range being converted to a standard deviation. Study of the relationship between the mean range of a number of samples and the standard deviation of the distribution shows that the best conversion is given by the equation

$$\bar{w} = d\sigma$$

where $\bar{w}$ is the mean range of the samples, σ is the standard deviation, and d, a constant depending only on the sample size, is given in the following table:

Sample size	2	3	4	5	6
d	1.13	1.69	2.06	2.33	2.53

It should be remembered that this does not give the exact value of σ, but only an estimate of it. However it is an adequate estimate for the purpose of control charts.

Note also that the mean value of the sample readings is an estimate of μ. It is usually given the name *grand average* and is denoted by $\bar{\bar{x}}$.

Example 4

Quality control is to be applied to the production of medicinal tablets by controlling the weight of the tablets. Ten samples, each containing four tablets were taken at random, with weights as follows.

	weight of tablets (grains)									
Sample No.	1	2	3	4	5	6	7	8	9	10
	3.5	3.1	3.5	3.1	3.5	3.3	3.8	3.3	3.1	3.2
	3.2	3.6	3.7	3.4	2.9	3.0	3.1	2.9	3.4	3.4
	3.3	3.3	3.8	3.6	3.2	3.1	3.7	3.7	3.6	2.8
	3.2	3.0	3.2	3.5	3.0	3.9	3.0	3.2	3.1	3.4

By calculating the means and ranges of the samples, find the action and warning limits for a quality control chart.

The mean weights and ranges for the samples are

Sample No.	1	2	3	4	5	6	7	8	9	10
Mean, $\bar{x}$	3.3	3.25	3.55	3.4	3.15	3.325	3.4	3.275	3.3	3.2
Range, w	0.3	0.6	0.6	0.5	0.6	0.9	0.8	0.8	0.5	0.6

The grand average

$$\bar{\bar{x}} = \frac{\Sigma \bar{x}}{10} = \frac{33.15}{10} = 3.315.$$

The mean range

$$\bar{w} = \frac{\Sigma w}{10} = \frac{6.2}{10} = 0.62.$$

For samples of size 4, $d = 2.06$. Thus

$$\sigma = \frac{\bar{w}}{d} = \frac{0.62}{2.06} = 0.30.$$

As $\bar{\bar{x}}$ is an estimate of μ, the warning limits are

$$\bar{\bar{x}} \pm \frac{1.96\sigma}{\sqrt{n}} = 3.315 \pm \frac{1.96 \cdot 0.30}{\sqrt{4}}$$

$$= 3.02 \quad \text{and} \quad 3.61.$$

The action limits are

$$\bar{\bar{x}} \pm \frac{3.09\sigma}{\sqrt{n}} = 3.315 \pm \frac{3.09 \cdot 0.30}{\sqrt{4}}$$

$$= 2.85 \quad \text{and} \quad 3.78.$$

Control limit factors. Writing $\bar{\bar{x}}$ for μ and $\bar{w}/d$ for σ, the warning limits $\mu \pm (1.96\sigma/\sqrt{n})$ become $\bar{\bar{x}} \pm (1.96\bar{w}/d\sqrt{n})$.

Similarly the action limits are $\bar{\bar{x}} \pm (3.09\bar{w}/d\sqrt{n})$.

The values $1.96/d\sqrt{n}$ and $3.09/d\sqrt{n}$ are tabulated in Table 2 on page 474 as A_1 and A_2 respectively, A_1 and A_2 being called *control limit factors*. The limits can therefore be calculated easily by substitution in the equation for the limits

$$\bar{\bar{x}} \pm A\bar{w}.$$

For example, if the grand average $\bar{\bar{x}} = 5.0$ and the mean range $\bar{w} = 2.0$, then for samples of size 6,

the upper action limit is $5.0 + 0.50 \cdot 2.0 = 6.00$,
the upper warning limit is $5.0 + 0.32 \cdot 2.0 = 5.64$,
the lower warning limit is $5.0 - 0.32 \cdot 2.0 = 4.36$, and
the lower action limit is $5.0 - 0.50 \cdot 2.0 = 4.00$.

Range control charts. The average control chart is used to indicate any change occurring in the setting of a machine or process. However, the setting may remain constant, whilst the number of items produced at some distance from the mean increases. That is, the spread of the distribution may change, whilst the mean remains unchanged. This is the sort of thing that occurs due to wear on parts of a machine. It will be some time before this change shows up on the average control chart. Thus in order to maintain complete control over the process it is essential:

(i) to control the setting by using average control charts,
(ii) to control the variability by using range control charts.

The ranges of the samples are plotted on range control charts which have warning and action limits ($\frac{1}{40}$ and $\frac{1}{1000}$ limits) as before.

The limits are given by $D\bar{w}$ where $\bar{w}$ is the mean sample range and D is obtained from Table 3 on page 474.

For example, if $\bar{w} = 2.0$ for samples of size 6, then

the upper action limit is $D_4\bar{w} = 2.22 \cdot 2.0 = 4.44$,
the upper warning limit is $D_3\bar{w} = 1.72 \cdot 2.0 = 3.44$,
the lower warning limit is $D_2\bar{w} = 0.42 \cdot 2.0 = 0.84$, and
the lower action limit is $D_1\bar{w} = 0.21 \cdot 2.0 = 0.42$.

The following example shows how the limits for average and range control charts may best be calculated from a number of samples.

Example 5

A machine was used to fill tins with a powder. The following samples of size 5 were obtained for weights of the powder in 60 of the tins.

Weights of powder (kilogram)											
Sample No. 1	2	3	4	5	6	7	8	9	10	11	12
6.15	5.98	6.09	5.92	5.91	6.13	5.89	6.06	5.86	6.13	5.95	6.00
5.97	6.12	6.08	5.81	6.03	5.94	6.13	6.04	6.02	5.92	6.06	5.86
6.10	6.01	5.89	5.99	6.17	6.01	5.94	6.01	6.10	5.98	6.05	5.92
6.00	6.08	6.02	6.05	5.96	6.03	5.93	5.97	6.02	6.00	5.98	6.04
5.87	6.03	5.94	6.02	5.96	6.09	6.20	6.05	6.08	6.17	6.02	5.99

Set up average and range control charts and plot the results from the 12 samples on the charts.

The mean weights and ranges for the samples are

Sample No.	1	2	3	4	5	6	7	8	9	10	11	12
Mean, $\bar{x}$	6.018	6.044	6.004	5.958	6.006	6.040	6.018	6.026	6.016	6.040	6.012	5.962
Range, w	0.28	0.14	0.20	0.24	0.26	0.19	0.31	0.09	0.24	0.25	0.11	0.18

The grand average

$$\bar{\bar{x}} = \frac{\Sigma \bar{x}}{12} = \frac{72.144}{12} = 6.0120.$$

The mean range

$$\bar{w} = \frac{\Sigma w}{12} = \frac{2.49}{12} = 0.2075.$$

The limits for the average control chart, using Table 2 on page 474 are,

warning limits, $\bar{\bar{x}} \pm A_1 \bar{w}$

$= 6.0120 \pm 0.38 \cdot 0.2075$

$= 5.933$ and 6.091,

action limits, $\bar{\bar{x}} \pm A_2 \bar{w}$

$= 6.0120 \pm 0.59 \cdot 0.2075$

$= 5.890$ and 6.134.

360

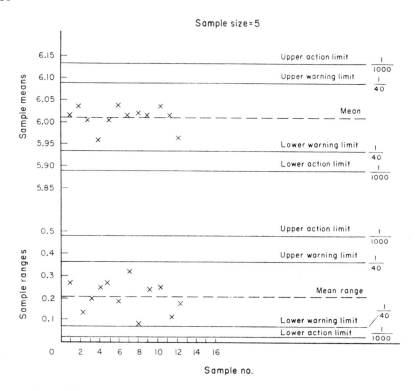

Fig. 16.5. Average and range control charts for weights of powder.

The limits for the range control chart, using Table 3 on page 474 are

the lower action limit $D_1 \bar{w} = 0.16 \cdot 0.2075 = 0.033$,
the lower warning limit $D_2 \bar{w} = 0.37 \cdot 0.2075 = 0.077$,
the upper warning limit $D_3 \bar{w} = 1.81 \cdot 0.2075 = 0.376$, and
the upper action limit $D_4 \bar{w} = 2.34 \cdot 0.2075 = 0.490$.

The average and range control charts may be drawn as shown in Fig. 16.5.

One reason for plotting the values for the 12 samples on the chart in Example 5, is to ensure that the machine is in statistical control. This is necessary, of course, if the charts are to be of any use. In this case all the points are well inside the warning limits so that the system would seem to be in control.

It may seem unnecessary to include lower limits on the range chart as values obtained outside these limits would not indicate a fault in the system, but in fact an improvement. They are usually included however for this very reason. If there is an improvement, due to some unknown cause, this will be shown on the chart, and having looked for the cause perhaps a permanent improvement can be obtained. Again, if a modification is made to the machine in order to improve the variability, then the chart will give a rough idea of whether the modification is successful.

It should be noted, that in practice the limits are calculated from about 20 samples of size 5, that is about 100 observations in all.

It must be remembered that these charts are to be used in order to give a good idea of how the machine or process is behaving. Any decisions that may be made about adjustment to the process will depend on these together with other considerations, such as tolerances. There is, of course, no relationship between tolerance limits and control chart limits. Tolerance limits state the requirements of the process, control chart limits give its capabilities. If the tolerances are small, it may be that any change in the setting etc. will require re-adjustment. If, however, the tolerances are much wider than the control limits, then a certain amount of deviation can be allowed without adjustment. Sometimes, in the latter case, modified control limits are used, which allow more deviation without re-assessment of limits or re-adjustment of the machine or process.

3. Number defective control charts

In the example considered in the previous section, the charts were used for a continuous variate. There are, however, many cases where it is practicable, and may only be possible, to say either that an article is satisfactory or that it is not satisfactory. This may occur, for example, when testing a valve to see that it operates correctly at a certain pressure, or when considering the finish on an article.

To control the process in these cases, samples are again taken at intervals, and the number of unsatisfactory articles or defectives in each sample are counted. Warning and action limits are determined as before where $\frac{1}{40}$ and $\frac{1}{1000}$ respectively of the points may be expected outside these limits.

To understand the basis of these charts, we note that after a fairly large number of articles have been taken it should be possible to say what proportion of articles are being produced as defectives while the process remains the same. We are therefore dealing with a binomial distribution in which values may be given to p and q. For example, if the process is producing 2% defectives, then $p = 0.02$ and $q = 0.98$. Now in most cases, p will be fairly small, while the sample size n is fairly large. Under these conditions the distribution approaches a Poisson, and assuming this is the case, the limits may be calculated more easily. Table 4 on page 475 gives the limits assuming a Poisson distribution.

If, for example, the numbers of defectives in 10 samples were 2, 4, 3, 0, 1, 1, 2, 0, 4, 1, the mean number of defectives per sample is $\frac{18}{10} = 1.8$. To find the upper action limit, refer to the appropriate column in Table 4. We see that, as 1.8 lies between the values 1.52 and 1.97 in the table, the action limit, given by the corresponding number in the control limit column is 8 defectives. The upper warning limit is similarly determined. Since 1.8 lies between 1.62 and 2.20, the upper warning limit is 6 defectives.

Note that no lower warning limit exists for a mean number of defectives less than 3.69, and no lower action limit for a mean less than 6.91.

The lower limits are included for the same reasons as in the range chart.

Although showing an improvement, if it is accidental, then investigation may lead to permanent improvement.

In the unusual event of having to deal with a mean greater than 8.96, the limits may be calculated assuming a normal distribution.

The limits are then given by $\lambda \pm 1.96\sqrt{\lambda}$ and $\lambda \pm 3.09\sqrt{\lambda}$.

Example 6

A new process is to be used in the manufacture of plastic beakers. Ten successive samples, each of size 80, contained the following number of beakers which were defective in some way.

Sample No.	1	2	3	4	5	6	7	8	9	10
No. of defectives	3	5	1	3	2	0	2	3	3	2

Use these results to set up a control chart and plot the ten results on the chart.

The mean number of defectives per sample is $\frac{24}{10} = 2.4$. From Table 4 on page 475, the upper action limit is 9, and the upper warning limit is 7. The control chart is drawn in Fig. 16.6.

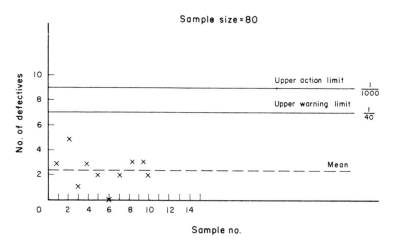

Fig. 16.6. Number defective control chart for plastic beakers.

Note that the sample size is not used to calculate the limits, but as the number of defectives obtained depends on the size of the sample, the limits do of course, depend on this too. The sample size is therefore noted on the chart to ensure that a constant size of sample is inspected. If it is impossible to use a constant sample size, then fraction defective charts can be used instead of number defective charts.

APPENDIX

Mean and variance of sample means

First we prove some general statistical theorems.

Consider a function z such that $z = x + y$ where x and y are independent variables.

Then the mean

$$\bar{z} = \frac{\Sigma f_i z_i}{\Sigma f_i}$$

$$= \frac{\Sigma f_i(x_i + y_i)}{\Sigma f_i}$$

$$= \frac{\Sigma f_i x_i}{\Sigma f_i} + \frac{\Sigma f_i y_i}{\Sigma f_i}$$

$$= \bar{x} + \bar{y}.$$

i.e.

$$\text{mean } z = \text{mean } x + \text{mean } y. \tag{1}$$

The variance of z, i.e.

$$\text{var}(z) = \frac{\Sigma f_i(z_i - \bar{z})^2}{\Sigma f_i}$$

$$= \frac{\Sigma f_i(x_i + y - \bar{x} - \bar{y})^2}{\Sigma f_i}$$

$$= \frac{\Sigma f_i[(x_i - \bar{x}) + (y_i - \bar{y})]^2}{\Sigma f_i}$$

$$= \frac{\Sigma f_i(x_i - \bar{x})^2 + \Sigma f_i(y_i - \bar{y})^2 + 2 \Sigma f_i(x_i - \bar{x})(y_i - \bar{y})}{\Sigma f_i}$$

Now $\Sigma f_i(x_i - \bar{x})(y_i - \bar{y}) = 0$ as x and y are independent. Thus

$$\text{var}(z) = \text{var}(x) + \text{var}(y). \tag{2}$$

Similarly, it can be shown that if $z = x - y$, then

$$\text{mean } z = \text{mean } x - \text{mean } y, \tag{3}$$

and

$$\text{var}(z) = \text{var}(x) + \text{var}(y). \tag{4}$$

Results (1), (2), (3), and (4) may be extended to several independent variables, $x_1, x_2, x_3, \ldots$, say.

If

$$z = x_1 \pm x_2 \pm x_3 \pm \ldots,$$

then

$$\bar{z} = \bar{x}_1 \pm \bar{x}_2 \pm \bar{x}_3 \pm \ldots, \tag{5}$$

and

$$\mathrm{var}(z) = \mathrm{var}(x_1) + \mathrm{var}(x_2) + \mathrm{var}(x_3) + \cdots. \tag{6}$$

If the functions z and x are such that

$$z = ax \quad (a \text{ constant})$$

then

$$\bar{z} = \frac{\Sigma f_i z_i}{\Sigma f_i} = \frac{\Sigma f_i a x_i}{\Sigma f_i} = a\bar{x}, \tag{7}$$

and

$$\mathrm{var}(z) = \frac{\Sigma f_i (z_i - \bar{z})^2}{\Sigma f_i} = \frac{\Sigma f_i a^2 (x_i - \bar{x})^2}{\Sigma f_i} = a^2 \, \mathrm{var}(z). \tag{8}$$

Consider the distribution of the means of samples of size n which are taken at random from a population which has a mean value μ and a standard deviation σ.

Suppose that a sample consists of $x_1, x_2, \ldots, x_n$ random independent values, then the sample mean $\bar{x}$ is given by

$$\bar{x} = \frac{x_1 + x_2 + \cdots + x_n}{n},$$

and from equation (5)

$$\mathrm{mean}\ \bar{x} = \frac{\mathrm{mean}\ x_1 + \mathrm{mean}\ x_2 + \cdots + \mathrm{mean}\ x_n}{n}$$

$$= \frac{\mu + \mu + \cdots + \mu}{n}$$

$$= \frac{n\mu}{n} = \mu.$$

i.e. the mean of the distribution of means is again μ.

From equation (6), the variance of $\bar{x}$ is given by

$$\mathrm{var}(\bar{x}) = \mathrm{var}\left(\frac{x_1}{n}\right) + \mathrm{var}\left(\frac{x_2}{n}\right) + \cdots + \mathrm{var}\left(\frac{x_n}{n}\right).$$

From equation (8), as n is constant,

$$\text{var}(\bar{x}) = \frac{1}{n^2}\,\text{var}(x_1) + \frac{1}{n^2}\,\text{var}(x_2) + \cdots + \frac{1}{n^2}\,\text{var}(x_n)$$

$$= \frac{1}{n^2}\,(\sigma^2 + \sigma^2 + \cdots + \sigma^2)$$

$$= \frac{\sigma^2}{n}\,.$$

i.e. the variance of the distribution of means is σ^2/n.

The relationship between sample and population variance estimates. We now prove a formula which is used in Chapter 17.

Consider a sample of n values with mean $\bar{x}$, taken at random from a population of size N and with mean μ.

Let s^2 denote the variance of the sample, and σ^2 denote the variance of the population.

Then by definition

$$s^2 = \frac{1}{n} \sum_{i=1}^{n} (x_i - \bar{x})^2$$

$$\sigma^2 = \frac{1}{N} \sum_{i=1}^{N} (x_i - \mu)^2.$$

In general, σ^2 must be estimated from the information obtained from the sample, $\bar{x}$ being the best estimate of μ.

The best estimate of σ^2 is given by

$$\frac{1}{n} \sum_{i=1}^{n} (x_i - \mu)^2.$$

Now

$$(x_i - \mu) = (x_i - \bar{x}) + (\bar{x} - \mu)$$

$$\therefore\ (x_i - \mu)^2 = (x_i - \bar{x})^2 + 2(x_i - \bar{x})(\bar{x} - \mu) + (\bar{x} - \mu)^2.$$

Summing over all values of x_i in a sample of size n

$$\sum_{i=1}^{n} (x_i - \mu)^2 = \sum_{i=1}^{n} (x_i - \bar{x})^2 + 2 \sum_{i=1}^{n} (x_i - \bar{x})(\bar{x} - \mu) + \sum_{i=1}^{n} (\bar{x} - \mu)^2,$$

and since for any given sample, $(\bar{x} - \mu)$ is a constant and

$$\sum_{i=1}^{n} (x_i - \bar{x}) = 0, \qquad \sum_{i=1}^{n} (x_i - \mu)^2 = \sum_{i=1}^{n} (x_i - \bar{x})^2 + \sum_{i=1}^{n} (\bar{x} - \mu)^2$$

$$= \sum_{i=1}^{n} (x_i - \bar{x})^2 + n(\bar{x} - \mu)^2.$$

If the calculations are repeated over a large number of samples, the mean value of $\Sigma_{i=1}^{n}(x_i - \mu)^2$ will tend, by definition, to $n\sigma^2$ and similarly the mean value of $n(\bar{x} - \mu)^2$ tends to n times the variance of $\bar{x}$, viz. $n(\sigma^2/n)$.

Thus we can write,

$$n\sigma^2 \rightarrow \sum_{i=1}^{n} (x_i - \bar{x})^2 + n\left(\frac{\sigma^2}{n}\right),$$

or in the limit

$$\sigma^2 = \frac{1}{n-1} \sum_{i=1}^{n} (x_i - \bar{x})^2.$$

This expression is exact only when the number of samples is infinite and the population is infinite, but in general the best estimate of σ^2, denoted by $\hat{\sigma}^2$, calculated from a sample will be given by

$$\hat{\sigma}^2 = \frac{1}{n-1} \sum_{i=1}^{n} (x_i - \bar{x})^2 = \frac{ns^2}{n-1}.$$

As

$$\Sigma (x_i - \bar{x})^2 = \Sigma (x_i^2 - 2x_i\bar{x} + \bar{x}^2)$$

$$= \Sigma x_i^2 - 2\bar{x} \Sigma x_i + \bar{x}^2 \Sigma 1$$

$$= \Sigma x_i^2 - 2(\Sigma x_i/n) \Sigma x_i + (\Sigma x_i/n)^2 n$$

$$= \Sigma x_i^2 - (\Sigma x_i)^2/n,$$

it is usually more convenient to calculate $\hat{\sigma}^2$ using the equation

$$\hat{\sigma}^2 = \frac{1}{n-1}\left(\sum_{i=1}^{n} x_i^2 - \frac{(\Sigma_{i=1}^{n} x_i)^2}{n} \right).$$

PROBLEMS

1. Articles are produced with a dimension having a mean value of 4.72 and a standard deviation of 0.24. What is the mean and the standard deviation of the distribution of means of samples of size 9?

2. Measurements follow a normal distribution with mean 58.65 units and standard deviation 2.42 units. If samples containing n measurements are taken, how large should n be to ensure that less than 2.5% of the means of these samples lie above 60.0 units?

3. A cement plant mixes cement by an automatic process. Quality control charts are to be used to control the compressive strength of the cement. If the plant is set up so that test cubes of the cement have a mean compressive strength of 90.5 kg

cm^{-2} and a standard deviation of 2.5 kg cm^{-2}, estimate warning and action limits for average and range quality control charts for samples of size 5, assuming results follow a normal distribution.

4. If, in the process explained in Question 3, the mean compressive strength of the cement changes to 92 kg cm^{-2}, whilst the standard deviation remains the same, what is the probability that the first sample mean is above the upper action limit?

5. Quality control charts are to be used in order to keep control on the octane number of a gasoline product. Fifteen samples of size 5 were taken as shown. Calculate action and warning limits for average and range control charts.

Sample No. 1	2	3	4	5	6	7	
82.5	88.2	89.4	87.1	87.5	85.4	87.6	
89.5	86.9	84.8	86.8	84.5	81.2	85.7	
87.1	87.4	83.7	84.6	89.3	88.3	90.1	
86.2	88.7	88.4	83.8	85.6	86.4	84.5	
83.0	83.9	84.0	87.9	85.6	86.4	85.2	

Sample No. 8	9	10	11	12	13	14	15
86.1	83.5	87.1	88.6	80.8	87.2	85.3	82.1
89.1	90.2	85.3	83.2	82.9	81.7	85.2	88.6
88.9	86.5	84.1	83.4	86.3	85.0	87.8	87.3
87.7	85.3	82.5	84.3	86.0	84.8	83.8	85.4
86.6	84.7	86.0	85.4	85.0	86.7	84.9	84.0

6. Calculate control limits for average and range control charts from the following data of the specific gravity of a powder. Draw the charts, plotting the ten values from the samples on each chart.

	Specific gravity of powder									
Sample No. 1	2	3	4	5	6	7	8	9	10
1.62	1.54	1.69	1.64	1.76	1.64	1.63	1.65	1.57	1.72
1.68	1.62	1.67	1.55	1.74	1.69	1.72	1.74	1.63	1.60
1.63	1.71	1.53	1.76	1.66	1.60	1.72	1.66	1.79	1.58
1.63	1.71	1.63	1.70	1.68	1.67	1.65	1.70	1.65	1.73
1.70	1.67	1.74	1.66	1.66	1.60	1.68	1.55	1.68	1.65
1.80	1.64	1.67	1.64	1.68	1.70	1.62	1.69	1.67	1.69

7. In the problem dealt with in Question 6, subsequent samples were obtained, giving the results below. Plot the sample means and ranges on the control charts of Question 6, and state the point at which you consider a change has occurred, and whether it is in mean or variability. Calculate new limits from the remaining samples.

Sample No.	11	12	13	14	15	16	17	18	19	20
				Specific gravity of powder						
	1.59	1.61	1.77	1.80	1.68	1.65	1.77	1.75	1.65	1.68
	1.65	1.65	1.73	1.71	1.71	1.69	1.72	1.74	1.70	1.71
	1.69	1.76	1.65	1.85	1.77	1.77	1.70	1.77	1.64	1.73
	1.61	1.72	1.82	1.65	1.63	1.62	1.73	1.74	1.68	1.78
	1.70	1.69	1.69	1.75	1.80	1.69	1.68	1.68	1.71	1.74
	1.64	1.68	1.70	1.73	1.67	1.70	1.70	1.85	1.78	1.76

8. If the mean range $\bar{w}$ of a number of samples is 2.5, what is the minimum sample size if an increase of 0.6 in the grand average $\bar{\bar{x}}$ results in over 2.5% of the sample means appearing above the upper action limit?

9. The number of accidents resulting in injuries were recorded for men working on a section of a plant. The results for 18 months were

Month	1	2	3	4	5	6	7	8	9
No. injured	7	5	6	9	6	3	8	11	7

Month	10	11	12	13	14	15	16	17	18
No. injured	9	3	15	4	6	9	7	12	8

Calculate warning and action limits for a number defective control chart.

10. Calculate limits for a control chart for the following number of defectives in ten samples of hydrometers, each of size 60.

Sample No.	1	2	3	4	5	6	7	8	9	10
No. of defectives	0	1	0	0	2	0	1	0	1	1

Draw a number defective control chart, plotting the sample values on the chart.

11. Subsequent samples for the system in Question 10 gave the following results:

Sample No.	11	12	13	14	15	16	17	18	19	20
No. of defectives	1	2	0	1	4	3	3	2	5	2

Plot these on the chart obtained previously. At what point do you consider a change has occurred in the process? Calculate new limits for the remaining data.

12. Suppose that for the system considered in Question 10, it was decided to double the sample size. What would the limits be, based on the results for the ten samples?

CHAPTER 17

Significance Tests

A general problem which exists in the chemical industry is the comparison of different products and different methods of analysis. An attempt to solve the problem will involve the comparison of data obtained from each product, or using each method. Some variability will be expected, even within data obtained under apparently identical conditions; this will be due to all the uncontrollable factors grouped under the heading of experimental error. Because of this inherent variability a parameter estimated from a random sample will not in general be identical with the true value for the population from which the sample was taken.

In order to make an exact comparison we require an infinite amount of data from each source and this of course is unobtainable. There will always be some limitations on the number of results. It may be that the test or form of analysis uses up material and there is a limit to the amount of material we can afford to waste. There will also be some limitation on the time which can be spent on the investigation. It is clear that the reliability of any parameters estimated will increase as the amount of data from which they are calculated increases, but there will always be the need for an economic balance between the precision required and the experimental facilities available.

In this chapter (excepting one type of test) we will consider comparisons between only two groups of data. What we aim to do, using statistical tests of significance, is to decide whether the groups (samples) have originated from the same, or from different sources. The statistical procedure puts a common sense assessment on a numerical basis.

Suppose, for example, it was thought that the use of a different catalyst would give an increased yield of product from a certain reaction. Some variation in yield from one reaction to another will be expected, without changing the type of catalyst. We have to decide how large the yield in an experiment, or small group of experiments, must be, before we can confidently assert that the use of the new type of catalyst increases yield. Without using statistical procedures, if the increase is very large we will make the assertion; if it is small we will check by further experiments. However, it is unlikely that everone would use the same dividing line for their decisions and it will be difficult to state just how certain one is that the correct conclusion has been drawn. Using statistical procedures we can draw conclusions and also state their reliability.

We first adopt a *null hypothesis* (H_0) i.e. make the assumption that the parameter being considered does not differ from some specified value, and then determine the probability of obtaining the observed (or more extreme) values, if this hypothesis is true.

Using the above example the null hypothesis would be, that changing the catalyst does not affect the yield. This implies that the two populations of yields coincide, so that yields obtained using the new catalyst are assumed to be a sample from the 'old' population. The position now, is that our sample has an average higher than that of the population from which it is assumed to have been drawn. We next calculate the probability of obtaining a sample as good, or better than this if the null hypothesis (H_0) is true.

Suppose this probability is very small, say $\frac{1}{1000}$. There are two alternative conclusions which can be drawn, either: (i) The $\frac{1}{1000}$ chance has occurred and H_0 is true; or (ii) H_0 is false.

If we conclude (ii) we know we have a $\frac{1}{1000}$ chance of being wrong.

The risk we are prepared to take of wrongly asserting a difference should be decided *before the experiment is carried out*. This is then called the *level of the significance test*, and is denoted by α. If the probability of obtaining the observed results, or better, when H_0 is true is less than or equal to α, we reject H_0 and assert a real difference.

If the probability of obtaining the observed results is high, say $\frac{2}{5}$, it means that results as good as, or better than this could be obtained with the first catalyst as often as 40% of the time, so H_0 has not been disproved.

The choice of significance level depends upon the nature and importance of the particular investigation being carried out, but the following table gives the generally accepted divisions:

Probability	Conclusion	Symbol
Greater than 0.1	NOT SIGNIFICANT, H_0 accepted	NS
Between 0.1 and 0.05	POSSIBLY SIGNIFICANT, some doubt is cast on H_0 but further evidence sought before rejection	PS
Between 0.05 and 0.01	SIGNIFICANT, H_0 rejected, if the result is very important confirm with further evidence	S
Between 0.01 and 0.001	HIGHLY SIGNIFICANT, H_0 confidently rejected	S*
Less than 0.001	VERY HIGHLY SIGNIFICANT, H_0 rejected and it is very unlikely that the conclusion is incorrect	S**

Summarizing, the purpose of statistical tests of significance is to separate differences which could not easily have occurred by chance from those which could. A difference which could not easily have occurred by chance is said to be statistically significant. To determine significance we first make an assumption of no difference and then calculate the probability that the observed difference could be equalled, or exceeded, by chance, if the null hypothesis is true. Having decided upon a level of significance we can then make the appropriate acceptance/rejection decision.

Before considering the different forms of significance test we must introduce the term *degrees of freedom*, which is used to express the reliability of any statistic calculated. Each parameter calculated from a group of data 'uses up' one degree of freedom. With n observations we have n degrees of freedom. If a total or mean value is calculated from these observations then one degree of freedom is lost, so that the variance will be based on $(n - 1)$ degrees of freedom.

If any statistic is calculated involving a variance estimate, the reliability of that statistic will depend upon the reliability of the variance and will be based on the same number of degrees of freedom.

The following table summarizes the statistical notation used:

Symbol	Interpretation of symbol
n	number of observations in sample
μ	population mean $= \dfrac{1}{N} \sum\limits_{i=1}^{N} x_i$
$\bar{x}$	sample mean $= \dfrac{1}{n} \sum\limits_{i=1}^{n} x_i$
σ^2	population variance $= \dfrac{1}{N} \sum\limits_{i=1}^{N} (x_i - \mu)^2$
$\hat{\sigma}^2$	the best estimate of population variance $= \dfrac{1}{n-1} \sum\limits_{i=1}^{n} (x_i - \bar{x})^2$ calculated from sample
ϕ	the number of degrees of freedom on which a given statistic is based.

1. Comparison of variability, F-test

It is known that if a given population variance is estimated independently from two random samples, drawn from that population, then the ratio of the two variance estimates will be distributed in a particular way, following what is known as the F-distribution. The F-test uses this distribution to compare estimates of population variance and is designed to test whether one population is more variable than another.

Suppose we have two variance estimates, $\hat{\sigma}_1^2$ and $\hat{\sigma}_2^2$ based on ϕ_1 and ϕ_2 degrees of freedom respectively. We wish to determine whether or not σ_1^2 is greater than σ_2^2.

Following the general procedure for significance tests, we first adopt a null hypothesis. This will be the assumption that the samples were drawn from populations with the same variance, i.e.

$$H_0 \equiv \sigma_1^2 = \sigma_2^2.$$

If $\hat{\sigma}_1^2 < \hat{\sigma}_2^2$ then clearly we cannot assert $\sigma_1^2 > \sigma_2^2$ and there is no need for a statistical test. Let us suppose $\hat{\sigma}_1^2 > \hat{\sigma}_2^2$. We then calculate

$$F = \frac{\text{larger variance estimate}}{\text{smaller variance estimate}}$$

$$= \frac{\hat{\sigma}_1^2}{\hat{\sigma}_2^2}.$$

If the whole population had been used to calculate each estimate, and H_0 was true, then F would be equal to unity. Since we always calculate F equal to the larger estimate over the smaller, F must always be equal to, or greater than unity. Obviously the greater the value of F, the less likely it is that H_0 is true, but the significance of the result cannot be assessed on the magnitude of F alone. We must take into account the reliability of the estimates, by considering the number of degrees of freedom upon which each is based.

Tables have been drawn up giving the value of F which will be exceeded with stated probability for various degrees of freedom (Table 5). We therefore determine, from the tables, the value of F, based on $\phi_1 = (n_1 - 1)$ and $\phi_2 = (n_2 - 1)$ degrees of freedom (ϕ_1 and ϕ_2 corresponding to $\hat{\sigma}_1^2$ and $\hat{\sigma}_2^2$ respectively) which will be exceeded with probability equal to our chosen level of significance. If the calculated value of F is greater than this, then the probability of H_0 being true is less than the risk we are prepared to take of wrongly rejecting it. We would assert therefore that the samples were not drawn from populations with equal variance and that $\sigma_1^2 > \sigma_2^2$.

Single-sided and double-sided tests. It is important to note that in the above example, the alternative to the null hypothesis is that $\sigma_1^2 > \sigma_2^2$, i.e.

$$H_0 \equiv \sigma_1^2 = \sigma_2^2$$
$$H_1 \equiv \sigma_1^2 > \sigma_2^2.$$

This is a *single-sided test.*

If, from technical knowledge, there is no indication, before the trials are carried out, which is likely to be the more variable source, then there are two possible alternatives to H_0: either $\sigma_1^2 > \sigma_2^2$ or $\sigma_1^2 < \sigma_2^2$. The test in this case would be for *any* difference in variability, rather than for a difference in a particular direction and we must make what is called a *double-sided test*. Thus for a double-sided test we have

$$H_0 \equiv \sigma_1^2 = \sigma_2^2$$
$$H_1 \equiv \sigma_1^2 \neq \sigma_2^2.$$

As noted at the beginning of the section the F-tables only give values of $F \geqslant 1$ and are designed for single-sided tests. To make a double-sided test we double the probability in the tables.

The degrees of freedom associated with an F value should always be quoted. $F(10, 24)$ indicates that the larger variance estimate is based on 10 degrees of freedom and the smaller estimate on 24.

When values are quoted from the tables a probability used as suffix denotes that value of the statistic which is exceeded with the stated probability, i.e. the proportion of the total area cut off by that value in one tail of the distribution.

Thus from Table 5

$$F_{0.05}(10, 24) = 2.25$$
$$F_{0.10}(10, 24) = 1.88.$$

Using a percentage as suffix indicates the appropriate level of significance for the particular test, e.g.

a single-sided test $F_{5\%}(10, 24) = 2.25$

a double-sided test $F_{10\%}(10, 24) = 2.25$.

The procedure is illustrated in the following example.

Example 1

Two packaging machines weigh out and pack a nominal 50 kg of a powdered fertilizer. For both machines a simple adjustment can correct for consistent under, or over, weight. It is suspected that the packs from machine A are more variable in weight than those from B; could this be concluded from the following data which give the checked weight (in kg) of 10 packs from each machine? The significance level is to be taken at $\alpha = 0.05$.

| m/c A | 50.8 | 51.0 | 49.5 | 52.1 | 51.8 | 47.4 | 51.5 | 48.2 | 49.0 | 48.0 |
| m/c B | 49.3 | 48.9 | 49.2 | 50.0 | 48.8 | 49.5 | 49.2 | 49.6 | 48.8 | 47.5. |

We first adopt the appropriate null hypothesis. This will be that there is no difference between the machines with respect to variability. The alternative hypothesis is that A is more variable than B. We are therefore looking for a difference in a particular direction and the test will be a single-sided test:

$$H_0 \equiv \sigma_A^2 = \sigma_B^2$$
$$H_1 \equiv \sigma_A^2 > \sigma_B^2.$$

Since subtracting a constant from each value in a group of data will not affect the variance calculation we can do this to simplify the calculations.

Machine A, subtracting 50 from each result:

| x_A | 0.8 | 1.0 | −0.5 | 2.1 | 1.8 | −2.6 | 1.5 | −1.8 | −1.0 | −2.0 |
| x_A^2 | 0.64 | 1.00 | 0.25 | 4.41 | 3.24 | 6.76 | 2.25 | 3.24 | 1.00 | 4.00 |

Hence

$$\Sigma x_A = -0.7$$
$$\Sigma x_A^2 = 26.79.$$

Then using the formula,

$$\hat{\sigma}^2 = \frac{1}{n-1}\left(\Sigma x^2 - \frac{(\Sigma x)^2}{n}\right)$$

we have,

$$\hat{\sigma}_A^2 = \frac{1}{9}\left(26.79 - \frac{(-0.7)^2}{10}\right)$$
$$= \tfrac{1}{9} \times 26.741 = 2.971.$$

Machine B, subtracting 49 from each result:

x_B	0.3	−0.1	0.2	1.0	−0.2	0.5	0.2	0.6	−0.2	−1.5
x_B^2	0.09	0.01	0.04	1.00	0.04	0.25	0.04	0.36	0.04	2.25

Hence

$$\Sigma x_B = 0.8$$
$$\Sigma x_B^2 = 4.12.$$

and

$$\hat{\sigma}_B^2 = \frac{1}{9}\left(4.12 - \frac{(0.8)^2}{10}\right)$$

$$= \tfrac{1}{9} \times 4.056$$

$$= 0.451.$$

Then

$$F = \frac{\hat{\sigma}_A^2}{\hat{\sigma}_B^2}\,(\phi_A, \phi_B)$$

$$= \frac{2.971}{0.451}\,(9, 9)$$

$$F = 6.58 \ (9, 9).$$

Referring to Table 5, $F_{(5\%)}(9, 9) = 3.18$.

Since the calculated value is greater than this, H_0 is rejected, the result judged significant and the conclusion drawn that machine A is more variable than B.

Suppose the question had not indicated that it was suspected that A was more variable but had been phrased, 'it is required to determine which machine is the more consistent'. The alternative to the null hypothesis will now be *either*, A is more variable than B *or*, B is more variable than A, and we require a double-sided test. The appropriate values of F for significance are obtained by doubling the probabilities in the F-tables, i.e.

$$F_{10\%}(9, 9) = 3.18$$
$$F_{2\%}(9, 9) = 5.35.$$

A higher value of F is required for significance since we are now including the possibility that σ_B^2 could be greater than σ_A^2 but by chance $\hat{\sigma}_A^2$ is greater than $\hat{\sigma}_B^2$.

Example 2

Over a very long period the weight of packs from machine A have been normally distributed with variance 1.5 kg^2. It is suggested that the machine has now become more variable. Can this be asserted from the data obtained in the previous example? ($\alpha = 0.05$)

The question indicates that the value 1.5 kg^2 has been calculated from a very large amount of data, it is therefore assumed to be an exact estimate of that population variance, and thus based on an infinite number of degrees of freedom.

We have

$$H_0 \equiv \sigma_A^2 = 1.5.$$
$$H_1 \equiv \sigma_A^2 > 1.5.$$

Then

$$F = \frac{\hat{\sigma}_A^2}{1.5} \, (\phi_A, \infty)$$

$$= \frac{2.971}{1.5}$$

$$= 1.97 \, (9, \infty).$$

From Table 5, $F_{5\%}(9, \infty) = 1.88$.

We therefore reject H_0 and assert that machine A has become more variable.

2. Comparisons of means, u- and t-tests

When comparing mean values we shall be testing either:

(a) whether a sample with mean value $\bar{x}$ could have been drawn from a given population with mean μ_0, i.e. comparing a sample mean with a standard value, or,

(b) whether two samples with means $\bar{x}_1, \bar{x}_2$ respectively could have originated from the same population, i.e. comparing two sample means.

There are two forms of test to be considered for case (a): the u-test based on the normal distribution and the t-test based on the t-distribution. The former is used when we know the standard deviation of the given population and are justified, from technical knowledge, in making the assumption that there will be no significant change in variability. When we either have no estimate for the population, or do not consider the assumption of no change in variability to be justified, then an estimate of the standard deviation must be calculated from the sample. The reliability of the estimate is taken into account, in assessing the significance of the result by using the t-test.

(a) (i) *Comparison of sample mean with standard value, standard deviation known (u-test).* Suppose we have a known population $N(\mu_0, \sigma_0)$, i.e. a normal distribution with mean μ_0 and standard deviation σ_0. The mean values of samples of size n drawn at random from this population will be distributed about μ_0 with standard deviation $\sigma_0/\sqrt{n}$.

The null hypothesis we make is that the population from which the sample is drawn is coincident with the known population. The alternative hypothesis will take one of two possible forms, either we are testing for any difference in mean value, or, for a difference in a particular direction.

Suppose we are interested in a real increase, i.e.

$$H_0 \equiv \mu = \mu_0 \qquad H_1 \equiv \mu > \mu_0$$

where the population mean μ is estimated by the sample mean $\bar{x}$.

N

We shall not be surprised if $\bar{x}$ is not exactly equal to μ_0; H_0 implies that if sampling is repeated indefinitely, then the average value of $\bar{x}$, i.e. μ, will be μ_0. To determine whether the observed value is consistent with this hypothesis we calculate the probability that a sample with mean as great as $\bar{x}$ could have been drawn from a population with mean μ_0.

In Chapter 15 we saw that any normal distribution can be transformed to the standard form by a suitable change of unit. For the population we are testing the transformation will be

$$u = \frac{\bar{x} - \mu_0}{\text{S.E.}(\bar{x})}.$$

From standard normal tables (Table 1) we obtain the probability of exceeding the value u. This will be the probability of exceeding the difference $\bar{x} - \mu_0$ and hence the probability that H_0 is true.

Knowing $\bar{x}$ and S.E.$(\bar{x})$ we can calculate the limits within which we expect the true value of μ (the mean of the population from which the given sample was drawn) to be. These are called *confidence limits*.

Suppose we are prepared to take a risk α that the true value of μ lies outside the calculated limits. In general this risk is divided into two parts: a risk $\frac{1}{2}\alpha$ that μ is above the upper limit and a risk $\frac{1}{2}\alpha$ that it is below the lower limit. Let $u_{\alpha/2}$ denote that value of u which is exceeded with probability $\frac{1}{2}\alpha$ (using a probability as suffix denotes that value of the statistic which cuts off an area equal to the given probability in one tail of the distribution). If a sample is drawn from a normal distribution, with mean μ, there is a probability $\frac{1}{2}\alpha$ that the sample mean will exceed the value $\bar{x}_1$, where $\bar{x}_1$ is given by (see Fig. 17.1)

$$u_{\alpha/2} = \frac{\bar{x}_1 - \mu}{\text{S.E.}(\bar{x}_1)}$$

i.e.

$$\bar{x}_1 = \mu + u_{\alpha/2} \cdot \text{S.E.}(\bar{x}_1).$$

Similarly a value less than $\bar{x}_2$ will be obtained with probability $\frac{1}{2}\alpha$ where $\bar{x}_2$ is given by (see Fig. 17.2)

$$-u_{\alpha/2} = \frac{\bar{x}_2 - \mu}{\text{S.E.}(\bar{x}_2)}$$

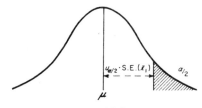

Fig. 17.1.

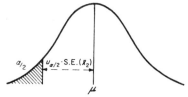

Fig. 17.2.

i.e.

$$\bar{x}_2 = \mu - u_{\alpha/2} \cdot \text{S.E.}(\bar{x}_2).$$

Therefore there is a probability α, that $\bar{x}$ lies outside the limits

$$\mu \pm u_{\alpha/2} \cdot \text{S.E.}(\bar{x}).$$

Conversely if $\bar{x}$ is known and is used to estimate μ there is a probability α that μ will lie outside the limit

$$\bar{x} \pm u_{\alpha/2} \cdot \text{S.E.}(\bar{x}).$$

These limits are called the $100(1 - \alpha)\%$ confidence limits. Let $\frac{1}{2}\alpha = 0.05$, then the 90% confidence limits on the true value of μ estimated by $\bar{x}$, will be

$$\bar{x} \pm u_{0.05} \cdot \text{S.E.}(\bar{x})$$

i.e.

$$\bar{x} \pm u_{0.05} \frac{\sigma_0}{\sqrt{n}}.$$

The above is not a full proof but illustrates the theory already used in the calculation of limits for quality control charts.

Example 3

The average daily scrap from a certain manufacturing process is 25.5 kg with standard deviation 1.6 kg. A modification is suggested in an attempt to reduce the scrap.

It can be assumed from the nature of the modification that day to day variability in the amount of scrap will not be affected.

The following data were obtained from a fortnight's trial of the modification:

	Mon.	Tues.	Wed.	Thurs.	Fri.	Sat.
1st week	23.8	24.5	25.0	26.1	22.8	25.0
2nd week	21.9	24.0	23.5	25.2	22.0	23.0

Is the modification effective? ($\alpha = 0.05$)

Put 95% confidence limits on the expected average daily scrap, using the modification.

$$H_0 \equiv \mu = \mu_0 = 25.5 \text{ kg}$$
$$H_1 \equiv \mu < 25.5 \text{ kg}.$$

To simplify the calculations subtract 24 from each value and multiply by 10.

$10(x - 24)$		-2	5	10	21	-12	10	-21	0	-5	12	-20	-10

$$10 \, \Sigma \, (x - 24) = -12$$

$$\therefore \; \bar{x} = 24.0 + \left(\frac{-12}{12}\right)\frac{1}{10}$$

$$= 23.9 \text{ kg.}$$

It is given that $\sigma = 1.6$ kg, therefore

$$\text{S.E.}(\bar{x}) = \sigma/\sqrt{n}$$

$$= 1.6/\sqrt{(12)}.$$

We calculate

$$u = \frac{\bar{x} - \mu_0}{\text{S.E.}(\bar{x})}$$

$$= \frac{23.9 - 25.5}{1.6/\sqrt{(12)}}$$

$$= -3.464.$$

Since the distribution is symmetrical about zero, the probability of a value of u less than $-x$ is the same as the probability of a value greater than $+x$. From Table 1 the probability that $u \geqslant 3.3$ is 0.000483. This is less than the chosen significance level ($\alpha = 0.05$), H_0 will be rejected and the alternative hypothesis, that the modification reduces scrap, accepted.

The best estimate available of the daily average using the modification is $\bar{x} = 23.9$ kg.

Given $\sigma = 1.6$ kg we have calculated the standard error of this estimate to be $1.6/\sqrt{(12)} = 0.462$, i.e.

$$\bar{x} = 23.9 \text{ kg}$$

$$\text{S.E.}(\bar{x}) = 0.462 \text{ kg.}$$

The 95% confidence limits on the true daily average will thus be given by,

$$\bar{x} \pm u_{0.025} \cdot \text{S.E.}(\bar{x}) = 23.9 \pm 1.96 \cdot 0.462$$

$$= 23.9 \pm 0.905$$

i.e. approximately 23.0 to 24.8 kg.

By carrying out the u-test we determine whether or not the decrease is real and by calculating the confidence limits we can determine whether or not it is of practical importance.

(a) (ii) Comparison of sample mean with standard value, standard deviation unknown (t-test).   The same null hypothesis is being tested but if the standard deviation σ is unknown a value must be estimated from the sample. The calculation of u assumed an exact value of σ, if this assumption is incorrect, u is not a standard normal deviate and the test of significance cannot be made using the normal distribution. We must take into account the number of degrees of freedom on which the estimate is based.

The statistic calculated, using standard notation is

$$t = \frac{\bar{x} - \mu_0}{\hat{\sigma}/\sqrt{n}}.$$

The distribution of t can be determined provided the data are drawn from an approximately normal distribution. The shape of the *t-distribution* will depend upon ϕ, the degrees of freedom on which $\hat{\sigma}$ is based and as ϕ tends to infinity so the t-distribution tends to the normal distribution, with which it is coincident when $\phi = \infty$.

Table 6 gives the values of t for various degrees of freedom, which will be exceeded in *absolute* value, with stated probability. The values in the table are therefore appropriate for a double-sided test, the percentages quoted being the sum of the areas in the two tails of the distribution, outside the limits $\pm t$. If we are interested in a difference in a specified direction, the probability in the tables must be halved.

Thus, from Table 6 if t is based on 24 degrees of freedom there is a 10% chance that it will fall outside the limits ± 1.71; and hence a 5% chance that it will be above $+1.71$.

Example 4

Test the data from Example 3 if the assumption that the modification will not affect the variability cannot be made.

As before

$$H_0 \equiv \mu = \mu_0 = 25.5 \text{ kg}$$

$$H_1 \equiv \mu < 25.5 \text{ kg}$$

From Example 3 we have,

$10(x - 24)$	-2	5	10	21	-12	10	-21	0	-5	12	-20	-10
$\{10(x - 24)\}^2$	2	25	100	441	144	100	441	0	25	144	400	100

$$10 \, \Sigma(x - 24) = -12$$

$$100 \, \Sigma(x - 24)^2 = 1924.$$

$$\hat{\sigma}^2 = \frac{1}{100} \times \frac{1}{11} \left(1924 - \frac{(-12)^2}{12} \right)$$

$$= 1.738 \text{ kg}^2.$$

$$\therefore \quad \frac{\hat{\sigma}}{\sqrt{n}} = \sqrt{\frac{1.738}{12}}$$

$$= 0.380(5) \text{ kg}.$$

We now calculate

$$t = \frac{\bar{x} - \mu_0}{\hat{\sigma}/\sqrt{n}} = \frac{23.9 - 25.5}{0.380(5)} = -4.205 \quad (\phi = 11)$$

From Table 6 we find that when t is based on 11 degrees of freedom, 0.1% of the distribution lies outside the limits ±4.44.

Since we are interested only in a decrease in mean we are concerned with only one tail of the t-distribution, i.e. we require a single-sided test and so halve the probability given in the tables.

Thus the appropriate significance points for this test are

$$t_{0.25\%}(\phi = 11) = 3.50$$

$$t_{0.05\%}(\phi = 11) = 4.44.$$

The probability that μ is not less than 25.5 kg is therefore only about 0.05% and we reject H_0 and assert a real decrease.

The t-test and not the u-test was used because the standard deviation was not exact, but an estimate based on $\phi = 11$. Similarly in calculating the confidence limits we must use the t-distribution. Thus the $100(1 - \alpha)\%$ confidence limits on the true daily average will be given by

$$\bar{x} \pm t_{\alpha/2(\phi)} \cdot \text{S.E.}(\bar{x})$$

where $t_{\alpha/2(\phi)}$ is that value of t, based on ϕ degrees of freedom, which cuts off an area $\frac{1}{2}\alpha$ in one tail of the distribution. As the tables are drawn up for a double-sided test, it will be that value of t which is exceeded in *absolute value* with probability α. The 95% confidence limits will be,

$$\bar{x} \pm t_{0.025(\phi=11)} \cdot \text{S.E.}(\bar{x})$$

i.e.

$$23.9 \pm 2.20 \times 0.38$$

i.e.

$$23.9 \pm 0.8.$$

It is important to note that since t is dependent upon the reliability of the S.E.$(\bar{x})$ it is always based on the same number of degrees of freedom as this estimate.

(b) (i) *Comparison of two samples when the data arise naturally in pairs* (*t-test*). One of the fundamental ideas in the statistical design of experiments, is that when comparisons are being made between certain conditions all other conditions should be maintained constant. This of course is an ideal which can only be approximated in practice. Suppose for example, we wish to determine whether dyeing affects the strength of a certain type of thread, the strength test involving breaking the thread. Ideally the test would be carried out first on the normal thread and then after dyeing. Since the test is destructive this will be impossible and the best we can do is to use two pieces of thread which initially, are as alike as possible. We could take a length of thread, halve it, and dye one half. The two halves are more likely to be similar than two pieces selected from two distant sections of the thread.

To analyse an experiment involving n lengths we can work in terms of the variable D_i, where D_i is the difference in strength between the two halves of the ith length. The null hypothesis is that dyeing does not affect the strength, and the alternative hypothesis is that it does, i.e.

$$H_0 \equiv \mu_D = 0$$

$$H_1 \equiv \mu_D \neq 0$$

where μ_D is the true average difference between dyed and undyed halves and is estimated by

$$\bar{D} = \frac{1}{n} \sum_{i=1}^{n} D_i.$$

The test now reduces to the previous form where the sample average was compared with a standard value and we calculate

$$t = \frac{\bar{D} - 0}{\hat{\sigma}_D/\sqrt{n}} \quad \phi = (n - 1)$$

where $\hat{\sigma}_D$ is the estimate of σ_D obtained from the sample and t is based on $\phi = (n - 1)$ degrees of freedom.

It is important to note that such an analysis is only justifiable if there is a practical reason for pairing the results. Here we have two groups of results, dyed and normal and a particular subsample in one group has an identifiable partner in the other.

Example 5

Six lengths of a certain type of synthetic fibre are halved, and one half of each pair (selected at random) dyed. The following data give the coded values of a strength test:

Sample	1	2	3	4	5	6
A	14.6	12.1	13.4	14.0	11.5	14.4
B (dyed)	13.8	12.5	11.6	12.0	10.8	13.6

Determine whether or not the strength is affected by dyeing if the level of significance is to be taken at $\alpha = 0.05$.

Put 95% confidence limits on the expected change in strength due to dyeing.

The results in this example fall naturally in pairs. Comparing A and B within each sample will remove the obvious variation from one sample to another:

$$H_0 \equiv \mu_D = 0.$$

$$H_1 \equiv \mu_D \neq 0.$$

The first step is to obtain the difference between the dyed and undyed half in each sample.

Sample	1	2	3	4	5	6
$D = A - B$	0.8	−0.4	1.8	2.0	0.7	0.8
D^2		0.64	0.16	3.24	4.00	0.49	0.64

$$\Sigma D = 5.7 \quad \therefore \bar{D} = 0.95$$

$$\Sigma D^2 = 9.17$$

$$\therefore \hat{\sigma}_D^2 = \frac{1}{5}\left(9.17 - \frac{(5.7)^2}{6}\right)$$

$$= 0.751.$$

We then calculate,

$$t = \frac{\bar{D} - 0}{\hat{\sigma}_D / \sqrt{n}}$$

$$= \frac{0.95}{\sqrt{(0.751/6)}} \quad (\phi = 5)$$

$$= 2.68 \quad (\phi = 5).$$

The question implied a double-sided test, since it did not specify a difference in a particular direction. From Table 6, $t_{5\%(\phi=5)} = 2.57$ (i.e. this value is exceeded in absolute value with probability 0.05). Thus H_0 is rejected and a real difference asserted.

The 95% confidence limits on the expected change in strength are

$$\bar{D} \pm t_{0.025(\phi=5)} \cdot \text{S.E.}(\bar{D}) = 0.95 \pm 2.57\sqrt{\frac{0.751}{6}}$$

$$= 0.95 \pm 0.91$$

i.e. 0.04 to 1.86.

We are therefore 95% confident that dyeing decreases the strength of this particular fibre by between 0.04 and 1.86 units.

(b) (ii) *Comparison of two samples when the data do not occur in pairs (t-test).* If there is no relationship between particular values in the two groups then there is no valid reason for considering the results in pairs. Suppose an experiment was carried out with the same objective as above but the experimental procedure was to take two lengths of thread sufficient for n_1, n_2 test pieces respectively. We now have two samples of results, n_1 in the first sample which are normal threads and n_2 dyed in the second.

Each sample will yield an estimate of the mean and variance of the population from which it was drawn. We wish to determine whether they could have been drawn from populations with the same mean. To do this we must make the assumption that the populations have the same variance. An F-test will verify this assumption, and only if the F-test gives a non-significant result are we justified in using the t-test to test for a difference in mean. If two populations showed significantly different variability, then even if they had exactly the same mean value, the population with greater variability could yield samples with means very different from the other population and thus a t-test may lead to incorrect conclusions.

Having shown that the two variances yielded by the respective samples could be estimates of the same value, a better estimate of that value will be obtained by combining the two, i.e. if $\hat{\sigma}_1^2$ is the estimate calculated from the n_1 results in sample 1 and $\hat{\sigma}_2^2$ is the estimate calculated from the n_2 results in sample 2, then

$$\hat{\sigma}^2 = \frac{(n_1 - 1)\hat{\sigma}_1^2 + (n_2 - 1)\hat{\sigma}_2^2}{(n_1 - 1) + (n_2 - 1)}$$

$$= \frac{(n_1 - 1)\hat{\sigma}_1^2 + (n_2 - 1)\hat{\sigma}_2^2}{n_1 + n_2 - 2}$$

and will be based on $(n_1 + n_2 - 2)$ degrees of freedom.

Again, the null hypothesis is that dyeing does not affect strength and the alternative hypothesis is that it does, i.e.

$$H_0 \equiv (\mu_1 - \mu_2) = 0$$

$$H_1 \equiv (\mu_1 - \mu_2) \neq 0$$

where μ_1 and μ_2 are estimated by $\bar{x}_1$ and $\bar{x}_2$ respectively.

Since we are asserting that $\hat{\sigma}^2$ is a better estimate of both population variances than either $\hat{\sigma}_1^2$ or $\hat{\sigma}_2$, we have

$$\text{var}(\bar{x}_1) = \frac{\hat{\sigma}^2}{n_1} \quad \text{and} \quad \text{var}(\bar{x}_2) = \frac{\hat{\sigma}^2}{n_2}.$$

$$\therefore \text{var}(\bar{x}_1 - \bar{x}_2) = \frac{\hat{\sigma}^2}{n_1} + \frac{\hat{\sigma}^2}{n_2}$$

(see Chapter 16 (appendix)), i.e.

$$\text{S.E.}(\bar{x}_1 - \bar{x}_2) = \hat{\sigma}\sqrt{\left(\frac{1}{n_1} + \frac{1}{n_2}\right)}.$$

We then calculate,

$$t = \frac{(\bar{x}_1 - \bar{x}_2) - 0}{\text{S.E.}(\bar{x}_1 - \bar{x}_2)}$$

$$= \frac{(\bar{x}_1 - \bar{x}_2) - 0}{\hat{\sigma}\sqrt{\left(\frac{1}{n_1} + \frac{1}{n_2}\right)}}.$$

Since $\hat{\sigma}$ is based on $(n_1 + n_2 - 2)$ degrees of freedom t is also based on $\phi = (n_1 + n_2 - 2)$.

The $100(1 - \alpha)\%$ confidence limits on the expected difference between the two means will be given by,

$$(\bar{x}_1 - \bar{x}_2) \pm t_{\alpha/2\,(\phi = n_1 + n_2 - 2)} \cdot \text{S.E.}(\bar{x}_1 - \bar{x}_2).$$

The u-test is obviously a special case of the t-test where the estimate of standard deviation is based on an infinite number of degrees of freedom.

If the null hypothesis is always expressed in the form of an estimate equal to a standard value (e.g. $\mu = \mu_0$; $\mu_1 - \mu_2 = 0$) then the formula for t can be remembered in a simple form, viz. t is equal to the estimate minus the standard value, divided by the standard error of the estimate.

Example 6

If the results in Example 5 were obtained by making six tests on each of the fibres A and B, B only being dyed, can it be asserted that dyeing affects strength? ($\alpha = 0.05$)

A	14.6	12.1	13.4	14.0	11.5	14.4
B(dyed)	13.8	12.5	11.6	12.0	10.8	13.6

There is now no relationship between a particular result on A and any result on B. Therefore they cannot be considered in pairs. The method of analysis will be to compare the averages $\bar{x}_A$ and $\bar{x}_B$ in relation to the variability within A and B.

$$H_0 \equiv (\mu_A - \mu_B) = 0$$

$$H_1 \equiv (\mu_A - \mu_B) \neq 0.$$

We first verify that the results within A and within B indicate approximately the same variability and then calculate,

$$t = \frac{(\bar{x}_A - \bar{x}_B) - 0}{\text{S.E.}(\bar{x}_A - \bar{x}_B)}, \quad \phi = (n_A + n_B - 2)$$

$(x_A - 13)$	1.6	−0.9	0.4	1.0	−1.5	1.4
$(x_A - 13)^2$	2.56	0.81	0.16	1.00	2.25	1.96

$$\Sigma(x_A - 13) = 2.0 \quad \therefore \bar{x}_A = 13.33$$

$$\Sigma(x_A - 13)^2 = 8.74.$$

$$\therefore \hat{\sigma}_A^2 = \frac{1}{5}\left(8.74 - \frac{(2.0)^2}{6}\right)$$

$$= \frac{8.073}{5}$$

$$= 1.615.$$

$(x_B - 12)$	1.8	0.5	−0.4	0	−1.2	1.6
$(x_B - 12)^2$	3.24	0.25	0.16	0	1.44	2.56

$$\Sigma(x_B - 12) = 2.3 \quad \therefore \bar{x}_B = 12.38$$

$$\Sigma(x_B - 12)^2 = 7.65.$$

$$\therefore \hat{\sigma}_B^2 = \frac{1}{5}\left(7.65 - \frac{(2.3)^2}{6}\right)$$

$$= \frac{6.768}{5}$$

$$= 1.354.$$

Hence

$$F = \frac{1.615}{1.354} \ (5, 5)$$

$$= 1.19.$$

Since, from Table 5, this result is clearly not significant, the populations can be assumed to have the same variance, a better estimate of this variance being given by

$$\hat{\sigma}^2 = \frac{(n_A - 1)\hat{\sigma}_A^2 + (n_B - 1)\hat{\sigma}_B^2}{n_A + n_B - 2}, \quad \phi = (n_A + n_B - 2)$$

$$= \frac{8.073 + 6.768}{10}$$

i.e.

$$\hat{\sigma}^2 = 1.484 \quad (\phi = 10)$$

Hence

$$\text{var}(\bar{x}_A - \bar{x}_B) = \frac{1.484}{6} + \frac{1.484}{6}$$

and

$$\text{S.E.}(\bar{x}_A - \bar{x}_B) = \sqrt{\frac{1.484}{3}}$$

$$t = \frac{(13.33 - 12.38) - 0}{\sqrt{(1.484/3)}}$$

$$= 1.35 \quad (\phi = 10).$$

From Table 6, for a double-sided test, the value of t required for significance is 2.23, i.e.

$$t_{5\%(\phi = 10)} = 2.23.$$

Thus the result will be judged not significant and the hypothesis that dyeing does not affect strength is not disproved.

Comparing the conclusions drawn in Examples 5 and 6 we can see how the test is made more sensitive by considering the results in pairs. This of course can only be done when there is practical justification.

Example 7

The following data give yields from two chemical processes:

| Process A | 65 | 72 | 68 | 74 | 73 | 71 | 72 | 74 | 70 | kg/batch |
| Process B | 60 | 67 | 64 | 65 | 67 | 62 | kg/batch | | | |

Process A is both more lengthy and more costly than process B. It is considered worthwhile using process A only if the average yield is more than 4 kg/batch higher than would be obtained using process B.

The probability of wrongly introducing process A is to be no more than 0.05. Should process A be recommended?

We are interested in a difference $(\mu_A - \mu_B) > 4$. We therefore determine the probability of obtaining a difference $(\bar{x}_A - \bar{x}_B)$ as great or greater than that observed, if $(\mu_A - \mu_B) = 4$. If this probability is no more than 0.05 then we shall assert that $(\mu_A - \mu_B) > 4$, i.e.

$$H_0 \equiv (\mu_A - \mu_B) = 4$$
$$H_1 \equiv (\mu_A - \mu_B) > 4$$

$(A - 70)$	-5	2	-2	4	3	1	2	4	0
$(A - 70)^2$		25	4	4	16	9	1	4	16	0

$\Sigma(A - 70) = 9, \quad \therefore \bar{x}_A = 71.00,$

$\Sigma(A - 70)^2 = 79,$

$\hat{\sigma}_A^2 = \dfrac{1}{8}\left(79 - \dfrac{9^2}{9}\right)$

$\quad = \dfrac{70}{8} \quad \therefore \hat{\sigma}_A^2 = 8.75.$

$(B - 65)$	-5	2	-1	0	2	-3
$(B - 65)^2$		25	4	1	0	4	9

$\Sigma(B - 65) = -5, \quad \therefore \bar{x}_B = 64.17,$

$\Sigma(B - 65)^2 = 43,$

$\hat{\sigma}_B^2 = \dfrac{1}{5}\left(43 - \dfrac{(-5)^2}{6}\right)$

$\quad = \dfrac{38.83}{5} \quad \therefore \hat{\sigma}_B^2 = 7.77.$

$F(8, 5) = \dfrac{\hat{\sigma}_A^2}{\hat{\sigma}_B^2}$

$\quad = \dfrac{8.75}{7.77}$

$\quad = 1.13.$

From Table 5 $F_{20\%}(8, 5) = 3.34$, so that the calculated value of F is not significant and we can assume that the samples are drawn from populations with equal variance, a better estimate of this variance being given by,

$$\hat{\sigma}^2 = \frac{70 + 38.83}{8 + 5}$$

$$\hat{\sigma}^2 = 8.37 \quad (\phi = 13).$$

Then

$$t = \frac{(\bar{x}_A - \bar{x}_B) - 4}{\text{S.E.}(\bar{x}_A - \bar{x}_B)}$$

$$= \frac{(71.00 - 64.17) - 4}{\sqrt{\{8.37(\frac{1}{9} + \frac{1}{6})\}}}$$

$$= \frac{2.83}{\sqrt{2.325}}$$

$$= 1.86 \quad (\phi = 13).$$

From tables, the positive value of $t(\phi = 13)$ which will be exceeded with probability 0.05 (i.e. single-sided test) is $t = 1.77$.

Thus the probability that $(\mu_A - \mu_B)$ is not greater than 4 kg/batch is less than 0.05. We therefore reject H_0 and assert that $(\mu_A - \mu_B) > 4$ kg/batch and thus recommend the use of process A.

By using F-, u-, and t-distributions we are making certain assumptions regarding the source of the data we are testing. Basically we assume that the observations are random values from normal populations. The validity of the tests is therefore dependent upon the assumptions:

(i) the error in one observation does not affect the error in any other observation;
(ii) the errors are normally distributed;
(iii) when comparing sample means the variance of the errors is approximately the same for each source.

Tests for comparing mean values are still valid when the departure from normality is quite large. The calculated probability will not be exact, but the conclusions drawn are unlikely to be misleading. The F-test is more sensitive and if serious departures from normality are suspected the data should be transformed (e.g. by taking logarithms or square roots) to give a closer approximation to a normal distribution.

Since in practice a population variance is invariably estimated from sample data, in further work we shall no longer distinguish between the true variance and its best estimate, i.e. σ^2 will denote the best estimate of σ^2.

3. Comparison of frequencies, χ^2-test

So far we have used significance tests to test for differences between groups of data arising from measurements. Similar tests have been devised to compare frequencies of occurrence. These tests are based on the χ^2-distribution.

The statistic *chi-square* is defined as the sum of the squares of a sample of n random values from a normal distribution with zero mean and unit standard deviation, i.e.

$$\chi^2 = \sum_{i=1}^{n} \frac{(x_i - \mu)^2}{\sigma^2}.$$

The χ^2-distribution will be formed by drawing an infinite number of such samples. The shape of the distribution will obviously depend upon n and tables give the value of χ^2 exceeded with stated probability and based upon a specific number of degrees of freedom.

If data can be classified into one of k classes and the probability of falling into the ith class is p_i (where $i = 1, 2, \ldots, k$), it can be shown by extending the binomial theory that the mean and variance of the resulting distribution will be np_i. To determine whether a particular event occurs more, or less frequently than expected, we can (provided the frequencies are not too small), approximate this distribution to a normal distribution. Then, if the observed frequency in the ith class is f_i

$$\sum_{i=1}^{n} \frac{(f_i - np_i)^2}{np_i}$$

will be distributed, approximately as χ^2, or

$$\chi^2 = \sum \frac{(\text{observed} - \text{expected})^2}{\text{expected}} \text{(approximately)}$$

$$= \sum \frac{(O - E)^2}{E}.$$

Since the difference between the observed and expected frequency is squared the same value of χ^2 will be obtained with either positive or negative differences and hence the table (Table 7) gives the appropriate values for a double-sided test.

The χ^2-distribution is a continuous distribution; by counting frequencies of occurrence we are dealing with discrete data. The approximation can in general be ignored but a correction is sometimes applied when χ^2 is based on only one degree of freedom. The correction consists of reducing the difference between the observed and expected frequencies by 0.5. If $O > E$, we calculate $(O - E - 0.5)^2$ and if $O < E$, calculate $(O - E + 0.5)^2$. The use of the correction is illustrated in the following example.

Example 8

A coin is tossed 100 times and shows heads 65 times. Are we justified in asserting that the coin is biased? ($\alpha = 0.01$)

With no bias, after an infinite number of trials we would expect an equal number of heads and tails, i.e.

$$H_0 \equiv \text{heads} = \text{tails}.$$

Therefore with 100 trials the best estimate of the expected frequencies is 50 heads and 50 tails, i.e.

$E_H = 50, \quad O_H = 65,$

$E_T = 50, \quad O_T = 35.$

$$\therefore \ \chi^2 = \frac{(65 - 50 - 0.5)^2}{50} + \frac{(35 - 50 + 0.5)^2}{50}$$

$$= 2\frac{(14.5)^2}{50}$$

$$= 8.41, \quad \phi = 1.$$

Since we know the total number of times the coin was tossed, quoting the number of heads automatically fixes the number of tails (i.e. only one of the frequencies can be arbitrarily chosen) and χ^2 is thus based on one degree of freedom.

Referring to the χ^2-tables (Table 7)

$$\chi^2_{1\%}(\phi = 1) = 6.64$$

$$\chi^2_{0.1\%}(\phi = 1) = 10.83.$$

We therefore conclude that the result is highly significant and the coin biased.

Example 9

The absenteeism, in lost man-hours, is recorded for five sections of a factory, over a given period. Is there a significant difference between the rates of absenteeism?

Section	Number employed	Lost man-hours
A	220	100
B	103	29
C	145	34
D	67	12
E	150	50
Total	685	225

From these data we can calculate the rate of absenteeism per employee for the period considered:

$$p = \frac{225}{685} = 0.328(5).$$

The null hypothesis to be tested is that p is the same for each section:

$$H_0 \equiv p_A = p_B = p_C = p_D = p_E = 0.328(5).$$

The expected absenteeism for each section is

$$E = 0.328(5) \times \text{number employed}.$$

Calculating this value for each section we have

Absenteeism

Section	O	E	$O - E$	$\dfrac{(O - E)^2}{E}$
A	100	72.3	27.7	10.61
B	29	33.8	−4.8	0.68
C	34	47.6	−13.6	3.89
D	12	22.0	−10.0	4.55
E	50	49.3	0.7	0.01
Total	225	225		19.74

The frequency for the fifth section is fixed when the other four are known. Thus

$$\chi^2 = 19.74, \quad \phi = 4.$$

From tables this value will be exceeded by chance (if H_0 is true) with probability less than 0.001. We therefore conclude that the rate of absenteeism is not the same for all sections.

If a more detailed investigation was required, we could proceed by considering the contributions made to χ^2 by each of the sections. The largest is made by A, which has the highest absenteeism. Omitting the data from this section and considering the other four we have

Section	Employed	O	E	$O - E$	$\dfrac{(O - E)^2}{E}$
B	103	29	27.7	1.3	0.06
C	145	34	39.0	−5.0	0.64
D	67	12	18.0	−6.0	2.00
E	150	50	40.3	9.7	2.33
Total	465	125	125		5.03

$p' = \frac{125}{465} = 0.269$

$H_0 \equiv p_B = \cdots = 0.269.$

The value $\chi^2 = 5.03$ ($\phi = 3$) will be exceeded more than 10% of the time if there is no difference between these sections with respect to absenteeism. There is thus insufficient evidence to reject the null hypothesis. The final conclusions drawn would be:

  (i) sections B, C, D, E are likely to have similar rates of absenteeism;
  (ii) section A has a higher rate of absenteeism than the other sections.

In general if there are n groups χ^2 will be based upon $(n - 1)$ degrees of freedom since having specified the total frequency of occurrence only $(n - 1)$ of the group frequencies can be arbitrarily chosen.

If there are two criteria of classification to be applied to the data the groups can be arranged in a two-dimensional array which is known as a *contingency table*.

Example 10

Three factories each produce four grades of a particular article. From the data given, is there sufficient evidence to assert a significant difference between the proportions produced by each factory in the four grades? ($\alpha = 0.05$)

Factory \ Grade	A	B	C	D	Total
1	11	25	17	13	66
2	7	14	7	12	40
3	13	14	15	14	56
Total	31	53	39	39	162

From the above table we calculate

Grade	A	B	C	D
Proportion of total in given grade	0.191	0.327	0.241	0.241

We then calculate the expected frequencies for each factory with the assumption that they all produce the same proportion in each grade.

Factory 1 produces a total of 66 articles
Factory 1 should therefore produce 66 x 0.191 in grade A
Factory 1 should therefore produce 66 x 0.327 in grade B
Factory 1 should therefore produce 66 x 0.241 in grade C
Factory 1 should therefore produce 66 x 0.241 in grade D

Similar calculations for the other factories yield the following table:

Factory \ Grade	A	B	C	D	Total
1	12.6	21.6	15.9	15.9	66
2	7.7	13.1	9.6	9.6	40
3	10.7	18.3	13.5	13.5	56
Total	31	53	39	39	162

We next determine the difference between the observed and expected frequency in each cell of the table and each will yield a contribution to the value of χ^2. In order to obtain the correct totals, two out of the three frequencies in any column and three out of the four in any row can be arbitrarily assigned. χ^2 therefore will be based upon $2 \times 3 = 6$ degrees of freedom. In general for an $m \times n$ table there will be $(m-1)(n-1)$ degrees of freedom.

From above

$$\chi^2 = \frac{(-1.6)^2}{12.6} + \frac{(3.4)^2}{21.6} + \frac{(1.1)^2}{15.9} + \frac{(-2.9)^2}{15.9} + \frac{(-0.7)^2}{7.7} + \frac{(0.9)^2}{13.1}$$

$$+ \frac{(-2.6)^2}{9.6} + \frac{(2.4)^2}{9.6} + \frac{(2.3)^2}{10.7} + \frac{(-4.3)^2}{18.3} + \frac{(1.5)^2}{13.5} + \frac{(0.5)^2}{18.5}$$

$$= 0.203 + 0.535 + 0.076 + 0.529 + 0.064 + 0.062 + 0.704$$

$$+ 0.600 + 0.494 + 1.010 + 0.167 + 0.019$$

$$\chi^2 = 4.463 \quad (\phi = 6).$$

This value will be exceeded by chance almost 50% of the time. The result therefore is judged not significant and the conclusion drawn is that there is insufficient evidence to assert a difference between the proportions made by the different factories in the four grades.

Care should be taken when interpreting the data if the calculated value of χ^2 is expected to be exceeded by chance a very large proportion of the time (say 0.99). This implies that there is a very small probability of obtaining a value less than that calculated, and means that the differences between the observed and expected frequencies are smaller than would be expected if H_0 is true.

The χ^2-test can also be used to test how well a set of data conforms to a theoretical distribution.

The procedure is to calculate, from the observations, the best estimates of the parameters defining the assumed distribution. From these values we calculate the theoretical frequencies for the distribution, and hence χ^2 from the difference between the observed and expected frequencies in each group.

It may be that there are few observations in some groups. In such a case, even a small difference from the expected frequency may have a large effect upon the value of χ^2. To avoid this disproportionate effect on χ^2, groups with frequencies less than 5 are usually combined with adjacent groups.

Example 11

Rubber cable was examined for faults in 100 m lengths. The number of faults per length ranged from 0 to 5 and the number of lengths containing the given number of faults recorded. Calculate the frequencies of the Poisson distribution with the same mean as this distribution and test the goodness of fit by a χ^2-test. ($\alpha = 0.05$)

Number of faults	0	1	2	3	4	5
Lengths with given number of faults	35	45	40	23	4	3

From the given data

$$\lambda = \frac{\Sigma f_i x_i}{\Sigma f_i}$$

$$= \frac{225}{150}$$

$$= 1.5.$$

We then calculate, assuming a Poisson distribution with mean $\lambda = 1.5$, the probability of obtaining each of the observed number of faults, i.e.

$$P_r = \frac{\lambda^r e^{-\lambda}}{r!}.$$

No. of faults	Proportion expected from Poisson ($\lambda = 1.5$)	No. of lengths expected from Poisson, E	No. of lengths observed, O	$\dfrac{(O - E)^2}{E}$
0	0.223	33.5	35	0.067
1	0.335	50.2	45	0.539
2	0.251	37.6	40	0.153
3	0.126	18.9	23	0.889
4	0.047	7.1 ⎫	4 ⎫ 7	0.800
5	0.014	2.1 ⎬ 9.8	3 ⎭	
6	0.003	0.4 ⎪		
7	0.001	0.2 ⎭		
Total		150	150	$2.448 = \chi^2$

$\chi^2 = 2.448$ is based on 3 degrees of freedom. Five groups contribute to the value of χ^2 but one degree of freedom is lost by making the mean of the Poisson distribution equal to the mean of the observed data and another by imposing the restriction that the total frequencies must be the same.

The calculated value $\chi^2 = 2.448 (\phi = 3)$ will be exceeded, by chance, about 50% of the time. There is therefore no reason to doubt the assumption that the observed data are drawn from a Poisson distribution.

Example 12

A sample of 100 items was measured and the results grouped in the form of a frequency table. The estimates of mean and standard deviation calculated from the data are 4.70 and 1.81 units respectively.

Determine whether or not the data can be assumed to be a random sample drawn from a normal distribution. ($\alpha = 0.05$)

Suppose we have a normal distribution with $\mu = 4.70$, $\sigma = 1.81$. By calculating $u = (x_i - \mu)/\sigma$ we can determine from Table 1, the probability of a random value from this distribution lying below any specified value x_i.

Upper group limit, x_i	$u = \dfrac{x_i - \mu}{\sigma}$	Probability of a value below x_i	Probability of a value between x_i and $x_{(i-1)}$	Expected frequency, E	Observed frequency, O	$O - E$	$\dfrac{(O-E)^2}{E}$
1	−2.04	0.0207	0.0207	2.07 ⎫ 6.81	1 ⎫ 7	0.19	0.01
2	−1.49	0.0681	0.0474	4.74 ⎭	6 ⎭		
3	−0.94	0.1736	0.1055	10.55	10	0.55	0.03
4	−0.39	0.3483	0.1747	17.47	19	1.53	0.13
5	0.17	0.5675	0.2192	21.92	22	0.08	0.00
6	0.72	0.7642	0.1967	19.67	21	1.33	0.09
7	1.27	0.8980	0.1338	13.38	9	4.38	1.43
8	1.82	0.9656	0.0676	6.76 ⎫	7 ⎫		
9	2.38	0.9913	0.0257	2.57 ⎬ 10.20	4 ⎬ 12	1.80	0.32
10	2.93	0.9983	0.0070	0.70	1		
above 10			0.0017	0.17 ⎭	0 ⎭		
				100	100		$\chi^2 = 2.01$
					100		

Let x_i, $i = 1, 2, \ldots, n$, take the values of the upper limits of the n groups in the frequency table. We can determine the probability of obtaining a measurement below any upper group limit x_i and hence the probability of a value falling between two limits x_i and $x_{(i-1)}$. Multiplying this probability by 100 gives the number of results expected, within each group, from a random sample of 100 measurements drawn from the hypothetical distribution. The observed and expected frequencies can then be compared using the χ^2-test.

The calculations are set out in the table shown on page 394. $\chi^2 = 2.01$ is based on 4 degrees of freedom. Three degrees of freedom are lost by imposing three restrictions upon the hypothetical distribution, the mean and standard deviation are made equal to the estimates calculated from the observed data and the totals of the observed and expected frequencies are made the same.

From Table 7, $\chi^2 = 2.01$ ($\phi = 4$) is clearly not significant and we would accept that the data are drawn from a normal distribution.

PROBLEMS

1. Two types of mixer, A and B are used to mix a powdered fertilizer. A mix consists of 500 kg, of which 57.5 kg is active ingredient. 10 samples are analysed from each of 2 mixes, one mixed by A, the other by B. The data give the percentage active ingredient.

Mixer A	11.4	11.7	11.1	11.6	11.5	11.7	11.3	11.7	11.8	11.2
Mixer B	11.3	11.4	11.5	11.5	11.7	11.6	11.6	11.3	11.4	11.7

Would you consider either mixer gives a more uniform mix than the other? ($\alpha = 0.05$).

2. Two trainee analysts, X and Y, made 10 repeat carbon analyses. Can it be asserted on the basis of the following data that the results by Y are significantly more variable than those by X? ($\alpha = 0.05$).

% carbon

X	1.24	1.23	1.25	1.26	1.27	1.25	1.23	1.26	1.25	1.25
Y	1.24	1.23	1.27	1.23	1.26	1.25	1.24	1.28	1.22	1.26

3. Over a long period, the average N_2 content of a chemical manufactured by a standard process is 8.40% and the standard deviation 0.85%. A new process is installed and 100 samples are analysed for N_2, giving a mean N_2 content, $\bar{x} = 8.62\%$. It can be assumed that the variability from one sample to another will remain unchanged.
   (i) Is the N_2 increase significant or can it be assumed to be due to random variation? ($\alpha = 0.05$).
   (ii) Put 95% confidence limits on the expected $N_2\%$ using the new process.

4. Effluent samples are tested from two different sites in a factory. One of the tests is the determination of the pH value.
   Given the following data, would you assert that the two sites differ with respect to pH value? ($\alpha = 0.05$)

Site A	6.9	7.0	6.6	7.0	7.2	6.9	7.9	7.4	6.6
Site B	6.5	6.2	5.9	7.4	6.3	6.2	7.0	6.8	6.5

5. Ten samples of a resin-treated fabric are each halved. One half of each pair is washed. The N_2 content is determined for each.
(i) Does washing reduce the N_2 content? ($\alpha = 0.05$). Data give $\%N_2$.

Sample	1	2	3	4	5	6	7	8	9	10
Before washing	2.28	2.35	0.81	2.52	1.91	2.11	2.06	1.55	3.21	1.87
After washing	2.06	2.08	0.75	2.34	1.27	1.25	1.32	0.98	2.69	1.57

(ii) Put 95% confidence units on the expected decrease in $\%N_2$ after washing.

6. Two laboratories determine the chlorine content of samples taken at the same time each day from the plant cooling water. Is there sufficient evidence in the following results to assert a bias between laboratories? ($\alpha = 0.05$). Data give chlorine in p.p.m.

Day	1	2	3	4	5	6	7
Lab. A	1.15	1.86	0.75	1.82	1.14	1.65	1.90
Lab. B	1.00	1.90	0.90	1.80	1.20	1.70	1.95

7. The following data give the rate of diffusion of CO_2 through two different soils. Can it be asserted that the two soils differ with respect to rate of diffusion of CO_2? ($\alpha = 0.05$)

Soil A	19	32	18	30	24	22	28	23	27	17	25		
Soil B	30	20	28	15	35	18	25	27	34	35	37	31	29

8. The following data give the number of α-particles emitted per unit time by a radioactive substance:

Particles emitted	0	1	2	3	4	5	6	7	8	9	10
No. of units of time when given no. are emitted	76	204	251	214	139	65	30	14	4	2	1

Can the data be assumed to be a random sample from a Poisson distribution?

9. Two types of spray were used to apply paint to 200 similar wooden panels. The panels were then inspected and assessed either as satisfactory or requiring a second application.
The following results were obtained

	Spray 1	Spray 2
Satisfactory	84	44
Requiring second application	40	32

Test the hypothesis that there is no significant difference between the two sprays.

10. Two drivers, A and B, each drove over eight different routes and their m.p.g. were recorded for each route. For practical purposes the cars can be assumed identical and each was filled with petrol of the same brand and octane. The two drivers covered the same route at approximately the same time. Do the following results indicate that one driver is better than the other with respect to petrol economy?

Route	1	2	3	4	5	6	7	8
Driver A	35.6	28.5	32.1	30.2	31.4	32.0	29.0	36.0
Driver B	30.4	29.0	31.0	28.8	30.0	32.5	28.5	34.0

11. Companies X and Y manufacture electrical fuses. Samples of 5 amp fuses were selected at random from the production lines of the companies; the fuses blown and the current, in amps, recorded as follows:

Company X	4.6	4.8	5.1	5.2	5.4	5.4	5.4	5.5	5.8	6.0
Company Y	5.2	5.4	5.4	5.5	5.5	5.5	5.6	5.8	5.9	6.2

Can the difference between the means of the two samples be attributed to random error?

12. Over a long period an upholstery manufacturer found that the covering material on one particular shape of footstool split within twelve months on 10% of the stools. A very thin layer of foam rubber was put under the cover and of the next 150 stools produced only 2 covers split in the first year. Would you accept that the foam rubber improved the durability of the covering material? Could you accept the manufacturer's claim that in future the percentage splitting would be down to 1%?

13. A particular road junction had a bad accident record for pedestrians. In an attempt to reduce accidents a panda crossing was put at the junction. To test the effectiveness of the crossing a survey of accidents at the junction was carried out before and after the crossing was in use. Surveys were also carried out at other points in the town to check the overall accident rate. On the basis of the following figures would you assert that the crossing was effective in reducing accidents?

	Number of Accidents	
	Before	After
Test junction	25	10
Rest of town	250	230

14. A factory operates three types of machine for three shifts each day. The following number of breakdowns for each type of machine is recorded for each shift for one year:

Shift \ Machine	1	2	3
1	38	20	14
2	25	11	10
3	15	16	14

Test the hypothesis that the breakdowns are independent of the shift.

CHAPTER 18

Analysis of Variance

In the previous chapter we used a t-test to determine whether or not the mean values of two samples could be accepted as estimates of the same population mean. The test considered the magnitude of the difference between the mean values in relation to the variability between the individual results in each sample.

Suppose the problem was to compare a larger number of samples. It may seem logical, using a t-test, to consider the samples in pairs, but such a procedure is not justified. When we carry out a t-test and reject the null hypothesis there is always a small probability that we are wrong. We specify this probability of error by stating the level of significance for the test. If ten sample means were tested in pairs there would be $_{10}C_2 = 45$, t-tests to be carried out and for each test there would be the possibility of a wrong decision. Thus the risk of wrongly rejecting the null hypothesis would not be controlled and it would be difficult to draw precise conclusions from data analysed in this way.

The method used for comparing a number of samples is the *analysis of variance*. Using this procedure a comparison of variance estimates provides a comparison between mean values.

In any experiment certain experimental conditions will be deliberately varied and others will be held constant. There will also be factors affecting the results which cannot be controlled and whose effects cannot be separately identified. Variation caused by such factors must be grouped under the heading of experimental error. The experimental error is thus the inherent variability due to random fluctuations and will be measured by the variance of observations made under apparently identical conditions.

If there are a number of independent sources of variation, whose variances are $\sigma_1^2, \sigma_2^2, \ldots, \sigma_n^2$, operating simultaneously on a system of observations, the resulting variance of the observations, σ^2, is given by

$$\sigma^2 = \sigma_1^2 + \sigma_2^2 + \cdots + \sigma_n^2,$$

see Chapter 16, page 364.

The analysis of variance is an algebraic procedure which breaks down the total variation into its component parts.

1. One-factor analysis of variance

Consider an experiment in which there is only one assignable cause of variation affecting the results.

Suppose an analyst uses k different methods for determining a given constituent in a certain powder, making n repeat tests with each method. We wish to determine whether or not the methods can be assumed equivalent.

The n results obtained using any given method are repeat observations made under what are *apparently* identical conditions. There may be some variation, in the amount of the given constituent, from one test sample to another but in this example such variability cannot be identified separately from the inherent variability of the method. Thus the variability within each group of n tests will provide an estimate of the experimental error; the true value of σ_0^2, the error variance, would be the variance of an infinite number of tests by a given method. Each of the k methods will provide a separate estimate, based on $(n-1)$ degrees of freedom. These can be combined to give the best available estimate of error variance for the whole experiment, i,e. σ_0^2 based on $\phi = k(n-1)$. This assumes that the separate values can be considered estimates of the same population variance. When using the analysis of variance technique it is not usual to verify this, but if technical information indicates that the variability may differ from one group to another, then the assumption should be checked.

The only assignable cause of variation is the use of k different methods. If an infinite number of determinations were made using each method then each mean would be the exact mean value for that method and the variability between the k mean values would estimate the true variance between methods (σ_m^2). If groups of n randomly selected values are drawn from the infinite possible number by one method, the mean values of these groups will be distributed about the true mean with variance σ_0^2/n. Each mean value calculated from n determinations thus has a component of error variance σ_0^2/n. This variability will be independent of the variability due to the difference between methods and the calculated variance between the k mean values will be

$$\sigma_1^2 = \frac{\sigma_0^2}{n} + \sigma_m^2, \quad \phi_1 = (k-1).$$

If the methods are equivalent, i.e. if the true average for each method is the same, then the true value of σ_m^2 is zero.

We can calculate

(i) σ_0^2, $\quad \phi_0 = k(n-1)$

and

(ii) $\sigma_1^2 = \dfrac{\sigma_0^2}{n} + \sigma_m^2, \quad \phi_1 = k-1.$

Multiplying both sides of (ii) by n we have

$$n\sigma_1^2 = \sigma_0^2 + n\sigma_m^2.$$

Now if σ_m^2 is zero, $n\sigma_1^2$ is an estimate of σ_0^2.

The null hypothesis we adopt is that there is no real difference between methods,

i.e.

$$H_0 \equiv \sigma_m^2 = 0,$$

or

$$H_0 \equiv \frac{\sigma_0^2 + n\sigma_m^2}{\sigma_0^2} = \frac{\sigma_0^2}{\sigma_0^2} = 1.$$

We calculate

$$F = \frac{\sigma_0^2 + n\sigma_m^2}{\sigma_0^2} = \frac{n\sigma_1^2}{\sigma_0^2}, \quad (\phi_1, \phi_0),$$

and determine from F-tables the significance of the result and hence whether or not the observed data are consistent with the null hypothesis (H_0).

If H_0 is rejected it will be asserted that the methods differ and if required, the best available estimate of σ_m^2 can be obtained from (i) and (ii),

Example 1

An analyst uses four different methods to determine the amount of a given constituent in a certain powder. He makes five repeat tests with each method.

Can the methods be considered equivalent for this test? (Data give constituent in p.p.m.)

Method	A	B	C	D
	1	1	2	4
	3	1	4	5
	4	1	3	3
	2	2	5	2
	4	2	6	4

We have to determine:
  (i) the variability within methods, the experimental error variance, σ_0^2;
  (ii) The variability between methods, the variance between the calculated mean values, σ_1^2.

The null hypothesis we adopt is that there is no real difference between methods and $\sigma_m^2 = 0$, i.e.

$$H_0 \equiv \frac{n\sigma_1^2}{\sigma_0^2} = \frac{\sigma_0^2}{\sigma_0^2}.$$

We have

x_A	x_A^2	x_B	x_B^2	x_C	x_C^2	x_D	x_D^2
1	1	1	1	2	4	4	16
3	9	1	1	4	16	5	25
4	16	1	1	3	9	3	9
2	4	2	4	5	25	2	4
4	16	2	4	6	36	4	16

Σx	14		7		20		18
Σx^2		46		11		90		70
$\frac{1}{5}(\Sigma x)^2$		39.2		9.8		80.0		64.8
$\sum_{1}^{n}(x-\bar{x})^2 = \Sigma x^2 - \frac{1}{5}(\Sigma x)^2$		6.8		1.2		10.0		5.2

Then

$$\sigma_0^2 = \frac{1}{k(n-1)} \sum_{1}^{k} \left(\sum_{1}^{n}(x-\bar{x})^2 \right)$$

$$= \frac{6.8 + 1.2 + 10.0 + 5.2}{4(5-1)}$$

$$= 1.45, \quad \phi_0 = 16.$$

$\bar{x}_A = \frac{14}{5} = 2.8, \quad \bar{x}_C = \frac{20}{5} = 4.0,$

$\bar{x}_B = \frac{7}{5} = 1.4, \quad \bar{x}_D = \frac{18}{5} = 3.6.$

We now calculate σ_1^2, the variance between these means, i.e.

$\bar{x}$	$\bar{x}^2$
2.8	7.84
1.4	1.96
4.0	16.00
3.6	12.96
Totals 11.8	38.76

Hence

$$\sigma_1^2 = \frac{1}{k-1} \left(\Sigma \bar{x}^2 - \frac{1}{k}(\Sigma \bar{x})^2 \right)$$

$$= \frac{1}{3} \left(38.76 - \frac{(11.8)^2}{4} \right) = 1.317, \quad \phi_1 = 3$$

$$\therefore F = \frac{n\sigma_1^2}{\sigma_0^2} \quad (\phi_1, \phi_0) \qquad = \frac{5 \times 1.317}{1.45} = 4.54(3, 16).$$

From tables

$$F_{5\%}(3, 16) = 3.24$$

$$F_{1\%}(3, 16) = 5.29.$$

Thus, the result is significant and we reject H_0 and conclude that there is a real difference between the four methods.

Further analysis will depend upon whether the four methods:

(i) can be assumed to be a random sample from a 'population' of methods; or

(ii) are four specially selected methods.

If (i) then the next step will be to estimate σ_m^2, the variance of the population of methods.

If (ii) the calculation of σ_m^2 will have no practical meaning and the procedure will be to calculate confidence limits on the mean value for each method.

Using the given data

(i) $\sigma_0^2 = 1.45$

$$n\sigma_1^2 = \sigma_0^2 + n\sigma_m^2$$

$$= 6.58.$$

$$\therefore \ \sigma_m^2 = \frac{n\sigma_1^2 - \sigma_0^2}{\sigma_0^2} = \frac{6.58 - 1.45}{5}$$

$$= 1.03.$$

(ii) For any given method the mean value estimated from n observations will be distributed about the true value with variance σ_0^2/n. Thus using any specified method, the $100(1 - \alpha)\%$ confidence limits on the true mean value for that method will be given by

$$\mu = \bar{x} \pm t_{\alpha/2(\phi=\phi_0)} \frac{\sigma_0}{\sqrt{n}},$$

where $\bar{x}$ is the estimate from n observations and $t_{\alpha/2(\phi=\phi_0)}$ is that value of t (based on ϕ_0 degrees of freedom) which is exceeded with probability $\frac{1}{2}\alpha$ (i.e. α outside the limits $\pm t_{\alpha/2}$).

Therefore, the 90% confidence limits on the true mean for method A are given by

$$\bar{x}_A \pm t_{0.05(\phi=16)} \frac{\sigma_0}{\sqrt{5}}$$

i.e.

$$2.8 \pm 1.75 \sqrt{\frac{1.45}{5}}$$

i.e. approximately 1.9 to 3.7.

The calculations involved in carrying out the analysis of variance can be simplified by working with totals rather than means. The total variability can be divided into two parts, the error variability and the variability between groups and it is usually more convenient to calculate the total variability and the variability between groups and obtain the error by subtraction rather than calculating it directly.

This method is derived and then illustrated by re-working the above example.

To test H_0 we determined: (i) σ_0^2; (ii) $n\sigma_1^2 = \sigma_0^2 + n\sigma_m^2$.

Let

    k = the number of groups,

    n = the number of observations per group,

    kn = the total number of observations,

    x_i = an individual observation, $i = 1, 2, \ldots, kn$,

    T_j = the total of the jth group, $j = 1, 2, \ldots, k$,

    G = the grand total, i.e. the total over kn observations.

Then

$$T_j = \sum_{i=1}^{n} x_i$$

$$G = \sum_{j=1}^{k} T_j = \sum_{i=1}^{kn} x_i$$

The variance between all observations is

$$\sigma^2 = \frac{1}{kn-1} \left(\sum_{i=1}^{kn} x_i^2 - \frac{G^2}{kn} \right).$$

The variance between groups is

$$\sigma_1^2 = \frac{1}{k-1} \left(\sum_{j=1}^{k} \left(\frac{T_j}{n} \right)^2 - \frac{\left\{ \sum_{j=1}^{k} \left(\frac{T_j}{n} \right) \right\}^2}{k} \right)$$

$$\therefore\ n\sigma_1^2 = \frac{1}{k-1} \left(\frac{1}{n} \sum_{j=1}^{k} T_j^2 - \frac{\left(\sum_{j=1}^{k} T_j \right)^2}{kn} \right)$$

$$= \frac{1}{k-1} \left(\frac{1}{n} \sum_{j=1}^{k} T_j^2 - \frac{G^2}{kn} \right).$$

The error variance is

$$\sigma_0^2 = \frac{1}{k(n-1)} \left(\sum_{i=1}^{kn} x_i^2 - \sum_{j=1}^{k} \frac{T_j^2}{n} \right).$$

Hence dropping the suffices we have:

Total sum of squares $= (kn - 1)\sigma^2$

$$= \Sigma x^2 - \frac{G^2}{kn} \, .$$

Sum of squares between groups $= n(k - 1)\sigma_1^2$

$$= \frac{1}{n} \Sigma T^2 - \frac{G^2}{kn} \, .$$

Error sum of squares $= k(n - 1)\sigma_0^2$

$$= \Sigma x^2 - \frac{1}{n} \Sigma T^2$$

and since

$$\Sigma x^2 - \frac{1}{n} \Sigma T^2 = \left(\Sigma x^2 - \frac{G^2}{kn} \right) - \left(\frac{1}{n} \Sigma T^2 - \frac{G^2}{kn} \right)$$

we have, error s.s. = total s.s. − s.s. between groups, where s.s. stands for sum of squares.

The following gives the routine procedure.

Calculate:

1. The total for each group $= T$.
2. The total over all observations $= G$.
3. The correction factor (CF) $= G^2/kn$.
4. The total sum of squares $= \Sigma x^2 - G^2/kn$.
5. The sum of squares between groups $= \dfrac{1}{n} \Sigma T^2 - \dfrac{G^2}{kn}$.
6. The error sum of squares by subtraction $= (4) - (5)$.
7. The mean square between groups $n\sigma_1^2 = \dfrac{1}{k-1} \left(\dfrac{1}{n} \Sigma T^2 - \dfrac{G^2}{kn} \right)$.
8. The error mean square $\sigma_1^2 = \dfrac{1}{k(n-1)} \left(\Sigma x^2 - \dfrac{1}{n} \Sigma T^2 \right)$.

The calculations are assembled to form the analysis of variance table:

Source of variation	Degrees of freedom (d.f.)	Sums of squares	Mean square (m.s.)	Expectation of the mean square (e.m.s.)
Between groups	$k - 1$	$\dfrac{1}{n} \Sigma T^2 - \dfrac{G^2}{kn}$	$\left(\dfrac{1}{n} \Sigma T^2 - \dfrac{G^2}{kn} \right) \Big/ (k - 1)$	$\sigma_0^2 + n\sigma_m^2$
Error	$k(n - 1)$	$\Sigma x^2 - \dfrac{1}{n} \Sigma T^2$	$\left(\Sigma x^2 - \dfrac{1}{n} \Sigma T^2 \right) \Big/ k(n - 1)$	σ_0^2
Total	$kn - 1$	$\Sigma x^2 - \dfrac{G^2}{kn}$		

We then proceed to test the significance of the mean square between groups by calculating

$$F = \frac{\sigma_0^2 + n\sigma_m^2}{\sigma_0^2}, \quad (\phi_1, \phi_0)$$

where

$$\phi_1 = (k-1), \quad \phi_0 = k(n-1).$$

Using the data given in Example 1:

1. $T_A = 14$, $T_B = 7$, $T_C = 20$, $T_D = 18$.
2. $G = 59$.
3. $CF = (59)^2/20 = 174.05$.
4. Total s.s. $= (1^2 + 3^2 + \cdots + 2^2 + 4^2) - CF = 42.95$
5. s.s. between groups $= \frac{1}{5}(14^2 + 7^2 + 20^2 + 18^2) - CF = 19.75$.
6. Error s.s. $= (4) - (5) = 23.20$.

The analysis of variance table is as follows:

Source of variation	d.f.	s.s.	m.s.	e.m.s.
Between groups (methods)	3	19.75	6.58	$\sigma_0^2 + 5\sigma_m^2$
Within groups (error)	16	23.20	1.45	σ_0^2
Total	19	42.95		

As before

$$F = \frac{\sigma_0^2 + 5\sigma_m^2}{\sigma_0^2} = \frac{6.58}{1.45}$$

$$= 4.54 \ (3, 16).$$

In the analysis of variance table the term mean square is used rather than variance, since every mean square is not an estimate of a true variance, e.g. the between groups mean square estimates $n\sigma_1^2$ not σ_1^2.

It is important that the above procedures should be clearly understood as they form the basis of the analysis of many statistically designed experiments. Having understood the theory behind the calculations then the calculations themselves are not difficult to remember if it is noted that:

(i) whenever a total is squared the resulting square is always divided by the number of observations making up that total, e.g.

G^2/kn, G is the total of all kn observations,

$\frac{1}{n}\Sigma T^2$, T is the total of a group of n observations,

Σx^2, x is an individual observation.

(ii) having specified a source of variation the appropriate sum of squares is obtained by summing the squares of the totals for each unit in that source and subtracting the square of the total over all units, each being divided as in (i) e.g. total s.s., each individual observation is a unit and the grand total is the total over all units, therefore calculate

$$\Sigma x^2 - G^2/kn.$$

s.s. between groups.   Each group is a unit, the grand total is again the total over all units. Thus calculate

$$\frac{1}{n} \Sigma T^2 - \frac{G^2}{kn}.$$

Error s.s., i.e. the variability within groups.   Each individual observation within a given group is a unit and the group total the total over all units. For any given group we have $\Sigma x^2 - T^2/n$. Thus, calculated over all groups this becomes

$$\Sigma x^2 - \frac{1}{n} \Sigma T^2.$$

It will often be convenient to work in units other than those of the given data. A constant may be subtracted or the data may be multiplied by a constant. Thus the calculations will be made using x' rather than x, where

$$x' = k(x - x_0)$$

x_0 and k being the suitably chosen constants.

Since the tests to determine significance are ratio tests we can work entirely in the units of the computing variable (x'). However, we must revert to the original units to determine variance components and to calculate confidence limits.

Example 2

Three methods of analysis are used to determine, in parts per million, the amount of a certain constituent in a given sample. Six analyses are made by each method. Determine whether or not the methods can be assumed equivalent and put 95% confidence limits on the mean value for each method.

Method	A	B	C
	0.16	0.12	0.19
	0.15	0.14	0.18
	0.15	0.16	0.20
	0.17	0.13	0.21
	0.17	0.10	0.19
	0.18	0.12	0.18

The null hypothesis to be tested is that there is no difference between the average result obtained by each method. If σ_M^2 denotes the variance between methods

$$H_0 \equiv \sigma_M^2 = 0.$$

To simplify the calculations we subtract 0.15 from the data and multiply by 100, i.e.

$$x' = (x - 0.15)100.$$

	A	B	C
	1	−3	4
	0	−1	3
	0	1	5
	2	−2	6
	2	−5	4
	3	−3	3
Totals	8	−13	25

1. Grand total $G = 8 - 13 + 25 = 20$
2. Correction factor $CF = (20)^2/18 = 22.22$
3. Total s.s. $= (1^2 + 0^2 + \cdots + 4^2 + 3^2) - CF = 155.78$
4. s.s. between methods $= \frac{1}{6}(8^2 + (-13)^2 + 25^2) - CF = 120.78$.

The analysis of variance table is:

Source of variation	d.f.	s.s.	m.s.	e.m.s.
Between methods	2	120.78	60.39	$\sigma_0^2 + 6\sigma_M^2$
Within methods (error)	15	35.00	2.33	σ_0^2
Total	17	155.78		

$$F = \frac{60.39}{2.33} = 25.9 \ (2, 15).$$

From tables $F_{1\%}(2, 15) = 6.36$, i.e. the result is highly significant.

Thus, H_0 is rejected and it is asserted that the methods are not equivalent but give different estimates of the content in the given sample.

Confidence limits. The mean value for each method is estimated from six results. Assuming a normal distribution, averages of six observations would be distributed about the true mean with variance $\sigma_0^2/6$.

Thus the 95% confidence limits, in the units of computation, are given by, method A:

$$\bar{x}'_A \pm t_{0.025(\phi=15)} \frac{\sigma_0}{\sqrt{6}} = 1.33 \pm 2.13 \sqrt{\frac{2.33}{6}}$$

method B:

$$\bar{x}'_B \pm t_{0.025(\phi=15)} \frac{\sigma_0}{\sqrt{6}} = -2.17 \pm 2.13 \sqrt{\frac{2.33}{6}}$$

o

method C:

$$\bar{x}'_C \pm t_{0.025 \ (\phi=15)} \frac{\sigma_0}{\sqrt{6}} = 4.17 \pm 2.13 \sqrt{\frac{2.33}{6}} \, .$$

In the original units,

$$x = \frac{x'}{100} + 0.15$$

and

$$\sigma_0^2 = 2.33 \times 10^{-4}.$$

Thus, the 95% confidence limits on

A are 0.163 ± 0.013 i.e. 0.150 to 0.176
B are 0.128 ± 0.013 i.e. 0.115 to 0.141
C are 0.192 ± 0.013 i.e. 0.179 to 0.205

It is important to note that such an experiment and its analysis reveals only that the methods are *biased* and the extent of their bias *with respect to one another*. The accuracy of each can be assessed only if the true value for the sample is known.

The analysis of variance procedures can be extended to cover any number of assignable causes of variation. All variation which cannot be attributed to a known source must be classed as experimental error.

2. Two-factor analysis of variance

Suppose in the previous example the six analyses by each method were made, one by each of six different analysts.

Analyst \ Method	A	B	C	Total
1	1	−3	4	2
2	0	−1	3	2
3	0	1	5	6
4	2	−2	6	6
5	2	−5	4	1
6	3	−3	3	3
Total	8	−13	25	20

Any differences which occur due to the analysts will affect the totals for each method, but provided each method is used by each analyst all method totals should be equally affected.

We can now identify two assignable causes of variation: (i) between methods; (ii) between analysts; and the variability due to each can be independently estimated by considering respectively the totals for each method and the totals for each analyst.

We calculate:

1. The correction factor $CF = (20)^2/18 = 22.22$
2. The total s.s. $= (1^2 + 0^2 + \cdots + 4^2 + 3^2) - CF = 155.78$
3. The s.s. between methods $= \frac{1}{6}\{8^2 + (-13)^2 + 25^2\} - CF = 120.78$
4. The s.s. between analysts $= \frac{1}{3}(2^2 + 2^2 + \cdots + 1^2 + 3^2) - CF = 7.78$
5. The error s.s. and d.f. by subtraction.

The analysis of variance table is:

Source of variation	d.f.	s.s.	m.s.	e.m.s.
Between methods	2	120.78	60.39	$\sigma_0^2 + 6\sigma_M^2$
Between analysts	5	7.78	1.56	$\sigma_0^2 + 3\sigma_A^2$
Residual (error)	10	27.22	2.72	σ_0^2
Total	17	155.78		

(i) Between methods

$$F = \frac{\sigma_0^2 + 6\sigma_M^2}{\sigma_0^2}$$

$$= \frac{60.39}{2.72} = 22.2 \quad (2, 10).$$

From tables

$$F_{1\%}(2, 10) = 7.56$$

and so the result is highly significant.

(ii) Between analysts

$$F = \frac{\sigma_0^2 + 3\sigma_A^2}{\sigma_0^2}$$

$$= \frac{1.56}{2.72} < 1 \quad (5, 10),$$

which is clearly not significant and the mean square between analysts could therefore be assumed to estimate σ_0^2.

The conclusions which can be drawn from the experiment are:
(i) the methods give different results (i.e. a bias between methods);
(ii) there is no difference between analysts.

This method of analysis assumes that any bias between analysts is independent of the method used and similarly that the differences between methods are independent of the analysts. If this assumption is incorrect there is said to be an *interaction* between methods and analysts. An interaction is denoted by a product of the factors involved, e.g., methods x analysts, or by using letters to denote these factors, e.g. $M \times A$ or MA.

The experimental error has been defined as the variability between observations made under apparently identical conditions. In the above data there is only one observation for each set of conditions, i.e. one observation per analyst per method. Thus the residual sum of squares estimates the true error plus the interaction. It is impossible to show whether or not an interaction exists unless there is an independent estimate of the experimental error, either from the experiment itself or from other work.

In the simple two-factor analysis of variance it is assumed that the residual estimates the experimental error variance and the significance of the other sources of variation are tested against it.

3. Two-factor analysis of variance with replicate tests

Consider an experiment similar to that above where each of four analysts carries out two analyses on a given sample, using each of three methods.

To simplify the calculations we subtract 0.15 from the data and multiply by 100.

Analyst \\ Method	A	B	C	Total
1	1 0	−1 0	4 3	7
2	2 2	−3 −1	5 5	10
3	1 3	−4 −2	4 5	7
4	1 2	0 −3	5 3	8
Total	12	−14	34	32

The identifiable sources of variation are:
    (i) between methods;
    (ii) between analysts;
    (iii) interaction, methods x analysts;
    (iv) experimental error.
The experimental error can now be independently estimated from the pairs of results obtained under apparently identical conditions.

We calculate:

1. correction factor $CF = (32)^2/24 = 42.7$
2. total s.s. $= (1^2 + 0^2 + \cdots + 5^2 + 3^2) - CF = 171.3$
3. s.s. between methods $= \frac{1}{8}\{12^2 + (-14)^2 + 34^2\} - CF = 144.3$
4. s.s. between analysts $= \frac{1}{6}(7^2 + 10^2 + 7^2 + 8^2) - CF = 1.0$
5. interaction s.s.

Let us consider this calculation in detail. If the pair of results in each 'cell' is summed we obtain the following table:

Method / Analyst	A	B	C	Total
1	1	−1	7	7
2	4	−4	10	10
3	4	−6	9	7
4	3	−3	8	8
Total	12	−14	34	32

This is identical in form to the data table in the previous examples and the interaction sum of squares can thus be calculated by subtracting the sum of squares between methods and the sum of squares between analysts from the sum of squares of the cell totals:

$$\text{s.s. between cells} = \tfrac{1}{2}(1^2 + 4^2 + \cdots + 9^2 + 8^2) - CF$$
$$= 156.3.$$

$\therefore$ interaction s.s. = s.s. between cells

$$- (\text{s.s. between analysts} + \text{s.s. between methods})$$
$$= 156.3 - (1.0 + 144.3)$$
$$= 11.0.$$

The number of degrees of freedom associated with the interaction is obtained by multiplying together the degrees of freedom for each of the factors involved, e.g. interaction $M \times A$, $\phi = 2 \times 3 = 6$.

For the general case let

T_M = method total summed over n_M results
T_A = analyst total summed over n_A results
T_C = cell total summed over n_C results
G = grand total summed over N results.

Then

(i) total s.s. = $\Sigma x^2 - G^2/N$

(ii) s.s. between methods = $\dfrac{1}{n_M} \Sigma T_M^2 - G^2/N$

(iii) s.s. between analysts = $\dfrac{1}{n_A} \Sigma T_A^2 - G^2/N$

(iv) interaction $(M \times A)$ s.s. = $\dfrac{1}{n_C} \Sigma T_C^2 - G^2/N$
$$- \left[\left(\frac{1}{n_M} \Sigma T_M^2 - \frac{G^2}{N} \right) + \left(\frac{1}{n_A} \Sigma T_A^2 - \frac{G^2}{N} \right) \right]$$

(v) error s.s. = $\Sigma x^2 - \dfrac{1}{n_C} \Sigma T_C^2 = \text{(i)} - \{\text{(ii)} + \text{(iii)} + \text{(iv)}\}$.

The analysis of variance table will be:

Source of variation	d.f.	s.s.	m.s.	e.m.s.
Between methods (M)	2	144.3	72.15	$\sigma_0^2 + 2\sigma_{MA}^2 + 8\sigma_M^2$
Between analysts (A)	3	1.0	0.33	$\sigma_0^2 + 2\sigma_{MA}^2 + 6\sigma_A^2$
Interaction $M \times A$	6	11.0	1.83	$\sigma_0^2 + 2\sigma_{MA}^2$
Error	12	15.0	1.25	σ_0^2
Total	23	171.3		

The expectation of each mean square can be identified by considering the variance between the appropriate averages and muliplying by the factor which makes the coefficient of σ_0^2 equal to unity.

(i) *Interaction*: since there are two observations in each cell, the component of σ_0^2 will be $\sigma_0^2/2$, the variance between cells will be $\sigma_0^2/2 + \sigma_{MA}^2$ and hence the e.m.s. will be $\sigma_0^2 + 2\sigma_{MA}^2$.

(ii) *Methods and analysts*: if an interaction exists it will affect the results for both methods and analysts. The mean for each method is based on 8 observations and 4 values which estimate σ_{MA}^2. Thus the variance of the method averages is $\sigma_0^2/8 + \sigma_{MA}^2/4 + \sigma_M^2$ and the e.m.s. will be $\sigma_0^2 + 2\sigma_{MA}^2 + 8\sigma_M^2$.

Similarly the mean for an analyst is based on 6 observations and 3 values estimating σ_{MA}^2 so that the variance is $\sigma_0^2/6 + \sigma_{MA}^2/3 + \sigma_A^2$ and the e.m.s. will be $\sigma_0^2 + 2\sigma_{MA}^2 + 6\sigma_A^2$.

The significance of the interaction mean square can be tested against the error mean square. If σ_{MA}^2 does not exist then the expected value of F is unity.

If the interaction mean square is shown to be not significant, we conclude that $\sigma_{MA}^2 = 0$ and the other mean squares can also be tested against the error mean square. If however the interaction is significant the mean squares for analysts and methods must be tested against the interaction mean square.

Thus

(a) Interaction:

$$F = \frac{\sigma_0^2 + 2\sigma_{MA}^2}{\sigma_0^2} = \frac{1.83}{1.25}$$

$$= 1.46 \quad (6, 12).$$

From tables,

$$F_{10\%}(6, 12) = 2.33.$$

Thus, the result is not significant and we assume that there is no interaction between methods and analysts and that the true value of σ_{MA}^2 is zero.

(b) Analysts: since $\sigma_{MA}^2 = 0$,

$$F = \frac{\sigma_0^2 + 6\sigma_A^2}{\sigma_0^2} = \frac{0.33}{1.25}$$

$$= 0.26 \quad (3, 12)$$

i.e. not significant.

(c) Methods: since $\sigma_{MA}^2 = 0$

$$F = \frac{\sigma_0^2 + 8\sigma_M^2}{\sigma_0^2} = \frac{72.15}{1.25}$$

$$= 57.8 \quad (2, 12).$$

From tables,

$$F_{1\%}(2, 12) = 6.93,$$

i.e. the result is highly significant.

The conclusions which can be drawn from the analysis of this experiment are therefore:

(i) there is a bias between methods;

(ii) there is no bias between analysts;

(iii) there is no interaction between methods and analysts, i.e. the difference between methods is independent of the analyst carrying out the test.

Without knowing the true value of the analysis it is impossible to determine the order of reliability for the methods. If we assume that each method is specially developed there is no practical meaning to the variance between methods but confidence limits can be put on the average for each method. These limits will indicate the range within which we would expect the average for the method to lie if an infinite number of replicate tests were carried out using that method.

Thus the 95% confidence limits on each average will be given by

$$\bar{x}' \pm t_{0.025(\phi=12)} \frac{\sigma_0}{\sqrt{n}},$$

i.e. reverting to the original units

$$\bar{x} \pm 2.18 \sqrt{\left(\frac{1.25}{8}\right)} \cdot 10^{-2}$$

or

$$\bar{x} \pm 0.0086.$$

The calculated mean values for the methods will be

$$\bar{x}_A = 0.165, \quad \bar{x}_B = 0.133, \quad \bar{x}_C = 0.193,$$

and hence the 95% confidence limits will be approximately (see Fig. 18.1)

A: 0.156 to 0.174

B: 0.124 to 0.142

C: 0.184 to 0.202

It can be seen that the 95% confidence limits do not overlap, emphasizing the conclusion that there is a significant difference between methods.

It should be noted that the above analysis considers the test results on only one sample. Obviously for a full evaluation of the methods a number of samples, with different nominal percentage content, should be tested by each method.

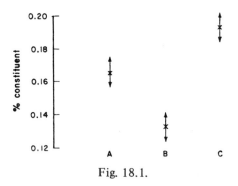

Fig. 18.1.

The analysis of variance procedures can be extended for any number of factors. Provided each level of each factor is tested over all levels of all other factors (i.e. the experimental design is symmetrical with respect to all factors), then the variability due to each identifiable source can be independently estimated. Non-symmetrical designs can be analysed, but not by the simple techniques considered in this chapter.

Suppose the previous example was repeated using more than one sample. There are now three main effects:

    (i) between methods;
    (ii) between analysts;
    (iii) between samples.

The analysis of such an experiment is given in the appendix at the end of the chapter.

In certain types of experimental work a calculated interaction can have no physical meaning. This will apply particularly in chemical sampling.

Suppose a certain chemical is delivered in batches and the object of the investigation is to determine whether the batches differ with respect to a given constituent (x). k batches are selected at random and n samples tested from each batch.

Let x_{ij} be the result on the jth sample from the ith batch, as follows:

Batch $\quad\quad\quad\quad i \rightarrow$							
Sample	1	2	.	.	.	k	
1	x_{11}	x_{21}	.	.	.	x_{k1}	
2	x_{12}	x_{22}	.	.	.	x_{k2}	
j .	.	.	.	.	.	.	
$\downarrow$.	.	.	.	x_{ij}	.	.	
.	.	.	.	.	.	.	
n	x_{1n}	x_{2n}	.	.	.	x_{kn}	
Total	B_1	B_2	.	.	.	B_k	G

There can be no interaction between batches and samples since a given sample from one batch cannot be identified with the same numbered sample from another. Similarly there is no meaning to a total for a given numbered sample calculated over all batches and hence the variability between samples must be considered within batches.

The total variability can be divided into two parts:

(i) between batches;

(ii) between samples within batches (written samples/batches).

The latter will include true sampling variability and the testing error. Without replicate tests on each sample the two sources cannot be estimated separately and the mean square will be the best available estimate of the experimental error variance σ_0^2. The analysis of variance table will therefore take the form:

Source of variation	d.f.	s.s.	e.m.s.
Between batches	$k - 1$	$\dfrac{1}{n} \Sigma B_i^2 - \dfrac{G^2}{kn}$	$\sigma_0^2 + n\sigma_B^2$
Between samples/batches	$k(n - 1)$	$\Sigma x_{ij}^2 - \dfrac{1}{n} \Sigma B_i^2$	σ_0^2
Total	$kn - 1$	$\Sigma x_{ij}^2 - \dfrac{G^2}{kn}$	

If the tests on each sample are replicated m times the variability between replicates will give an independent estimate of the testing error,

Source of variation	d.f.	s.s.	e.m.s.
Between batches	$k - 1$	$\dfrac{1}{nm} \Sigma B_i^2 - \dfrac{G^2}{knm}$	$\sigma_0^2 + m\sigma_S^2 + nm\sigma_B^2$
Between samples/batches	$k(n - 1)$	$\dfrac{1}{m} \Sigma S_f^2 - \dfrac{1}{nm} \Sigma B_i^2$	$\sigma_0^2 + m\sigma_S^2$
Testing error	$kn(m - 1)$	$\Sigma x_{ij}^2 - \dfrac{1}{m} \Sigma S_f^2$	σ_0^2
Total	$knm - 1$	$\Sigma x_{ij}^2 - \dfrac{G^2}{knm}$	

where S_f denotes the total of the m results on the fth sample.

In each case the sum of squares to estimate σ_0^2 is calculated by subtraction in the analysis of variance table.

Variability between samples will not in general affect each batch total to the same extent. Hence variability between samples will be included in the calculated mean square between batches.

Example 3

Three samples are taken from each of four deliveries of urea formaldehyde syrup. Two analyses to determine the percentage formaldehyde are made on each sample. Data given in units of 0.1% above 6%.

Sample \ Delivery	1	2	3	4
1	2	2	8	3
	1	4	5	2
2	3	3	6	1
	3	4	4	4
3	2	4	7	3
	4	5	3	1

(i) Does the % HCHO vary within a delivery?

(ii) Does the % HCHO vary between deliveries?

(iii) Put 95% confidence limits on the estimated % HCHO for each delivery.

correction factor, $CF = (84)^2/24 = 294$

total s.s. $= (2^2 + 1^2 + \cdots + 3^2 + 1^2) - CF$

$$= 368 - 294 = 74$$

s.s. between deliveries $= \frac{1}{6}(15^2 + 22^2 + 33^2 + 14^2) - CF$

$$= 332.3 - 294 = 38.3$$

s.s. between samples/deliveries $= \frac{1}{2}(3^2 + 9^2 + \cdots + 5^2 + 4^2)$

$$- \frac{1}{6}(15^2 + 22^2 + 33^2 + 14^2)$$

$$= 341 - 332.3 = 8.7$$

The analysis of variance table is as follows:

Source of variation	d.f.	s.s.	m.s.	e.m.s.
Between deliveries	3	38.3	12.8	$\sigma_0^2 + 2\sigma_S^2 + 6\sigma_D^2$
Between samples/deliveries	8	8.7	1.1	$\sigma_0^2 + 2\sigma_S^2$
Testing error	12	27.0	2.3	σ_0^2
Total	23	74.0		

(i) variability within a delivery (between samples/deliveries).

$$F = \frac{\sigma_0^2 + 2\sigma_S^2}{\sigma^2} = \frac{1.1}{2.3} = 0.48 \quad (8, 12),$$

i.e. clearly not significant and it can be assumed that a delivery is homogeneous with respect to % HCHO.

(ii) Variability between deliveries.

Having shown that σ_S^2 can be assumed zero, the mean square between deliveries will be expected to estimate $\sigma_0^2 + 6\sigma_D^2$. To assess the significance of σ_D^2 we therefore calculate,

$$F = \frac{\sigma_0^2 + 6\sigma_D^2}{\sigma_0^2}$$

$$= \frac{12.8}{2.3} = 5.56 \quad (3, 12).$$

From tables,

$$F_{5\%}(3, 12) = 3.49$$
$$F_{1\%}(3, 12) = 5.95.$$

The value of F is therefore significant and we conclude that there is real variation in % HCOH from one delivery to another.

Since σ_S^2 is assumed non-existent the sum of squares between samples/deliveries and the testing error sum of squares could be combined to give a more reliable estimate of σ_0^2, based on $\phi = 8 + 12 = 20$:

i.e.

$$\sigma_0^2 = \frac{8.7 + 27.0}{20} = 1.8.$$

The between deliveries mean square would then be tested against this value, i.e.

$$F = \frac{12.8}{1.8} = 7.1 \quad (3, 20)$$

and we again conclude a real difference between deliveries.

The procedure of combining non-significant sums of squares to give a more reliable estimate of the error variance will be of use when σ_0^2 is based on few degrees of freedom. However care should be taken if either:

(a) there are practical reasons which suggest the existence of effects which the analysis has shown non-significant; or
(b) the mean squares give an F value which approaches significance.
(iii) Taking the original estimate of $\sigma_0^2 = 2.3$ ($\phi = 12$) the 95% confidence limits on the mean for each delivery will be given by,

$$\bar{x}' \pm t_{0.025(\phi = 12)}\sigma/\sqrt{n}$$

i.e.

$$\bar{x}' \pm 2.18\sqrt{(2.3/6)}$$

i.e.

$$\bar{x}' \pm 1.35,$$

in units of 0.1%. We have

$$\bar{x}_1 = 6 + \tfrac{1}{10} \cdot \tfrac{15}{6} = 6.25\%$$

$$\bar{x}_2 = 6 + \tfrac{1}{10} \cdot \tfrac{22}{6} = 6.37\%$$

$$\bar{x}_3 = 6 + \tfrac{1}{10} \cdot \tfrac{33}{6} = 6.55\%$$

$$\bar{x}_4 = 6 + \tfrac{1}{10} \cdot \tfrac{14}{6} = 6.23\%$$

and the 95% confidence limits will be approximately

1. 6.25 ± 0.14 i.e. 6.11 to 6.39%
2. 6.37 ± 0.14 i.e. 6.23 to 6.51%
3. 6.55 ± 0.14 i.e. 6.41 to 6.69%
4. 6.23 ± 0.14 i.e. 6.09 to 6.37%.

APPENDIX

Suppose k samples are taken and each of q analysts makes n determinations with each of m methods. The analysis of variance table will take the form:

Source of variation		d.f.	e.m.s.
Main effects	Between samples (S)	$k - 1$	$\sigma_0^2 + n\sigma_{SAM}^2 + qn\sigma_{SM}^2 + mn\sigma_{SA}^2 + qmn\sigma_S^2$
	Between analysts (A)	$q - 1$	$\sigma_0^2 + n\sigma_{SAM}^2 + kn\sigma_{AM}^2 + mn\sigma_{SA}^2 + kmn\sigma_A^2$
	Between methods (M)	$m - 1$	$\sigma_0^2 + n\sigma_{SAM}^2 + kn\sigma_{AM}^2 + qn\sigma_{SM}^2 + kqn\sigma_M^2$
Interactions	$S \times A$	$(k-1)(q-1)$	$\sigma_0^2 + n\sigma_{SAM}^2 + mn\sigma_{SA}^2$
	$S \times M$	$(k-1)(m-1)$	$\sigma_0^2 + n\sigma_{SAM}^2 + qn\sigma_{SM}^2$
	$A \times M$	$(q-1)(m-1)$	$\sigma_0^2 + n\sigma_{SAM}^2 + kn\sigma_{AM}^2$
	$S \times A \times M$	$(k-1)(q-1)(m-1)$	$\sigma_0^2 + n\sigma_{SAM}^2$
	Error	$kqm(n-1)$	σ_0^2
Total		$kqmn - 1$	

The sum of squares for each source of variation is calculated as for only one or two factors by summing over all levels of the other factors.

Suppose there are 3 samples (with differing nominal % content) and 2 analysts making 2 determinations with each of 3 methods; the following results being obtained:

Sample	1		2		3		
Method \\ Analyst	I	II	I	II	I	II	Total
A	1 2	0 1	5 4	6 5	10 12	11 14	71
B	2 2	2 3	7 8	5 7	14 14	13 14	91
C	3 2	3 3	7 9	6 7	13 13	14 16	96
Total	12	12	40	36	76	82	
Total	24		76		158		258

Totals   Sample (T_S) 1 = 24   Method (T_M) A = 71   Analyst (T_A) I = 128

$\qquad\qquad\qquad$ 2 = 76 $\qquad\qquad\qquad$ B = 91 $\qquad\qquad\qquad\quad$ II = 130

$\qquad\qquad\qquad$ 3 = 158 $\qquad\qquad\qquad$ C = 96

$\qquad\qquad\quad$ CF = $(258)^2/36$ = 1849.

Total s.s.

$$= \Sigma x^2 - G^2/knmq = (1^2 + 2^2 + \cdots + 14^2 + 16^2) - 1849 = 821.0$$

s.s. between samples

$$= \frac{1}{mnq} \Sigma T_S^2 - \frac{G^2}{knmq} = \tfrac{1}{12}(24^2 + 76^2 + 158^2) - 1849 = 760.67$$

s.s. between analysts

$$= \frac{1}{kmn} \Sigma T_A^2 - \frac{G^2}{knmq} = \tfrac{1}{18}(128^2 + 130^2) - 1849 = 0.11$$

s.s. between methods

$$= \frac{1}{knq} \Sigma T_M^2 - \frac{G^2}{knmq} = \tfrac{1}{12}(71^2 + 91^2 + 96^2) - 1849 = 29.17$$

Interaction $S \times A$

Summing over methods

A \\ S	1	2	3
I	12	40	76
II	12	36	82

Let cell total = c, then

$$S \times A = \frac{1}{mn} \Sigma c^2 - \frac{G^2}{knmq}$$

$\quad$ — (s.s. between samples + s.s. between analysts)

$\quad = \frac{1}{6}(12^2 + 12^2 + 40^2 + 36^2 + 76^2 + 82^2) - 1849 - (760.67 + 0.11)$

$\quad = 4.22.$

Interaction $S \times M$

Summing over analysts

M \ S	1	2	3
A	4	20	47
B	9	27	55
C	11	29	56

Let cell total = d, then

$$S \times M = \frac{1}{qn} \Sigma d^2 - \frac{G^2}{knmq}$$

$\quad$ — (s.s. between samples + s.s. between methods)

$\quad = \frac{1}{4}(4^2 + 9^2 + 11^2 + 20^2 + 27^2 + 29^2 + 47^2 + 55^2 + 56^2)$

$\quad - 1849 - (706.67 + 29.17)$

$\quad = 0.66.$

Interaction $A \times M$

Summing over samples

A \ M	1	2	3
I	34	47	47
II	37	44	49

Let cell total = e, then

$$A \times M = \frac{1}{kn} \Sigma e^2 - \frac{G^2}{knmq}$$

$\quad$ — (s.s. between analysts + s.s. between methods)

$\quad = \frac{1}{6}(34^2 + 37^2 + 47^2 + 44^2 + 47^2 + 49^2) - 1849$

$\quad - (0.11 + 29.17)$

$\quad = 1.72.$

Interaction $S \times A \times M$

Summing over duplicate observations

S		1		2		3
	A					
	I	II	I	II	I	II
M						
A	3	1	9	11	22	25
B	4	5	15	12	28	27
C	5	6	16	13	26	30

Let cell total $= f$:

$$S \times A \times M = \frac{1}{n} \Sigma f^2 - \frac{G^2}{knmq} - \{\text{s.s.}(S) + \text{s.s.}(A) + \text{s.s.}(M) + \text{s.s.}(S \times A)$$

$$+ \text{s.s.}(S \times M) + \text{s.s.}(A \times M)\}$$

$$= \tfrac{1}{2}(3^2 + 4^2 + \cdots + 27^2 + 30^2) - 1849 - (760.67 + 0.11 + 29.17$$

$$+ 4.22 + 0.66 + 1.72)$$

$$= 7.45.$$

The error s.s. can then be obtained by subtraction; a check will be given by

$$\text{error s.s.} = \Sigma x^2 - \frac{1}{n} \Sigma f^2$$

$$= 2670 - 2653$$

$$= 17.00$$

or

$$\text{error s.s.} = \Sigma x^2 - \frac{1}{n} \Sigma f^2 = \tfrac{1}{2} \Sigma \alpha^2$$

(where $\alpha =$ difference between duplicate observations)

$$= \tfrac{1}{2} \times 34$$

$$= 17.00.$$

The analysis of variance table is as follows:

Source of variation		d.f.	s.s.	m.s.	e.m.s.
Main effects	S	2	760.67	380.34	$\sigma_0^2 + 2\sigma_{SAM}^2 + 4\sigma_{SM}^2$ $+ 6\sigma_{SA}^2 + 12\sigma_S^2$
	A	1	0.11	0.11	$\sigma_0^2 + 2\sigma_{SAM}^2 + 6\sigma_{AM}^2$ $+ 6\sigma_{SA}^2 + 18\sigma_A^2$
	M	2	29.17	14.59	$\sigma_0^2 + 2\sigma_{SAM}^2 + 6\sigma_{AM}^2$ $+ 4\sigma_{SM}^2 + 12\sigma_M^2$
Interactions	$S \times A$	2	4.22	2.11	$\sigma_0^2 + 2\sigma_{SAM}^2 + 6\sigma_{SA}^2$
	$S \times M$	4	0.66	0.17	$\sigma_0^2 + 2\sigma_{SAM}^2 + 4\sigma_{SM}^2$
	$A \times M$	2	1.72	0.86	$\sigma_0^2 + 2\sigma_{SAM}^2 + 6\sigma_{AM}^2$
	$S \times A \times M$	4	7.45	1.86	$\sigma_0^2 + 2\sigma_{SAM}^2$
Error		18	7.00	0.94	σ_0^2
Total		35	821.00		

Since the samples were chosen with different nominal % content the mean square between samples has no practical meaning.

We now proceed to test the significance of the effects starting with the highest order interaction.

The 3-factor interaction being 'not significant', the 2-factor interactions can be tested against the error mean square. None of the 2-factor interactions are significant so that the main effects can also be tested against the error mean square.

Clearly there is no significant difference between analysts, but for methods we have,

$$F = \frac{14.59}{0.94} = 15.5 \quad (2, 18).$$

From tables,

$$F_{1\%}(2, 16) = 6.23.$$

Hence the data indicate a real bias between methods. Which method is the most accurate can only be determined if the exact % content for each sample is known.

(The interaction and error mean squares could have been combined to give an estimate of $\sigma_0^2 = 1.04(\phi_0 = 30)$.)

PROBLEMS

1. Pellets are mass-produced by the automatic compression of a powder. The specific gravity is determined of each of a random sample of five pellets, from each of four machines. Is there a significant difference between machines with respect to the specific gravity of the pellets they produce?

Machine	1	2	3	4
	1.72	1.68	1.70	1.70
	1.70	1.69	1.72	1.69
	1.71	1.71	1.70	1.72
	1.69	1.70	1.71	1.69
	1.73	1.69	1.72	1.71

2. A sample of fertilizer, taken from each of six batches is divided into three parts and analysed for percent potash by three different analysts.

Batch / Analyst	1	2	3	4	5	6
A	15.1	15.3	14.9	15.5	15.4	14.8
B	14.7	15.6	15.4	15.6	14.8	15.1
C	15.2	15.5	15.3	15.6	15.2	15.2

(i) Is there any bias between analysts?
(ii) Does the potash content vary from one batch to another?
(iii) Put 95% confidence limits on the estimated potash content of batch 4.
(iv) What assumption must be made in order to analyse this experimental data?

3. To minimize the corrosion action in aqueous antifreeze solutions an inhibitor is incorporated in the concentrated antifreeze. The effect of four inhibitors is to be tested in three types of antifreeze. Metal samples are allocated at random, two to each antifreeze/inhibitor combination. All samples are immersed for an equal time and the corrosive action assessed by the loss in weight. The results are given in the following table in milligrams per square decimetre per day:

Inhibitor / Antifreeze	A	B	C	D
1	16	17	16	14
	15	18	16	15
2	16	15	13	17
	16	17	14	15
3	16	17	15	13
	15	18	14	14

What conclusions can be drawn from an analysis of the data?

4. Four samples are taken from each of three types of vulcanized rubber. The percentage of free carbon in each is determined and the results are given in the following table:

Sample \ Rubber	A	B	C
1	27.9	28.0	30.6
2	28.0	30.0	30.0
3	29.0	29.2	29.0
4	29.5	29.0	29.5

Is there sufficient evidence to assert that the amount of free carbon present differs for the three rubbers?

5. The compressive strength is to be determined for four kinds of Portland cement. Three samples are taken from a mix of each cement and two test cubes made from each sample. The following results were obtained. The data give strength in kg cm^{-2}.

Sample \ Cement	I	II	III	IV
1	58	74	69	59
	60	80	86	59
2	70	64	80	70
	68	66	75	67
3	66	70	65	65
	64	72	70	66

Does the compressive strength differ from one cement to another?

6. An isotope filter is made by brushing calcium fluoride powder on to strips of sintered nickel. An experiment is designed to investigate the effect of different grades of powder and different brush pressures, on the porosity of the filter. The porosity is measured in units of γ and the results given in the following table. Three tests are made for each powder/brush combination.

Brush pressure \ Powder	A	B	C
1	4.1	4.3	4.3
	4.2	4.4	4.2
	4.5	4.3	4.1
2	4.4	4.4	4.4
	4.6	4.5	4.5
	4.3	4.7	4.5
3	4.1	4.5	4.4
	4.6	4.2	4.4
	4.2	4.4	4.5

Is the porosity affected by changing:
 (i) the grade of calcium fluoride;
(ii) the brush pressure?

CHAPTER 19

Regression Analysis

So far we have considered the analysis of data which arises from the measurement of one particular characteristic. We have compared groups of such data to determine whether or not those groups differ with respect to that characteristic. Often the problem is not the analysis of one characteristic or parameter but the investigation of a possible functional relationship between two or more. Such relationships will be of two types:

   (i) those governed by exact physical or chemical laws, e.g.

       force = mass × acceleration

       $2H_2 + O_2 \rightarrow 2H_2O$

   (ii) those where the relationship is:

      (a) subject to experimental error, e.g. the temperature of a liquid is measured simultaneously during heating, by a centigrade and a fahrenheit thermometer, the measured temperatures will be related by the formula, $°F = \frac{9}{5} °C + 32$.

      (b) subject to experimental error and only an approximate relationship, e.g. the specific gravity (s) of an aqueous solution of HCl is dependent upon the amount of gas (v) dissolved in the water, the approximate relationship being, $s = A + Bv$, where A and B are constants.

Type (i) are not statistical problems. The relationship having been determined it is expected to be obeyed exactly. In type (ii) a relationship is determined from data available, but some variability is expected so that observed values will not in general coincide with those predicted from the relationship.

We shall be concerned only with problems of type (ii). They can be considered in two stages:

   1. deciding what form of relationship shall be fitted and determining the equation of that form which best fits the available data;

   2. deciding how well the data are represented by the fitted equation and putting confidence limits on predicted values.

These procedures are known as *regression analysis*, the fitted equation being called the *regression equation*.

By taking sufficient terms an equation can be determined which will exactly represent any given set of data. When the relationship between the variables is not

exact then, in general, if a further set of observations is taken they will not conform to the calculated equation.

In this chapter we shall sonsider the practical problem where an approximate, and
essentially simple equation relating the variables is required. The knowledge of such
a relationship will enable one parameter to be estimated (within known limits of
accuracy) from measurements of the other variables. This will be of use if, for
example, there is a simple relationship between two parameters, one of which is easy
to measure, the other difficult, or perhaps involving a destructive test. The second
can then be estimated from a measurement of the first.

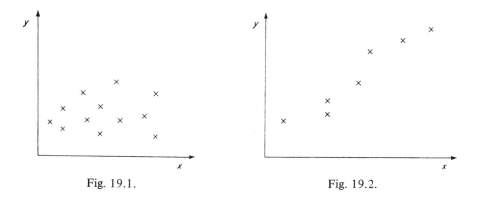

Fig. 19.1.   Fig. 19.2.

Suppose we are interested in a possible relationship between two variables x and
y. Assume that a number of pairs of observations (x_i, y_i) have been taken covering
the range of interest. The first step should always be to plot the available data in the
form of a *scatter diagram*.

In Fig. 19.1 it is clear that no simple relationship exists between the two variables
whereas in Fig. 19.2 an approximately linear relationship is indicated. Often the form
of the relationship will be indicated by practical knowledge.

1. Determination of the linear regression equation

Let us consider the simple case where a variable y is expected to be linearly dependent upon the variable x. We wish to determine the best linear equation to estimate
y for any measured value of x.

First we define the best equation as that which gives minimum error in prediction.
Any line through the average values $(\bar{x}, \bar{y})$ will give zero sum of errors; it is also
important to have consistency in prediction, i.e. to minimize the variability of the
y values about the fitted line.

Let the equation of the line be

$$y = a + bx$$

where a and b are constants.

Since the point $(\bar{x}, \bar{y})$ lies on this line the equation can be written

$$y - \bar{y} = b(x - \bar{x})$$

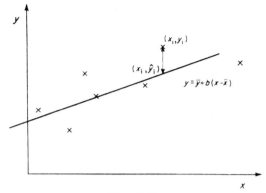

Fig. 19.3.

or

$$y = \bar{y} + b(x - \bar{x}).$$

We then determine the value of b (the slope of the line) which will minimize the standard deviation of the errors in prediction.

Let (x_i, y_i), $i = 1, 2 \ldots, n$, be any observed pair of values. Let $\hat{y}_i$ be the value of y, predicted from the equation, corresponding to x_i, i.e.

$$\hat{y}_i = \bar{y} + b(x_i - \bar{x}).$$

Then the error in prediction is $y_i - \hat{y}_i = y_i - \bar{y} - b(x_i - \bar{x})$.

The standard deviation of such errors will be minimized by minimizing the sum of squares:

$$U = \sum_{i=1}^{n} \{y_i - \bar{y} - b(x_i - \bar{x})\}^2. \tag{1}$$

This procedure is known as the *method of least squares.*

We have

$$U = \sum_{i=1}^{n} \{(y_i - \bar{y}) - b(x_i - \bar{x})\}^2.$$

For maximum or minimum U, $dU/db = 0$. Now

$$\frac{dU}{db} = -2 \sum_{i=1}^{n} \{(y_i - \bar{y}) - b(x_i - \bar{x})\}(x_i - \bar{x})$$

$$= -2 \left(\sum_{i=1}^{n} (x_i - \bar{x})(y_i - \bar{y}) - b \sum_{i=1}^{n} (x_i - \bar{x})^2 \right).$$

Thus, for maximum or minimum U

$$b = \frac{\sum_{i=1}^{n} (x_i - \bar{x})(y_i - \bar{y})}{\sum_{i=1}^{n} (x_i - \bar{x})^2}. \tag{2}$$

That this is a true minimum is verified since

$$\frac{d^2 U}{db^2} = 2 \sum_{i=1}^{n} (x_i - \bar{x})^2$$

which is positive.

By expanding and putting

$$\bar{x} = \frac{\sum_{i=1}^{n} x_i}{n}, \quad \bar{y} = \frac{\sum_{i=1}^{n} y_i}{n}$$

it can be shown that

$$b = \frac{\sum_{i=1}^{n} (x_i - \bar{x})(y_i - \bar{y})}{\sum_{i=1}^{n} (x_i - \bar{x})^2} = \frac{\sum_{i=1}^{n} x_i y_i - \frac{1}{n} \sum_{i=1}^{n} x_i \sum_{i=1}^{n} y_i}{\sum_{i=1}^{n} x_i^2 - \frac{1}{n} \left(\sum_{i=1}^{n} x_i \right)^2}$$

which in general gives a more convenient form for calculations.

Thus the equation of the line of best fit, i.e. the regression line of y on x, is given by

$$y = \bar{y} + b(x - \bar{x})$$

where

$$b = \frac{\sum_{i=1}^{n} x_i y_i - \frac{1}{n} \sum_{i=1}^{n} x_i \sum_{i=1}^{n} y_i}{\sum_{i=1}^{n} x_i^2 - \frac{1}{n} \left(\sum_{i=1}^{n} x_i \right)^2}$$

and is known as the *regression coefficient.*

2. Variation about the fitted line

Whatever the true functional relationship may be for any set of data we can determine the best linear equation to represent that data. The equation itself gives no indication as to how reliable we can expect the predicted y values to be. Thus having determined the equation we have to assess its use in the prediction of y values.

Suppose measurements are made of only the y characteristic. The best estimate of the mean of the y population will be given by $\bar{y}$, the mean of the n measured values. The variability of the population about the mean will be estimated by the sum of squares $\Sigma(y - \bar{y})^2$. In general we can assume that the distribution of the y values will be approximately normal, so that from these estimates confidence limits can be calculated within which further data, from the same population, will be expected to lie.

Now suppose we have the x value corresponding to each y and from this data we calculate the regression line of y on x. The variability of the predicted y values is now the variability about the fitted line, i.e. the sum of squares given by the minimum value of U.

From equation (1),

$$U = \Sigma(y_i - \bar{y})^2 - 2b\,\Sigma(x_i - \bar{x})(y_i - \bar{y}) + b^2\,\Sigma(x_i - \bar{x})^2$$

the summation extending from 1 to n.

Substituting for b from equation (2) gives

$$U = \Sigma(y_i - \bar{y})^2 - \frac{2\{\Sigma(x_i - \bar{x})(y_i - \bar{y})\}^2}{\Sigma(x_i - \bar{x})^2}$$

$$+ \left(\frac{\Sigma(x_i - \bar{x})(y_i - \bar{y})}{\Sigma(x_i - \bar{x})^2}\right)^2 \Sigma(x_i - \bar{x})^2$$

$$= \Sigma(y_i - \bar{y})^2 - \frac{\{\Sigma(x_i - \bar{x})(y_i - \bar{y})\}^2}{\Sigma(x_i - \bar{x})^2}$$

Thus by using the linear relationship between x and y the prediction range for y is reduced (see Fig. 19.4).

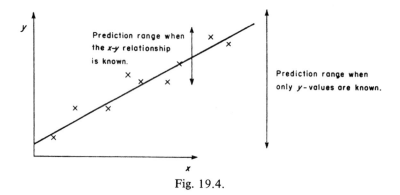

Fig. 19.4.

We wish to know whether or not it has been reduced by a significant amount. Summarizing, we have:

(i) Without the knowledge of the regression equation, a sample of n values has been drawn from the y population. The best estimate of any further value, drawn at random is $\bar{y}$. The sum of squares representing the error in prediction is:

the total sum of squares $= \Sigma(y_i - \bar{y})^2$.

(ii) The regression equation of y on x is known and having measured x_i, the best estimate of y_i is given by

$$\hat{y}_i = \bar{y} + b(x_i - \bar{x}).$$

The sum of squares now representing the error in prediction is the minimum value of U, i.e.

$$U = \Sigma(y_i - \bar{y})^2 - \frac{\{\Sigma(x_i - \bar{x})(y_i - \bar{y})\}^2}{\Sigma(x_i - \bar{x})^2}.$$

The difference between the sums of squares in (i) and (ii) represents the sums of squares accounted for by the regression, i.e. the sum of squares explained by the regression is

$$\frac{[\Sigma(x_i - \bar{x})(y_i - \bar{y})]^2}{\Sigma(x_i - \bar{x})^2}.$$

The significance of the regression can be assessed by using the analysis of variance procedures.

We calculate the total sum of squares and the sum of squares explained by the regression, obtaining by subtraction the sum of squares about the regression.

Thus

s.s. about regression = total s.s. − s.s. explained by regression.

Hence the analysis of variance table will be:

Source of variation	d.f.	s.s.	m.s.
Explained by regression	1	$\dfrac{\{\Sigma(x_i - \bar{x})(y_i - \bar{y})\}^2}{\Sigma(x_i - \bar{x})^2}$	M_1
About regression	$n - 2$	$\Sigma(y_i - \bar{y})^2 - \dfrac{\{\Sigma(x_i - \bar{x})(y_i - \bar{y})\}^2}{\Sigma(x_i - \bar{x})^2}$	M_2
Total	$n - 1$	$\Sigma(y_i - \bar{y})^2$	

Since one degree of freedom is 'lost' with each parameter calculated from a set of data, one degree of freedom will be used in calculating the slope and this degree of freedom is associated with the fitted line. (We have already allowed for the degree of freedom associated with the mean value.)

We now adopt the appropriate null hypothesis. That is, that there is no real linear relationship between the variables and hence that the mean square associated with the fitted line is not significantly greater than the mean square representing the variability about the line.

The hypothesis is tested by calculating

$$F = \frac{M_1}{M_2}, \quad (1, n - 2)$$

and referring to F-tables to assess its significance.

3. The standard error of the regression coefficient

If the equation of the line was estimated from an infinite number of pairs of observations the constants in that equation (i.e. $\bar{y}$ and b) would be exact. Since the equation will, in general, be determined from a limited amount of data, the parameters calculated will be subject to the experimental error variance of the observed y values. The best estimate of this error variance will be given by the mean square about the regression line, i.e. $\sigma_0^2 = M_2$.

We have calculated

$$b = \frac{\sum_{i=1}^{n}(x_i - \bar{x})(y_i - \bar{y})}{\sum_{i=1}^{n}(x_i - \bar{x})^2} .$$

This equation can be re-written

$$b = \frac{\sum_{i=1}^{n}(x_i - \bar{x})y_i - \bar{y}\sum_{i=1}^{n}(x_i - \bar{x})}{\sum_{i=1}^{n}(x_i - \bar{x})^2}$$

$$= \frac{\sum_{i=1}^{n}(x_i - \bar{x})y_i}{\sum_{i=1}^{n}(x_i - \bar{x})^2}$$

since

$$\sum_{i=1}^{n}(x_i - \bar{x}) = 0.$$

$$\therefore\ b = \frac{1}{\sum_{i=1}^{n}(x_i - \bar{x})^2}\left\{(x_1 - \bar{x})y_1 + (x_2 - \bar{x})y_2 + \cdots + (x_n - \bar{x})y_n\right\}$$

$$\therefore\ \text{var}(b) = \frac{1}{\{\sum_{i=1}^{n}(x_i - \bar{x})^2\}^2}\{(x_1 - \bar{x})^2\ \text{var}(y_1)$$

$$+ (x_2 - \bar{x})^2\,\text{var}(y_2) \qquad + \cdots + (x_n - \bar{x})^2\,\text{var}(y_n)\}.$$

Thus, since the experimental error variance of the observed y values is σ_0^2, we have

$$\text{var}(b) = \frac{1}{\{\sum_{i=1}^{n}(x_i - \bar{x})^2\}^2}\left(\sigma_0^2\sum_{i=1}^{n}(x_i - \bar{x})^2\right)$$

i.e.

$$\text{var}(b) = \frac{\sigma_0^2}{\sum_{i=1}^{n}(x_i - \bar{x})^2}$$

and

$$\text{S.E.}(b) = \frac{\sigma_0}{\sqrt{\{\sum_{i=1}^{n}(x_i - \bar{x})^2\}}}$$

The $100(1 - \alpha)\%$ confidence limits on the true value of b will be given by

$$b \pm t_{\frac{1}{2}\alpha(\phi=n-2)} \cdot \text{S.E.}(b)$$

i.e.

$$b \pm t_{\alpha/2(\phi=n-2)} \cdot \frac{\sigma_0}{\sqrt{\{\sum_{i=1}^{n}(x_i - \bar{x})^2\}}} .$$

4. The standard error of the estimated y

The estimate of y_i for any measured x_i is

$$\hat{y}_i = \bar{y} + b(x_i - \bar{x}),$$

and the variance of this estimate will be given by

$$\text{var}(\hat{y}_i) = \text{var}(\bar{y}) + (x_i - \bar{x})^2 \, \text{var}(b)$$

$$= \frac{\sigma_0^2}{n} + (x_i - \bar{x})^2 \frac{\sigma_0^2}{\sum_{i=1}^n (x_i - \bar{x})^2}$$

$$\therefore \text{S.E.}(\hat{y}_i) = \sigma_0 \sqrt{\left(\frac{1}{n} + \frac{(x_i - \bar{x})^2}{\sum_{i=1}^n (x_i - \bar{x})^2}\right)}$$

The standard error of the estimate is therefore a minimum when $x_i = \bar{x}$ and increases as $(x_i - \bar{x})^2$ increases.

The $100(1 - \alpha)\%$ confidence limits on the estimate will be given by

$$\hat{y}_i \pm t_{\alpha/2\,(\phi=n-2)} \cdot \sigma_0 \sqrt{\left(\frac{1}{n} + \frac{(x_i - \bar{x})^2}{\sum\limits_{i=1}^n (x_i - \bar{x})^2}\right)}$$

Example 1

Given the following data:

x	−1	0	1	2	2	3
y	0	1	1	2	3	3

(i) Determine the best linear equation to predict y from a given x.
(ii) Test the significance of the relationship.
(iii) Put 90% confidence limits on the expected value of y when $x = 1$.

(i) We first plot the scatter diagram (Fig. 19.5).

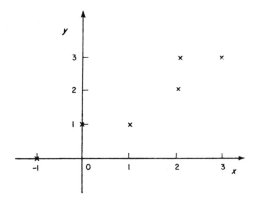

Fig. 19.5.

Since there appears to be an approximately linear relationship we proceed with the calculations. We require to calculate Σx, Σy, Σx^2, Σxy and Σy^2.

	x	y	x^2	xy	y^2
	-1	0	1	0	0
	0	1	0	0	1
	1	1	1	1	1
	2	2	4	4	4
	2	3	4	6	9
	3	3	9	9	9
Totals	7	10	19	20	24

$\bar{x} = \frac{7}{6} = 1.17$

$\bar{y} = \frac{10}{6} = 1.67.$

$$b = \frac{\Sigma x_i y_i - \dfrac{1}{n}\Sigma x_i \Sigma y_i}{\Sigma x_i^2 - \dfrac{1}{n}\left(\Sigma x_i\right)^2} = \frac{20 - \frac{1}{6}\cdot 7 \cdot 10}{19 - \frac{1}{6}(7)^2} = \frac{8.33}{10.33}$$

$$= 0.77.$$

Thus, the equation of the line of best fit is

$$y = \bar{y} + b(x - \bar{x})$$

i.e.

$$y = 1.67 + 0.77(x - 1.17)$$

i.e.

$$y = 0.77x + 0.77.$$

The line is shown on the scatter diagram in Fig. 19.6.

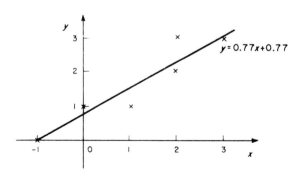

Fig. 19.6.

(ii) The total sum of squares

$$\Sigma(y_i - \bar{y})^2 = \Sigma y_i^2 - \frac{1}{n}\left(\Sigma y_i\right)^2 = 24 - \tfrac{1}{6}(10)^2$$

$$= 7.3$$

the sum of squares explained by the regression

$$= \frac{\{\Sigma(x_i - \bar{x})(y_i - \bar{y})\}^2}{\Sigma(x_i - \bar{x})^2}$$

$$= \frac{(8.33)^2}{10.83} = 6.4.$$

Source of variation	d.f.	s.s.	m.s.
Explained by regression	1	6.4	6.4
About regression	4	0.9	0.23
Total	5	7.3	

$$F = \frac{6.4}{0.23} = 27.8 \quad (1, 4).$$

From tables,

$$F_{1\%}(1, 4) = 21.2$$

i.e. the result is highly significant, indicating that the use of the relationship increases the accuracy in prediction.

(iii)

$$\text{var}(\bar{y}) = \frac{\sigma_0^2}{n} = \frac{0.23}{6} = 0.038,$$

$$\text{var}(b) = \frac{\sigma_0^2}{\Sigma(x_i - \bar{x})^2} = \frac{0.23}{10.83} = 0.021,$$

and since

$$y = \bar{y} + b(x - \bar{x})$$

$$\text{var}(y) = \text{var}(\bar{y}) + (x - \bar{x})^2 \, \text{var}(b).$$

Thus when $x = 1$, $y = 1.54$

$$\text{var}(y_{x=1}) = 0.038 + (0.17)^2(0.021)$$

$$\approx 0.039.$$

The 90% confidence limits on the estimate of y_i are

$$\hat{y}_i \pm t_{0.05(\phi=4)} \cdot \text{S.E.}(y_i)$$

and so 90% confidence limits on y when $x = 1$ are

$$1.54 \pm 2.13\sqrt{(0.039)}$$

or

$$1.12 \text{ to } 1.96.$$

Example 2

In a certain chemical process it is known that the temperature affects the amount of a given impurity A. Given the following data:

Temperature (°C)	81	84	83	88	85	90	87
Amount of A(%)	0.1	0.3	0.2	0.4	0.3	0.4	0.3

(i) Determine the best linear equation to assess the amount of A likely to be present at any given temperature between 80 °C and 90 °C.

(ii) Does the use of this relationship increase the accuracy of the estimated amount of A?

(iii) Calculate the 95% confidence limits on the fitted line.

The scatter diagram (Fig. 19.7) indicates that there may be an approximately linear relationship over the temperature range covered by the available data.

(i) To simplify the calculations let

$$x = (°C - 85) \quad \text{and} \quad y = 100A.$$

	x	y	x^2	xy	y^2
	−4	1	16	−4	1
	−1	3	1	−3	9
	−2	2	4	−4	4
	3	4	9	12	16
	0	3	0	0	9
	5	4	25	20	16
	2	3	4	6	9
Totals	3	20	59	27	64

$\bar{x} = \frac{3}{7} = 0.43$

$\bar{y} = \frac{20}{7} = 2.86$

$$b = \frac{27 - \frac{1}{7} \cdot 3 \cdot 20}{59 - \frac{1}{7} \cdot (3)^2} = \frac{18.43}{57.71} = 0.319.$$

Thus the equation of the line in computing units is

$$y = 2.86 + 0.319(x - 0.43)$$

i.e.

$$y = 0.319x + 2.72.$$

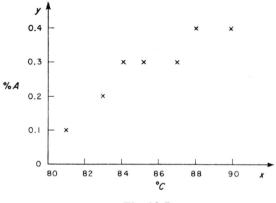

Fig. 19.7.

In the original units

$$10A = 0.319(°C - 85) + 2.72$$

i.e.

$$\%A = 0.0319(°C) - 2.44.$$

The line is plotted on the scatter diagram (Fig. 19.8).

(ii) Since the test of significance is a ratio test, the analysis of variance can be carried out in the units of computation

$$\text{total s.s.} = \Sigma y_i^2 - \frac{1}{n}\left(\Sigma y_i\right)^2 = 64 - \tfrac{1}{7}(20)^2 = 6.86$$

$$\text{s.s. explained by regression} = \frac{\left(\Sigma x_i y_i - \frac{1}{n}\Sigma x_i \Sigma y_i\right)^2}{\Sigma x_i^2 - \frac{1}{n}\left(\Sigma x_i\right)^2} = \frac{(18.43)^2}{57.71} = 5.88,$$

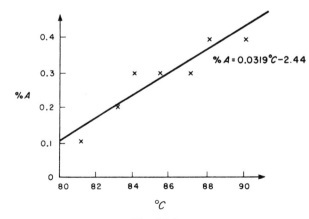

Fig. 19.8.

giving the analysis of variance table:

Source of variation	d.f.	s.s.	m.s.
Explained by regression	1	5.88	5.88
About regression	5	0.98	0.20
Total	6	6.86	

$$F = \frac{5.88}{0.20} = 29.4 \quad (1, 5).$$

From tables,

$$F_{1\%}(1, 5) = 16.3.$$

Hence the result is highly significant, indicating that an appreciable amount of variability is removed by fitting the line and hence the accuracy in the prediction of the amount of A is increased by the use of the relationship.

(iii) To calculate the confidence limits we must transform back to the original units.

In computing units $\sigma_0^2 = 0.20$, $\bar{x} = 0.43$. Thus in measured units $\sigma_0^2 = 0.0020$, $\bar{C} = 85.43$.

The standard error of any estimated A_i will be given by

$$\text{S.E.}(\hat{A}_i) = \sigma_0 \sqrt{\left(\frac{1}{n} + \frac{(C_i - \bar{C})^2}{\Sigma(C_i - \bar{C})^2}\right)}.$$

Since the computing unit x is C minus a constant

$$(C_i - \bar{C}) = (x_i - \bar{x})$$

and

$$\Sigma(C_i - \bar{C})^2 = \Sigma(x_i - \bar{x})^2 = 57.71.$$

$$\therefore \quad \text{S.E.}(\hat{A}_i) = \sqrt{\left\{0.002\left(\frac{1}{7} + \frac{C_i - 85.43}{57.71}\right)\right\}}$$

and when

$$C = 80, \quad \text{S.E.}(A_{C=80}) = 0.036$$
$$C = 85, \quad \text{S.E.}(A_{C=85}) = 0.017$$
$$C = 90, \quad \text{S.E.}(A_{C=90}) = 0.032.$$

The 95% confidence limits on each estimated A_i are given by

$$\hat{A}_i \pm t_{0.025(\phi=5)} \cdot \text{S.E.}(\hat{A}_i).$$

When $C = 80$

$$\%A = 0.11$$

and the 95% confidence limits on A are

$$0.11 \pm 2.57 \cdot 0.036 \quad \text{i.e. } 0.02 \text{ to } 0.20.$$

When $C = 85$

> $\%A = 0.27$

and the 95% confidence limits on A are

> $0.27 \pm 2.57 \cdot 0.017$ i.e. 0.23 to 0.31.

When $C = 90$

> $\%A = 0.43$

and the 95% confidence limits on A are

> $0.43 \pm 2.57 \cdot 0.032$ i.e. 0.35 to 0.51.

These limits are shown plotted about the regression line (Fig. 19.9).

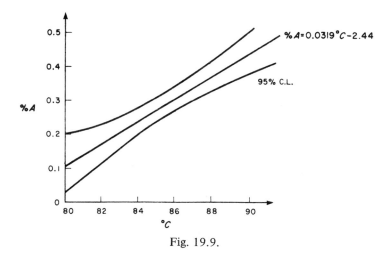

Fig. 19.9.

Although for clarity a separate diagram has been used at each stage in the above examples, in practice the regression line and its confidence limits would be plotted on the original scatter diagram.

So far we have made the asumption that the x variable can be accurately measured and all the variability is in the measurement of y.

Similar calculations would be made if the value of x was to be predicted from an accurately measured y. The equation to be fitted is then called the *regression line of x on y*. By symmetry the equation will be

> $x = \bar{x} + b'(y - \bar{y})$

where

$$b' = \frac{\Sigma(x_i - \bar{x})(y_i - \bar{y})}{\Sigma(y_i - \bar{y})^2} = \frac{\Sigma x_i y_i - \dfrac{1}{n}\Sigma x_i \, \Sigma y_i}{\Sigma y_i^2 - \dfrac{1}{n}\left(\Sigma y_i\right)^2},$$

P

and if $\sigma_0'^2$ is the error variance of the observed x values

$$S.E.(b') = \frac{\sigma_0'}{\{\sqrt{\Sigma(y_i - \bar{y})^2}\}},$$

and

$$S.E.(\bar{x}_i) = \sigma_0' \sqrt{\left(\frac{1}{n} + \frac{(y_i - \bar{y})^2}{\Sigma(y_i - \bar{y})^2}\right)}.$$

Example 3

Given the following data (as in Example 1):

y	0	1	1	2	3	3
x	-1	0	1	2	2	3

(i) Determine the regression line of x on y.
(ii) Test the significance of the regression.
(iii) Put 90% confidence limits on the expected value of x when $y = 1$.
(i) From Example 1 we have

$$\Sigma x_i = 7, \quad \Sigma y_i = 10, \quad \Sigma x_i^2 = 19, \quad \Sigma x_i y_i = 20, \quad \Sigma y_i^2 = 24.$$

$$\bar{x} = 1.17$$

$$\bar{y} = 1.67$$

$$\therefore b' = \frac{\Sigma x_i y_i - \dfrac{1}{n}\Sigma x_i \Sigma y_i}{\Sigma y_i^2 - \dfrac{1}{n}\left(\Sigma y_i\right)^2}$$

$$= \frac{20 - \frac{1}{6} \cdot 7 \cdot 10}{24 - \frac{1}{6} \cdot (10)^2} = \frac{8.33}{7.33} = 1.14.$$

Thus the equation of the regression line of x on y is

$$x = \bar{x} + b'(y - \bar{y})$$

i.e.

$$x = 1.17 + 1.14(y - 1.67)$$

i.e.

$$x = 1.14y - 0.73.$$

The two equations $y = f(x)$ and $x = g(y)$ are shown in Fig. 19.10. Note that both lines must pass through $(\bar{x}, \bar{y})$, but they will not in general be coincident since $y = f(x)$ is determined by minimizing the errors in the y direction and $x = g(y)$ by minimizing the errors in the x direction.

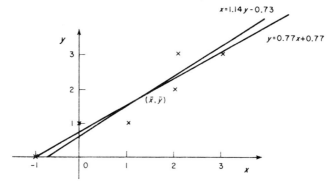

Fig. 19.10.

(ii)

$$\text{Total s.s.} = \Sigma x_i^2 - \frac{1}{n}\left(\Sigma x_i\right)^2 = 19 - \tfrac{1}{6}(7)^2 = 10.83$$

$$\text{s.s. explained by regression} = \frac{\Sigma x_i y_i - \dfrac{1}{n}\Sigma x_i \Sigma y_i}{\Sigma y_i^2 - \dfrac{1}{n}\left(\Sigma y_i\right)^2} = \frac{(8.33)^2}{7.33} = 9.47.$$

Source of variation	d.f.	s.s.	m.s.
Explained by regression	1	9.47	9.47
About regression	4	1.36	0.34
Total	5	10.83	

$$F = \frac{9.47}{0.34} = 27.8 \quad (1, 4).$$

From tables,

$$F_{1\%}(1, 4) = 21.2.$$

Hence the result is significant and use of the regression equation will increase the accuracy in estimating x.

(iii)

$$\text{S.E.}(\hat{x}_i) = \sigma_0 \sqrt{\left(\frac{1}{n} + \frac{(y_i - \bar{y})^2}{\Sigma(y_i - \bar{y})^2}\right)}.$$

When $y = 1$,

$$x = 0.41.$$

$$\therefore \text{S.E.}(\hat{x}_{y=1}) = \sqrt{\left\{0.34\left(\frac{1}{6} + \frac{(0.67)^2}{7.33}\right)\right\}}$$

i.e.

$$S.E.(\hat{x}_{y=1}) = 0.28.$$

The 90% confidence limits on any estimated x_i are given by

$$\hat{x}_i \pm t_{0.05(\phi=4)} \cdot S.E.(\hat{x})$$

and so when $y = 1$ the 90% confidence limits on the values of x are

$$0.41 \pm 2.13 \times 0.28$$

i.e.

$$-0.19 \text{ to } 1.01.$$

There may be practical reasons why the fitted line must pass through some fixed point other than $(\bar{x}, \bar{y})$. If this is the point (α, β) then the regression equation of y on x will be

$$y = \beta + b(x - \alpha)$$

where b is determined by minimizing

$$U = \Sigma[y_i - \beta - b(x_i - \alpha)]^2$$

i.e.

$$b = \frac{\Sigma(x_i - \alpha)(y_i - \beta)}{\Sigma(x_i - \alpha)^2}.$$

In the special case where the fixed point is the origin of measurement for both variables i.e. the point $(0, 0)$, the equation will be

$$y = bx$$

where

$$b = \frac{\Sigma x_i y_i}{\Sigma x_i^2}.$$

The expected relationship may be of the form

$$Y = AB^X$$

where A and B are constants.

By taking logarithms we have

$$\log_{10} Y = \log_{10} A + X \log_{10} B.$$

Putting $\log_{10} Y = y$, $\log_{10} A = a$ and $\log_{10} B = b$ the equation becomes

$$y = a + bX$$

which can be written

$$y = \bar{y} + b(X - \bar{X})$$

and the slope, b, determined as before.

Note that we minimize deviations not of Y but of $\log_{10} Y$. In general the errors introduced will be of negligible practical importance.

Similar transformations may be made to reduce other relationships to a linear form.

5. Curvilinear regression

To fit a curve whose equation is of a second or higher degree, say

$$y = a + bx + cx^2$$

the sum of squares of the errors in prediction will be

$$U = \sum_{i=1}^{n} (y_i - a - bx_i - cx_i^2)^2,$$

and the values of a, b and c are determined to give a minimum value of U.

We have

$$\frac{\partial U}{\partial a} = 0 = -2 \sum_{i=1}^{n} (y_i - a - bx_i - cx_i^2)$$

$$\frac{\partial U}{\partial b} = 0 = -2 \sum_{i=1}^{n} (y_i - a - bx_i - cx_i^2)(x_i)$$

$$\frac{\partial U}{\partial c} = 0 = -2 \sum_{i=1}^{n} (y_i - a - bx_i - cx_i^2)(x_i^2).$$

Giving

$$\Sigma y_i - an - b\,\Sigma x_i - c\,\Sigma x_i^2 = 0$$
$$\Sigma x_i y_i - a\,\Sigma x_i - b\,\Sigma x_i^2 - c\,\Sigma x_i^3 = 0$$
$$\Sigma x_i^2 y_i - a\,\Sigma x_i^2 - b\,\Sigma x_i^3 - c\,\Sigma x_i^4 = 0,$$

the summations extending from 1 to n.

The sum of squares explained by the regression will be given by

$$b\,\Sigma(x_i - \bar{x})(y_i - \bar{y}) + c\,\Sigma(x_i^2 - \bar{x}^2)(y_i - \bar{y}),$$

and the total sum of squares can be partitioned as before.

Example 4

It is expected that a value y can be predicted from the knowledge of a variable x, using a relationship of the form

$$y = a + bx + cx^2.$$

Given the following pairs of observations:

x	-2	-1	0	1	2
y	2	1	3	5	9

(i) Determine the constants a, b, c using the method of least squares.
(ii) Test the significance of the regression.

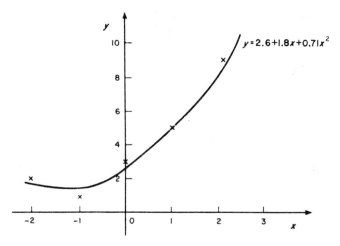

Fig. 19.11.

The scatter diagram is plotted in Fig. 19.11.

x	y	x^2	x^3	x^4	xy	x^2y	y^2
-2	2	4	-8	16	-4	8	4
-1	1	1	-1	1	-1	1	1
0	3	0	0	0	0	0	9
1	5	1	1	1	5	5	25
2	9	4	8	16	18	36	81
Totals 0	20	10	0	34	18	50	120

We have

$$\Sigma y_i - an - b \Sigma x_i - c \Sigma x_i^2 = 0,$$
$$\Sigma x_i y_i - a \Sigma x_i - b \Sigma x_i^2 - c \Sigma x_i^3 = 0,$$
$$\Sigma x_i^2 y_i - a \Sigma x_i^2 - b \Sigma x_i^3 - c \Sigma x_i^4 = 0,$$

i.e.

$$20 - 5a - 10c = 0, \tag{3}$$
$$18 - 10b = 0, \tag{4}$$
$$50 - 10a - 34c = 0. \tag{5}$$

From equation (4) $b = 1.8$, from equations (3) and (5):

$$10a + 20c = 40,$$
$$10a + 34c = 50.$$

$$\therefore \quad c = 0.71$$

and $a = 2.6$. Thus the regression equation is

$$y = 2.6 + 1.8x + 0.71x^2$$

and is shown plotted in Fig. 19.11.

$$\text{The total s.s.} = \Sigma y_i^2 - \frac{1}{n}\left(\Sigma y_i\right)^2 = 120 - \frac{(20)^2}{5}$$

$$= 40.$$

The s.s. explained by the regression

$$= b\ \Sigma(x_i - \bar{x})(y_i - \bar{y}) + c\ \Sigma(x_i^2 - \bar{x}^2)(y_i - \bar{y})$$

$$= b\left(\Sigma x_i y_i - \frac{1}{n}\Sigma x_i\ \Sigma y_i\right) + c\left(\Sigma x_i^2 y_i - \frac{1}{n}\Sigma x_i^2\ \Sigma y_i\right)$$

$$= 1.8(18 - 0) + 0.71(50 - \tfrac{1}{5} \cdot 10 \cdot 20)$$

$$= 39.5.$$

We now construct the analysis of variance table:

Source of variation	d.f.	s.s.	m.s.
Explained by regression	2	39.5	19.75
About regression	2	0.5	0.25
Total	4	40.0	

Two degrees of freedom are now associated with the sum of squares explained by the regression, one with the calculation of the slope (b) and the other with the calculation of the 'curvature' (c).

To test the significance of the regression we calculate

$$F = \frac{19.75}{0.25} = 79.00\ (2, 2).$$

From tables,

$$F_{5\%}(2, 2) = 19.0$$
$$F_{1\%}(2, 2) = 99.0$$

i.e. the value of F, and hence the regression, is significant.

In some cases where the form of the functional relationship between the variables cannot be anticipated from practical knowledge, the scatter diagram may leave some doubt as to whether a quadratic equation should be fitted or whether a linear equation would be adequate over the required range.

This can be determined by fitting both forms of equation to the data, calculating the sum of squares explained by the regression in both cases and assessing whether the quadratic equation accounts for a significantly greater part of the variability.

The procedure can be extended to assess the value of the addition of terms of higher power to the equation.

The disadvantage of increasing the degree of the fitted equation in this way is that the coefficients must be re-calculated.

6. Multiple regression

Any form of polynomial, involving any number of variables can be fitted to a given set of data, the method of least squares being used to determine the constants in that equation.

Suppose we expect the data to be adequately represented by an equation of the form,

$$z = a + bx + cy$$

where x and y are the independent variables.

The sum of squares of the errors in prediction will be given by

$$U = \Sigma(z_i - a - bx_i - cy_i)^2$$

and the values of a, b, and c to give the minimum value of U are obtained by solving the following equations:

$$\Sigma z_i - an - b\,\Sigma x_i - c\,\Sigma y_i = 0,$$
$$\Sigma x_i z_i - a\,\Sigma x_i - b\,\Sigma x_i^2 - c\,\Sigma x_i y_i = 0,$$
$$\Sigma y_i z_i - a\,\Sigma y_i - b\,\Sigma x_i y_i - c\,\Sigma y_i^2 = 0.$$

Hence the regression equation.

The sum of squares explained by fitting this equation will be given by

$$b\,\Sigma(x_i - \bar{x})(z_i - \bar{z}) + c\,\Sigma(y_i - \bar{y})(z_i - \bar{z})$$

and the significance of the regression can be tested by contructing the analysis of variance table.

PROBLEMS

1. A spectrochemical analysis was made of the calcium oxide in standard samples of a refractory material. The logarithm of the weight of calcium oxide in the standards was plotted against the logarithm of the relative energy of the calcium 3179 Å line, the following results being obtained:

log(micrograms CaO)	1.8	2.0	2.5	3.0	3.2	3.5	4.0
log (energy) (arbitrary units)	4	7	13	16	20	22	26

Determine the best linear equation to estimate the amount of calcium oxide present, in similar samples, from the spectrochemical analysis.

2. The voltage across a two-ohm resistance is measured by a voltmeter for the following known currents:

i (amp)	1	2	4	6	8	10
v (volt)	1.8	3.7	8.2	12.0	15.8	20.2

Determine a linear calibration curve for the instrument.

3. The following data show the effect of temperature on the loss in weight of chrysolite asbestos fibres

°F	500	600	700	800	900
% loss	0.5	0.9	1.7	2.2	2.8

It is known that there is a rapid increase in the rate of loss above 1000 °F but the relationship is expected to be approximately linear over the range 500 °F to 900 °F.
 (i) Determine the regression equation to give the percentage loss in weight over the temperature range given.
 (ii) Test the significance of the regression using analysis of variance.

4. A number of simultaneous measurements are made of two variables x and y, between which there is a linear relationship of the form

$$y = a + bx.$$

Explain the difference between the regression line of y on x and the regression line of x of y.
 Theoretically there is a relationship between two variables x and y of the form

$$y = Mx^2 + K.$$

The x value can be accurately measured.
 (i) Using the method of least squares, determine the values of the constants M and K, given the following observed values:

x	1	2	3	4	5
y	21	24	30	38	47

 (ii) Put 95% confidence limits on the estimated value of y when $x = 3.6$.

5. The following observations give the temperature during a chemical experiment, measured at five minute intervals:

Time (min)	5	10	15	20	25	30	35	40	45	50	55
Temp. (°C)	100.2	100.7	100.9	101.2	101.3	101.4	101.4	101.3	101.0	100.8	100.4

Fit an equation of the form

$$T = a + bt + ct^2$$

where T denotes temperature and t denotes time. Use the equation to estimate the time at which the temperature will be a maximum.

6. The following table gives the molar-heat capacity of steam at one atmosphere pressure in cal $°C^{-1}$:

°C	100	400	700	1000	1300
cal $°C^{-1}$	8.85	8.65	9.05	9.89	11.15

Fit a parabolic regression equation to the data.

CHAPTER 20

Design of Experiments

Within a single chapter we can only indicate the basic principles to be considered in the design and analysis of experiments and give a brief outline of the most useful of the simple designs.

The purpose of an experiment is to answer a question or questions. An efficient design and its appropriate analysis will enable the questions to be answered with a minimum of experimental effort. Thus before attempting to design an experiment, its objectives must be clearly stated and this means that the technical details of the problem must be clearly understood.

In general we are investigating the effect of changing one or more of the experimental conditions. Any given feature of the experimental conditions is known as a *factor*. The particular condition or value of the factor used in a trial is called the *level* of that factor. The numerical result of the trial is called the *response* and the *effect* of a factor is the change in response due to a change in factor level.

First we must ensure that the experiment will cover the factor of interest. We must also consider, when designing the experiment, the analysis which will be used and whether the use of the significance tests involved will be justified.

In using statistical tests of significance we assume that:

(i) The errors are independent of one another.
(ii) The error variance within each group of observations is approximately the same.
(iii) The errors are normally distributed.

Let us consider the importance of these assumptions in tests comparing estimates of mean values and of variances.

Assumption (i) is very important for both t- and F-tests. The experimental error arises from variation in factors which are not controlled. Provided this is random variation, the error in any one observation is independent of the error in any other. If however there is some form of trend in the experimental conditions the errors will not be independent.

Suppose a strength testing machine is subject to slight fatigue, such that if the same sample could be tested at the beginning and end of a day it would appear to have increased in strength during the day. If, to compare the strength of two groups of samples, one group was tested in the morning and the other in the afternoon, then with no real difference between the two, the second group would appear to have the greater average strength. We may know, or suspect, the existence of this fatigue but it may be impracticable to adjust the machine between tests. By testing

448

the samples in random order the experimental error will be slightly increased but the comparison between groups will be independent of the trend and we shall draw valid conclusions from the significance test.

This procedure of randomizing the allocation of experimental material and order of testing should always be followed as a precaution against an unsuspected trend.

When the nature of the variability in the uncontrolled experimental conditions is known, e.g. variation between batches of raw material we may use what is known as the *block principle*. This involves dividing the experimental material into units, within which the variability is expected to be less than within the whole. These units are termed *blocks*. The factor levels are compared within blocks so that the comparisons are independent of any differences between blocks. The process of randomization is used within each block as a safeguard against a trend within the block.

The second assumption, that there is similar variation within groups is not very important for tests comparing mean values, provided the number of observations is approximately the same within each group. If it is suspected that the standard deviation is proportional to the mean value then it may be made independent by a suitable transformation of the data.

If $\sigma \propto f(\mu)$ then the observed data may be transformed to the variable x by the equation,

$$x = \int \frac{1}{f(\mu)} \, d\mu.$$

Suppose we are comparing proportions of occurrences, e.g. comparing samples from a binomial distribution, then the variance is equal to $np(1 - p)$ (using previous notation) so that $\sigma \propto \sqrt{\{p(1 - p)\}}$. A suitable transformation will be given by,

$$x = \int \frac{1}{\sqrt{(p - p^2)}} \, dp \quad \text{or} \quad x = \sin^{-1} \sqrt{p}$$

(omitting the factor of 2). The x values can then be compared using a u-test.

The assumption that the errors are normally distributed is most important when comparing variance estimates. If this assumption is not justified an empirical transformation of the data can be made using normal probability paper. The values of the variable are plotted against their cumulative probability of occurrence. The points from a normal distribution with the same mean and variance will lie along a straight line. Suppose the points representing the observed data give the cumulative curve shown in Fig. 20.1. The probability axis is on a logarithmic scale the x-axis is linear.

The observed value x_1 is transformed to the value x_2 where x_2 is the value from the normal distribution corresponding to the same probability as x_1.

A great deal of information from past experimental work is filed away and will never be referred to again. The important information which should always be recorded is the experimental error associated with any test. A knowledge of the error variance is of great value when designing an experiment involving similar tests. It enables us to determine the size of the experiment required in order to draw our conclusions with the required precision.

450

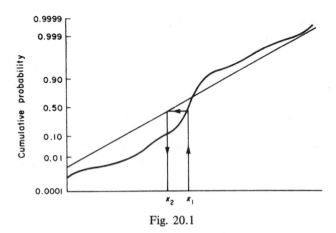

Fig. 20.1

1. Size of experiment

There are two ways in which an error can be made when a conclusion is drawn from a test of significance, viz.:

(i) The risk of an *error of the first kind* (α) is the risk of wrongly rejecting the null hypothesis.

(ii) The risk of an *error of the second kind* (β) is the risk of wrongly accepting the null hypothesis.

The risk α is the level of significance for the test.

Suppose the null hypothesis is that the mean of the population from which the observed results are drawn is μ_0 and let δ represent the minimum increase which is of practical importance. Thus if μ exceeds μ_0 by an amount as great as δ it is important that this should be detected.

Let the risk we are prepared to take of wrongly asserting the change be α and the risk of failing to detect such a change be β.

We have to determine the size of sample (n) and the limiting mean value of that sample ($\bar{X}'$), such that, with the given risks of error the decision can be made either

$$\mu = \mu_0$$

or

$$\mu = \mu_0 + \delta.$$

Let the known experimental error variance be σ^2.

Consider the distribution of the means of samples of n observations when $\mu = \mu_0$ and when $\mu = \mu_0 + \delta$ (Fig. 20.2).

If the sample mean is greater than $\bar{X}'$ we assert that $\mu = \mu_0 + \delta$ and the probability that $\mu = \mu_0$ is α.

If the sample mean is less than $\bar{X}'$ we accept that $\mu = \mu_0$ and the probability that $\mu = \mu_0 + \delta$ is β.

From Fig. 20.2

$$\delta = u_\alpha \frac{\sigma}{\sqrt{n}} + u_\beta \frac{\sigma}{\sqrt{n}}$$

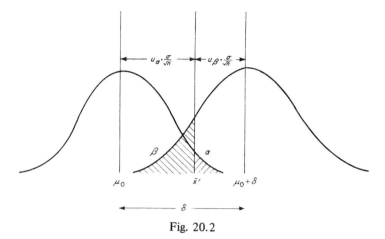

Fig. 20.2

where u_α and u_β denote the values of the standard normal deviate which cut off 'tails' with areas α and β respectively. Therefore

$$\delta = \frac{\sigma}{\sqrt{n}} (u_\alpha + u_\beta)$$

$$\therefore n = \left(\frac{u_\alpha + u_\beta}{\delta/\sigma}\right)^2.$$

If we are interested in a difference $\pm \delta$ an error of the first kind can be made in two ways; we can assert that $\mu = \mu_0 + \delta$ when $\mu = \mu_0$, or that $\mu = \mu_0 - \delta$ when $\mu = \mu_0$. Hence for a total risk of error α we allow a risk of $\frac{1}{2}\alpha$ for each wrong assertion and

$$n = \left(\frac{u_{\alpha/2} + u_\beta}{\delta/\sigma}\right)^2.$$

To compare two sample means, since both estimates will be subject to error, the size of each sample should be, for a one-sided test

$$n = 2\left(\frac{u_\alpha + u_\beta}{\delta/\sigma}\right)^2$$

and for a two-sided test

$$n = 2\left(\frac{u_{\alpha/2} + u_\beta}{\delta/\sigma}\right)^2.$$

This theory is dependent upon a reliable estimate of σ. If the standard deviation differs from the value assumed the risks of error will not be controlled.

Example 1

A reaction process is known to give an average yield of 70% with standard deviation 3%. It is suggested that a certain modification will increase the yield. It is required to detect an increase as great as 5% with probability 0.95 but since the modification should not be made unnecessarily α is set at 0.025 (see Fig. 20.3).

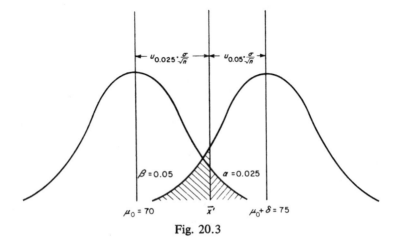

Fig. 20.3

(i) How many experimental runs should be made to decide whether or not the modification should be introduced?

(ii) What is the minimum average yield which must be attained from the trials before it is accepted that the modification should be introduced?

(i)

$\delta = 5, \quad \alpha = 0.025,$

$\sigma = 3, \quad \beta = 0.05.$

$$n = \left(\frac{u_\alpha + u_\beta}{\delta/\sigma}\right)^2$$

$$= \left(\frac{1.96 + 1.645}{5/3}\right)^2$$

$$\simeq 5.$$

(ii)

$$\bar{x}' = \mu_0 + u_{0.025}\frac{\sigma}{\sqrt{n}}$$

$$= 70 + 1.96\frac{3}{\sqrt{5}}$$

$$= 72.6.$$

Thus five trial runs will be made using the modified process and if the average yield is greater than 72.6% the modification is to be accepted.

2. Randomized block designs

The simple *randomized block design* enables us to make comparisons within one set of treatments or factor levels. (The term treatment is used in a general sense to define the particular factor level or combination of factor levels used in a given trial.) The experimental conditions are divided into blocks and all treatments are tested within each block, the treatments being allocated at random within each block. The analysis is then a simple two-factor analysis of variance, the factors being blocks and treatments. Provided we can assume that there is no interaction between blocks and treatments the residual sum of squares can be used to estimate the error variance. If tests are replicated an independent estimate of error can be calculated.

Example 2

An experiment is designed to compare the resultant yields from a reaction process using four different catalyst concentrations (A, B, C, D). A batch of material from the previous stage in the process is sufficient for four trials. The investigation is designed as a randomized block experiment. Four batches are used in the experiment, each being divided into four parts and the catalyst concentrations allocated at random to those parts as follows:

Block (batch)	Treatment (catalyst concentration)			
1	A	C	D	B
2	B	C	A	D
3	C	D	B	A
4	A	B	C	D

Is there a difference in yield for different catalyst concentration?
The yields (above 70%) from the experiment are given in the following table:

Batch \ Catalyst	A	B	C	D	Total
1	−3	−4	−1	−1	−9
2	−1	−2	2	0	−1
3	−1	0	1	2	2
4	2	0	3	2	7
Total	−3	−6	5	3	−1

$$CF = (-1)^2/16 = 0.06$$

$$\text{total s.s.} = [(-3)^2 + (-1)^2 + \cdots + 2^2 + 2^2] - CF = 58.94$$

s.s. between batches
$$= \tfrac{1}{4}\{(-9)^2 + (-1)^2 + 2^2 + 7^2\} - CF = 33.69$$

s.s. between catalysts
$$= \tfrac{1}{4}\{(-3)^2 + (-6)^2 + 5^2 + 3^2\} - CF = 19.69.$$

The analysis of variance table is:

Source of variation	d.f.	s.s.	m.s.
Between batches	3	33.69	11.23
Between catalysts	3	19.69	6.56
Error	9	5.56	0.62
Total	15	58.94	

An F-test shows a highly significant difference between catalysts and if required, confidence limits could be calculated on the expected average yield for each catalyst concentration, viz.

$$\bar{Y} \pm t_{\alpha/2(\phi=9)} \cdot \text{S.E.}(\bar{Y})$$

where

$$\text{S.E.}(\bar{Y}) = \sqrt{\frac{0.62}{4}}.$$

The highly significant value of F when testing the mean square between batches indicates that the use of blocks has been effective in removing an appreciable amount of variation from the comparisons between catalysts.

3. Latin squares

The randomized block design enables comparisons to be made between a given set of treatments under similar experimental conditions. It may be that the experimental conditions can be subdivided in more than one way. A *Latin square design* is used when there is a possible two-way division. As its name implies, the design is in the form of a square. One set of conditions (factor) is allocated to the rows of the square and the other to the columns. The treatments are then allocated to the cells of the square such that each occurs once, and once only, in each row and in each column. A possible design for a 3 x 3 Latin square, testing treatments A, B and C would be:

Row \ Column	1	2	3
1	A	B	C
2	B	C	A
3	C	A	B

A Latin square design can be used *only* if we are justified in the assumption of no interactions between factors.

We are also limited to testing the same number of levels of each factor (i.e. no. of rows = no. of columns = no. of letters).

Example 3

Suppose in Example 2, it was required to complete the investigation within a limited time and to do this it was necessary to use four different reaction vessels.

A suitable design if we can assume no interactions would be a 4 x 4 Latin square. The four batches and four reactors can be allocated at random to the rows and columns respectively.

Reactor / Batch	2	1	4	3	Total
3	$A: -3$	$B: -2$	$C:\ 0$	$D: -1$	-6
2	$B: -3$	$A: -2$	$D:\ 0$	$C:\ 1$	-4
1	$C:\ 2$	$D:\ 4$	$A: -1$	$B:\ 0$	5
4	$D:\ 2$	$C:\ 4$	$B:\ 0$	$A:\ 2$	8
Total	-2	4	-1	2	3

$$\begin{array}{ccccc} & A & B & C & D \\ \text{Treatment totals} & -4 & -5 & 7 & 5 \end{array}$$

$CF = (3)^2/16 = 0.56.$

Total s.s. $= \{(-3)^2 + (-3)^2 + \cdots + 0^2 + 2^2\} - CF = 72.44$

s.s. between columns (reactors)

$$= \tfrac{1}{4}\{(-2)^2 + 4^2 + (-1)^2 + 2^2\} - CF = 5.69$$

s.s. between rows (batches)

$$\tfrac{1}{4}\{(-6)^2 + (-4)^2 + 5^2 + 8^2\} - CF = 34.69$$

s.s. between letters (catalysts)

$$= \tfrac{1}{4}\{(-4)^2 + (-5)^2 + 7^2 + 5^2\} - CF = 28.19.$$

Source of variation	d.f.	s.s.	m.s.	F
Between reactors	3	5.69	1.90	2.92
Between batches	3	34.69	11.56	17.8
Between catalysts	3	28.19	9.40	14.5
Error	6	3.87	0.65	
Total	15	72.44		

From tables

$$F_{10\%}(3, 6) = 3.29$$
$$F_{1\%}(3, 6) = 9.79.$$

Thus again a highly significant difference in yield is shown for different catalyst concentrations (the nature of this could be further investigated).

A significant amount of variation has been removed from the comparison by the use of blocks but there is no significant difference between reaction vessels.

Provided that it can be assumed that there is negligible interaction, the above procedures can be extended to cover further factors by introducing a further set of letters (Graeco-Latin squares) or by working in further dimensions (Latin cubes).

In each design, each source of variation is isolated by summing over all levels of all other factors.

Designs beyond that of the Latin square will rarely be used in chemical investigations since the assumption of no interactions will not be justified. A simple design, such as a Latin square, may however be of great use at the beginning of an investigation, where preliminary trials are to be made, seeking large changes; and interactions, if they do exist will be negligible compared with the main effects.

4. Factorial designs

A factorial design is a design which tests all combinations of all levels of all factors, enabling all main effects and all interactions to be estimated. Such a design will be of great use in chemical investigations provided the number of factors and factor levels does not make the number of trials prohibitive. The analysis of such an experiment has been considered in the appendix to Chapter 18.

If a large number of factors are to be investigated, a useful design, particularly for any preliminary trials, is the factorial design with each factor tested at two levels. With n factors this is referred to as a 2^n factorial design.

A particular notation is used for such experiments. The factors are denoted by capital letters, the upper level of the factor by the corresponding small letter and the lower level by (1) (e.g. $A, a, (1)$). The treatment (set of factor levels) to be used in any given trial is denoted by including the appropriate small letter for each factor which occurs at its upper level.

Suppose we have a 2^3 experiment, with factors A, B, C. The treatment with all factors at their lower level will be denoted by (1).

The treatment with A at its upper level and B and C at their lower levels will be denoted by a.

The treatment with A, B and C each at their upper levels will be denoted by abc.

Thus the set of 8 treatments for the full experiment will be $(1), a, b, ab, c, ac, bc, abc$. The treatments are always written in standard order, that order being obtained by considering the addition of the factors one at a time. As each factor is added the previous treatments are multiplied by the small letter corresponding to the added factor to give the additional treatments to be tested. For example

    factor A, treatments $(1), a$

    add factor B, add treatments b, ab

    add factor C, add treatments c, ac, bc, abc.

The main effect of any given factor is the difference between the average response at its upper and lower levels. Noting that the upper level of a factor is indicated by the inclusion of the appropriate small letter in the treatment and the lower level by the absence of the small letter, we have

main effect $A = \frac{1}{4}(a + ab + ac + abc) - \frac{1}{4}((1) + b + c + bc)$.

Treating a, b, c and (1) as algebraic symbols this can be written

$A = \frac{1}{4}(a - 1)(b + 1)(c + 1)$.

Similarly

$B = \frac{1}{4}(a + 1)(b - 1)(c + 1)$

and

$C = \frac{1}{4}(a + 1)(b + 1)(c - 1)$.

The interaction between two factors is *defined* as half the difference between the effect of the one factor at the upper and lower level of the second, e.g. interaction AB.

The treatments which test B at its upper level are

b, ab, bc, abc.

Thus the effect of A when B is at its upper level is

$\frac{1}{2}(ab + abc) - \frac{1}{2}(b + bc)$.

Similarly the effect of A when B is at its lower level is

$\frac{1}{2}(a + ac) - \frac{1}{2}((1) + c)$

and the interaction AB will be given by

$AB = \frac{1}{2}[\{\frac{1}{2}(ab + abc) - \frac{1}{2}(b + bc)\} - \{\frac{1}{2}(a + ac) - \frac{1}{2}((1) + c)\}]$

and treating a, b, c and (1) as algebraic symbols this can be written

$AB = \frac{1}{4}(a - 1)(b - 1)(c + 1)$.

Similarly

$AC = \frac{1}{4}(a - 1)(b + 1)(c - 1)$

and

$BC = \frac{1}{4}(a + 1)(b - 1)(c - 1)$.

The interaction ABC will be given by half the difference between the effects of AB at the upper and lower level of C

interaction AB at the upper level of $C = \frac{1}{2}\{(abc - bc) - (ac - c)\}$

interaction AB at the lower level of $C = \frac{1}{2}\{(ab - b) - (a - (1))\}$

$\therefore \ ABC = \frac{1}{2}\{\frac{1}{2}(abc - bc - ac + c) - \frac{1}{2}(ab - b - a + (1))\}$

$\qquad = \frac{1}{4}(a - 1)(b - 1)(c - 1)$.

Similarly with n factors $A, B, \ldots, P$:

$$A = \frac{1}{2^{n-1}}\{(a-1)(b+1)\ldots(p+1)\}$$

$$AB = \frac{1}{2^{n-1}}\{(a-1)(b-1)\ldots(p+1)\}$$

$$AP = \frac{1}{2^{n-1}}\{(a-1)(b+1)\ldots(p-1)\}$$

$$AB\ldots P = \frac{1}{2^{n-1}}\{(a-1)(b-1)\ldots(p-1)\}.$$

This algebraic notation when expanded will give the appropriate sign to attach to each treatment response in order to estimate any given effect.

For any given set of factors we can draw up the appropriate *table of signs*, the treatments being written in standard order. For three factors A, B, C, the table will take the form:

Treatment \ Effect	Total	A	B	C	AB	AC	BC	ABC
(1)	+	−	−	−	+	+	+	−
a	+	+	−	−	−	−	+	+
b	+	−	+	−	−	+	−	+
ab	+	+	+	−	+	−	−	−
c	+	−	−	+	+	−	−	+
ac	+	+	−	+	−	+	−	−
bc	+	−	+	+	−	−	+	−
abc	+	+	+	+	+	+	+	+

For each main effect (A, B, C) a positive sign is attached to each treatment which contains the appropriate small letter (i.e. the factor at its upper level) and a negative sign to those which do not. The signs attached to the treatments estimating each interaction can be obtained by multiplying the signs for the main effects involved in that interaction.

Thus ignoring the divisor

A	−(1)	+a	−b	+ab	−c	+ac	−bc	+abc
AB	(1)	−a	−b	+ab	+c	−ac	−bc	+abc

When each factor is tested at two levels only, the calculation of the sum of squares for the effects can be simplified.

Suppose we have n factors, i.e. 2^n observations. Each effect is estimated from the difference between two groups of 2^{n-1} observations. Thus, if for any given effect, the sum over all treatment responses with a positive sign attached is denoted by U and the sum over those with a negative sign is denoted by L, we have:

(i) the estimated effect is

$$\frac{1}{2^{n-1}} [U - L]$$

(ii) the sum of squares between the levels of the factor is

$$= \frac{1}{2^{n-1}} [U^2 + L^2] - \frac{[U + L]^2}{2^n}$$

$$= \frac{1}{2^{n-1}} [U^2 + L^2] - \frac{1}{2^n} [U^2 + L^2 + 2UL]$$

$$= \frac{1}{2^n} [U - L]^2.$$

Example 4

It is required to investigate the effect on the yield of a reaction process using two different grades of raw material, either one or two hours stirring time and two different catalysts.

	Upper level	Lower level
A, grade of material	grade α	grade β
B, time of stirring	2 hours	1 hour
C, catalyst	catalyst p	catalyst q

It is considered very unlikely that there will be any interaction between factors A and B or between A and C. The experiment is designed as a 2^3 factorial experiment and the following results obtained. Data give yield above 60%.

Treatment	(1)	a	b	ab	c	ac	bc	abc
Response	2	8	2	6	1	4	5	9

Using the table of signs for 3 factors we have:

Response	2	8	2	6	1	4	5	9	
$4A$:	−	+	−	+	−	+	−	+ = 17
$4B$:	−	−	+	+	−	−	+	+ = 7
$4C$:	−	−	−	−	+	+	+	+ = 1
$4AB$:	+	−	−	+	+	−	−	+ = −1
$4AC$:	+	−	+	−	−	+	−	+ = −3
$4BC$:	+	+	−	−	−	−	+	+ = 11
$4ABC$:	−	+	+	−	+	−	−	+ = 3

Sums of squares:

main effect $A: (17)^2/8 = 36.125$
$B: (7)^2/8 = 6.125$
$C: (1)^2/8 = 0.125$

interactions AB: $(-1)^2/8 = 0.125$
AC: $(-3)^2/8 = 1.125$
BC: $(11)^2/8 = 15.125$
ABC: $(3)^2/8 = 1.125$

Total   59.875.

Since each factor is at two levels there is only one degree of freedom for each effect and hence the mean square is equal to the sum of squares in each case.

It is unlikely that compensating errors will be made in these calculations so that an adequate check is provided by an independent calculation of the total sum of squares. This should agree (within calculation approximations) with the total obtained by summing the sums of squares for each effect.

$$\text{Total sum of squares} = (2^2 + 8^2 + \cdots + 5^2 + 9^2) - \frac{(37)^2}{8}$$

$$= 59.875$$

therefore calculations can be assumed correct.

The analysis of variance table is:

Source of variation	d.f.	m.s.	F
Main effect A	1	36.125	45.7
B	1	6.125	
C	1	0.125	
Interactions AB	1 ⎫	0.125 ⎫	
AC	1 ⎬ 3	1.125 ⎬ 0.79	19.2
BC	1	15.125	
ABC	1 ⎭	1.125 ⎭	
Total	7	57.875	

In the absence of replication the highest order interaction must be used as the estimate of experimental error. Since the interaction mean squares AB and AC, which were not expected to exist, are comparable with the ABC mean square, the three mean squares should be combined to give an estimate of the error variance.

$$\sigma_0^2 = \frac{0.125 + 1.125 + 1.125}{3} = 0.79 \ (\phi = 3).$$

As the interaction BC is shown to be significant the main effect B and C should be tested against the BC mean square.

From tables,

$$F_{5\%}(1, 3) = 10.1$$
$$F_{1\%}(1, 3) = 34.1.$$

The conclusions which can be drawn from this experiment are:
(i) Grade α material is superior to grade β.

(ii) Although neither the difference between catalysts nor between stirring times is shown significant when considered over all levels of the other two factors, there is a significant interaction between catalyst and stirring time. This means the results for the two catalysts should be examined at each level of stirring time.

Summing over the levels of factor A we have

C \ B	(1)	b
(1)	$\dfrac{(1)+a}{10}$	$\dfrac{b+ab}{8}$
c	$\dfrac{c+ac}{5}$	$\dfrac{bc+abc}{14}$

indicating that catalyst q gives a higher response when stirring for one hour but catalyst p increases the response with two hours stirring.

It would therefore be recommended that grade α material and catalyst p should be used with stirring time of two hours.

Whenever an interaction is shown to be significant, it is important that the effect of each factor involved should be examined at each level of the other factors involved.

5. Confounding in factorial designs

When an additional factor is introduced the number of experimental trials required in a 2^n design will be doubled. If all the treatments cannot be tested under similar conditions it may be necessary to divide the experimental conditions into blocks, testing some of the treatments in each block. This will mean that certain comparisons will be confused or '*confounded*' with the differences between blocks. It is important that the experiment should be designed so that only relatively unimportant comparisons are lost in this way.

The process of dividing the treatments into blocks, such that the important effects can be estimated independently of inter-block differences, while the unimportant comparisons are confused is known as *confounding*.

To confound between two blocks the treatments must be divided into two groups such that the difference between those groups measures the effect to be confounded.

Suppose we have a 2^3 experiment, factors A, B, C, and only four tests can be made under similar conditions. It is therefore necessary to use two sets of conditions (blocks) and it is decided to confound interaction ABC between blocks. We have

$$ABC = \tfrac{1}{4}(a - 1)(b - 1)(c - 1).$$

The treatments with a positive sign will be tested in one block and those with a negative sign in the other, i.e.

Block (1) (+)	Block (2) (−)
a	(1)
b	ab
c	ac
abc	bc

We can verify that all other effects can be determined independently of any difference between blocks. Let block 1 give a response α units higher than block 2. Then the measured responses will be

$$(a + \alpha, b + \alpha, c + \alpha, abc + \alpha)((1), ab, ac, bc).$$

Main effect $A = \frac{1}{4}\{-(1) + (a + \alpha) - (b + \alpha) + ab - (c + \alpha) + ac - bc + (abc + \alpha)\}$

$\qquad\qquad = \frac{1}{4}\{-(1) + a - b + ab - c + ac - bc + abc\}$

$\qquad\qquad = \frac{1}{4}(a - 1)(b + 1)(c + 1)$

i.e. independent of α.

Similarly B, C, AB, AC, and BC also can be shown to be independent of any difference between the blocks. To confound between blocks in a 2^n experiment there must always be 2^p blocks where $p < n$ is integral. Confounding between 2^p blocks confounds $(2^p - 1)$ effects, p of which can be chosen at random and the remaining $(2^p - 1 - p)$ are automatically confounded.

Suppose we test factors A, B, C in four blocks each of two observations and we chose to confound AB and BC between blocks, $(n = 3, p = 2)$.

The division of treatments will be:

	1		2	
(1)	abc	a	bc
ab	c	c	ac
	3		4	

Then

$\qquad (1 + 3) - (2 + 4)$ estimates AB,

$\qquad (1 + 2) - (3 + 4)$ estimates BC.

So that AB is confounded between 'columns' and BC is confounded between 'rows'.

The other indpendent comparison which can be made between the four blocks is $(1 + 4) - (2 + 3)$, i.e.

$\qquad ((1) + abc + b + ac) - (a + bc + ab + c)$

i.e.

$\qquad (a - 1)(b + 1)(c - 1)$ which estimates AC.

Thus the third interaction which is confounded is AC.

To simplify the division of treatments into blocks two rules can be followed:

(i) Given the p chosen interactions to be confounded between 2^P blocks, the remaining $(2^P - 1 - p)$ can be obtained from the products of the p interactions taken in all possible ways, i.e. two at a time, three at a time, up to p at a time, and when a factor appears as a squared term, replace it by unity, i.e.

$$AB \times BC = AB^2 C = AC.$$

(ii) When dividing the treatments into two groups to confound a given interaction, one group consists of the treatments with an even number of letters corresponding to the confounded interaction and the other group the treatments with an odd number of corresponding letters.

Example 5

To investigate five factors A, B, C, D, and E, each at two levels, the experiment is confounded between eight blocks. It is decided to confound interactions ABC, CDE, BCD.

What other interactions will automatically be confounded?

We have a 2^5 experiment to be confounded in 2^3 blocks, so that 3 of the $(2^3 - 1)$ confounded interactions can be chosen and the remaining 4 will automatically be confounded.

Taking the given interactions two at a time gives

$$ABC^2DE, AB^2C^2D, BC^2D^2E$$

i.e.

$$ABDE, AD, BE$$

and three at a time gives

$$AB^2C^2CD^2E,$$

i.e.

$$ACE.$$

Thus the confounded interactions will be,

$$AD, BE, ABC, BCD, CDE, ACE, ABDE.$$

Example 6

Four factors A, B, C, D are tested in four blocks confounding interactions AB, BCD, ACD. Determine the treatments tested in each block.

Writing the treatments in standard order we have

$$(1), a, b, ab, c, ac, bc, abc, d, ad, bd, abd, cd, acd, bcd, abcd.$$

To confound AB the division into two groups will be

1 (even) $(1), ab, c, abc, d, abd, cd, abcd.$

2 (odd) $a, b, ac, bc, ad, bd, acd, bcd.$

Subdividing [1] to confound BCD.

   $1a$ (even) $(1), abc, abd, cd$.

   $1b$ (odd) $ab, c, d, abcd$.

Subdividing [2] to confound ACD

   $2a$ (even) b, ac, ad, bcd.

   $2b$ (odd) a, bc, bd, acd.

Thus the four blocks will be

Block 1	$(1),$	$abc,$	$abd,$	$cd;$
2	$a,$	$bc,$	$bd,$	$acd;$
3	$b,$	$ac,$	$ad,$	$bcd;$
4	$ab,$	$c,$	$d,$	$abcd.$

6. Fractional factorial designs

A full factorial design will enable us to estimate all the effects with the maximum precision, all the observations being used to estimate each of the effects. However, we have seen that with a large number of factors or factor levels it may become impracticable to carry out all the necessary trials.

Often we may know from practical considerations that it is impossible that certain interactions could exist. We may be looking for very large main effects so that by comparison all interactions will be negligible. It may be that the experiment is to be a preliminary trial and a high degree of precision in the estimation of the effects is not required.

Under any of these circumstances it may be possible to obtain the required information by carrying out only part of the full factorial design. That part must however be carefully chosen depending upon the effects it is required to estimate. Such an experiment is called a *fractional factorial design* and may be obtained by selecting any block from a suitably confounded full factorial design. Alternatively we can select a full factorial design which tests a smaller number of factors and use the comparisons which estimate unwanted effects to estimate the additional factors.

We shall consider the procedures in relation to a 2^n design although designs involving a larger number of factor levels can be derived in a similar way.

Example 7

An experiment is to be designed to investigate the main effects of three factors A, B, C in four observations. Determine a suitable set of treatments.

   Method (i).   The full design to estimate all effects will be treatments

   $(1), a, b, ab, c, ac, bc, abc$.

Confounding interaction ABC between 2 blocks by dividing the treatments into

1. those which have an even number of letters in common with ABC

2. those which have an odd number of letters in common with *ABC* gives

Block 1:   (1)   *ab*   *ac*   *bc*

Block 2:   *a*   *b*   *c*   *abc*.

With four observations, three independent comparisons can be made and either of the above blocks will estimate the three main effects A, B, C. For example, block 1 gives

$$A = \tfrac{1}{2}(ab + ac) - \tfrac{1}{2}((1) + bc)$$
$$B = \tfrac{1}{2}(ab + bc) - \tfrac{1}{2}((1) + ac)$$
$$C = \tfrac{1}{2}(ac + bc) - \tfrac{1}{2}((1) + ab).$$

Method (ii).   A full design testing four treatments can estimate two factors and their interaction (i.e. a 2^2 design). Let the two factors be A and B, then the treatments will be

$$(1), a, b, ab.$$

The table of signs showing the signs attached to the treatments to estimate each effect is:

Trial	Treatment	A	B	AB
y_1	(1)	−	−	+
y_2	a	+	−	−
y_3	b	−	+	−
y_4	ab	+	+	+

Provided we can assume that AB does not exist the comparisons estimating this interaction can be used to estimate C.

Re-writing the table of signs and putting $C = AB$ we have:

Trial	A	B	C	Treatment tested
y_1	−	−	+	c
y_2	+	−	−	a
y_3	−	+	−	b
y_4	+	+	+	abc

Since a positive sign indicates a factor at its upper level and a negative sign, the factor at its lower level we see that C is tested at its upper level in trials y_1 and y_4 and at its lower level in trials y_2 and y_3. Hence the treatments tested in the experiment will be

$$c, a, b, abc.$$

Alternatively we could put $C = -AB$, giving

Trial	A	B	C	Treatment
y_1	—	—	—	(1)
y_2	+	—	+	ac
y_3	—	+	+	bc
y_4	+	+	—	ab

The two possible designs are therefore

> 1: $a, b, c, abc,$

or

> 2: $(1), ac, bc, ab.$

as obtained using method (i).

Example 8

Five factors A, B, C, D and E are to be investigated in eight observations. Determine a suitable set of treatments to estimate the effects A, B, C, D, E, AB and AC.

Eight observations represent a full design investigating three factors (2^3 design). Suppose these are the factors A, B, C. The effects which will be estimated are

> $A, B, C, AB, AC, BC, ABC.$

Thus the required design will use the comparisons estimating BC and ABC to estimate the added factors D and E.

Trial	Treatment with factors A, B, C	A	B	C	AB	AC	$D = BC$	$E = ABC$	Treatment tested
y_1	(1)	—	—	—	+	+	+	—	d
y_2	a	+	—	—	—	—	+	+	ade
y_3	b	—	+	—	—	+	—	+	be
y_4	ab	+	+	—	+	—	—	—	ab
y_5	c	—	—	+	+	—	—	+	ce
y_6	ac	+	—	+	—	+	—	—	ac
y_7	bc	—	+	+	—	—	+	—	bcd
y_8	abc	+	+	+	+	+	+	+	$abcde$

The treatment tested is determined by adding to the treatment with factors A, B, C, the appropriate small letter when D and E occur at their upper level.

There are four possible designs for the experiment obtained from $D = \pm BC$ and $E = \pm ABC$. Together, these will give the four blocks in the full factorial design for factors $A, B, C, D,$ and E, confounding $BCD, ADE, ABCE$ between blocks. The interactions confounded in a given design are called the *defining contrasts*. Adding the term I (unity) gives the *group of defining contrasts*.

When a fractional design is obtained by equating an additional factor to an interaction the defining contrast is obtained by multiplying both sides of the equation by the added factor, e.g.

$$D = + BC$$

gives

$$D^2 = + BCD$$

i.e.

$$I = + BCD, \tag{1}$$

using the rule

$$A^2 = B^2 = \text{etc.} = I.$$

Similarly

$$E = + ABC$$

gives

$$I = + ABCE. \tag{2}$$

Multiplying equations (1) and (2) gives the last member of the group. Thus the group of defining contrasts will be

$$I, BCD, ABCE, ADE.$$

If further factors are introduced, multiplying all equations of the form $I = inter$-*action*, in all possible ways, will identify the whole group of defining contrasts.

It is essential that we should be able to identify the effects measured by each comparison in order to avoid confusion of important effects. This can be done by multiplying the group of defining contrasts by each effect in turn, using the rule $A^2 = B^2 = \text{etc.} = I$, until all comparisons have been identified. The set of effects estimated by the same comparison are called *aliases*. By including the appropriate signs in the group of defining contrasts we can determine whether a sum or difference of effects is measured.

Thus in the above design the group of defining contrasts, with appropriate signs, is

$$I, +BCD, +ABCE, +ADE$$

and the seven independent comparisons between the eight observations will estimate

$$
\begin{array}{llll}
A + ABCD + BCE & + DE \\
B + CD & + ACE & + ABDE \\
C + BD & + ABE & + ACDE \\
D + BC & + ABCDE & + AE \\
E + BCDE & + ABC & + AD \\
AB + ACD & + CE & + BDE \\
AC + ABD & + BE & + CDE.
\end{array}
$$

To analyse a fractional design investigating n factors in 2^p observations (where $p < n$) it is convenient to first analyse the data as if it was a full design testing p factors and then to identify the aliases.

Example 9

An experiment is designed to investigate five factors in eight observations. The effects of interest are A, B, C, D, E, AB and AD. It can be assumed that no other interactions exist.

The following data are obtained:

Treatment	e	a	bde	abd	cd	$acde$	bc	$abce$
Response	2	1	3	1	5	5	2	2

(i) Verify that this is a suitable design, obtained by putting $D = -BC$, $E = AC$ in a 2^3 design with factors A, B, C.

(ii) Identify the aliases and test the significance of the effects. It can be assumed that the error variance is $\sigma_0^2 = 0.2$.

(i)

Treatment with A, B, C	A	B	C	AB	$E = AC$ AC	$D = -BC$ BC	ABC	Treatment tested
(1)	−	−	−	+	+	+	−	e
a	+	−	−	−	−	+	+	a
b	−	+	−	−	+	−	+	bde
ab	+	+	−	+	−	−	−	abd
c	−	−	+	+	−	−	+	cd
ac	+	−	+	−	+	−	−	$acde$
bc	−	+	+	−	−	+	−	bc
abc	+	+	+	+	+	+	+	$abce$

(ii) Initially we analyse the 8 observations assuming they represent the full design investigating A, B and C. The aliases can then be introduced in the final stage of the analysis, from the group of defining contrasts,

$$I, -BCD, ACE, -ABDE.$$

Thus using the above table of signs and applying the appropriate signs to the responses gives:

Aliases	Effect assumed estimated	Degrees of freedom	Total	Mean square
$A - ABCD + CE - BDE$	A	1	−3	1.125
$B - CD + ABCE - ADE$	B	1	−5	3.125
$C - BD + AE - ABCDE$	C	1	7	6.125
$AB - ACD + BCE - DE$	AB	1	−1	0.125
$AC - ABD + E - BCDE$	E	1	3	1.125
$BC - D + ABE - ACDE$	$-D$	1	−7	6.125
$ABC - AD + BE - CDE$	ABC	1	1	0.125
Total		7		17.875

From tables $F_{5\%}(1, \infty) = 3.84$ and hence effects A, B, C, E and D are significant. Each total gives four times the estimated effect and hence each effect and its confidence limits could be calculated.

Note, that if σ_0^2 had not been given, an estimate would be obtained from the AB and ABC mean squares and the appropriate significance point would then be

$$F_{5\%}(1, 2) = 18.5.$$

The factors involved in chemical experiments will often have real interactions, so that experiments which represent a very small fraction of a full design are unlikely to be used. Fractional designs will be most useful for preliminary experiments when either, large main effects are expected or a large number of factors are being investigated, only a few of which are likely to have real effects.

To design an efficient experiment it is important that both the technical and the statistical aspects of the problem are appreciated. The objectives of the investigation must be stated clearly and the validity of any assumptions made should be considered.

The appropriate statistical design and its correct analysis will always be the most efficient design possible for any investigation. Conclusions can be drawn from a minimum of experimentation and the reliability of those conclusions can be assessed.

PROBLEMS

1. Samples from the cooling water of a chemical plant are to be analysed daily to determine the amount of chlorine present. The sampling and testing error variance can be assumed to be $\sigma_0^2 = 0.0625$ (p.p.m.)^2. It is required to detect with probability 0.99 if the amount of chlorine present is as great as 2 p.p.m. If it is no greater than 1.5 p.p.m. the probability of asserting a value of 2 p.p.m. should be no more than 0.05. How many samples should be analysed each day?

2. An experiment was designed to determine the effect of ageing on the compressive strength of cement. The experiment used only one mix of cement but the test samples were moulded in three different moulds. The following randomized block design was used:

Mould 1	A	B	D	C
2	C	A	B	D
3	B	C	D	A

where $A, B, C,$ and D denote the times of ageing and the design shows the order in which the samples from each mould were allocated to the times of ageing.

Mould \ Ageing Time	A	B	C	D
1	58	80	96	100
2	74	81	100	105
3	54	83	110	120

The table gives the results obtained in (kg cm^{-3}).
 (i) Draw up the analysis of variance table for the experiment.
 (ii) What conclusion can be drawn if A, B, C, D represent increasing ageing times?

3. Nylon filaments are produced by extruding the molten polymer through fine orifices. After cooling, the filaments are twisted to give greater cohesion and are then cold drawn to four times their original length. Four machines are used for the cold drawing.

It is required to compare the resulting strength of the filaments when drawn in four different organic liquids, A, B, C and D.

Since there may be a difference between the filaments from different extrusions and also from the different drawing machines, the experiment is designed as a 4 x 4 Latin square, the extrusions and machines being allocated at random to the rows and columns respectively.

The yield strength in grams per denier is given in the following table:

Extrusion \ Machine	1	2	3	4
1	A: 0.44	B: 0.45	C: 0.46	D: 0.42
2	B: 0.45	A: 0.47	D: 0.43	C: 0.45
3	C: 0.47	D: 0.44	A: 0.45	B: 0.44
4	D: 0.44	C: 0.46	B: 0.45	A: 0.46

(i) What conclusions can be drawn from an analysis of the data?
(ii) Put 95% confidence limits on the strength of filament to be expected using liquid C.

4. An experiment is designed to investigate the production conditions of small, ceramic, dielectric tubes. The experiment covers four different mixes of slip, two extrusion speeds and three levels on the firing trolley.

The table gives the puncture voltages of tubes made under all combinations of these conditions.

Level on trolley \ Mix	Extrusion speed	1		2		3		4	
		1	2	1	2	1	2	1	2
Top		470	470	480	470	480	470	490	470
Middle		480	470	470	480	460	480	470	470
Bottom		450	440	460	450	470	470	450	460

Draw up the analysis of variance table for the experiment and test the significance of the results.

5. A 2^4 experiment with factors A, B, C and D is to be tested in four blocks. It is decided to confound interactions ABC and ABD between blocks.
(i) What other interaction will automatically be confounded between blocks?
(ii) Divide the treatments into the appropriate blocks.

6. An experiment is designed to investigate the effect of different annealing conditions on the ultimate tensile strength of a titanium alloy. Each factor is to be considered at two levels as given below.

Factor	+	−
A: Annealing time	2 hours	$1\frac{1}{2}$ hours
B: Annealing temperature	745 °C	735 °C
C: Furnace	F_1	F_2
D: Position in furnace	Inner	Outer
E: Level in furnace	Upper	Lower

A preliminary experiment is carried out and since it is considered unlikely that any of the factors will interact, the five factors are examined in eight observations using a fractional factorial design. The design is obtained by taking the full design which investigates factors A, B and C, and using the comparisons which estimate ABC and AC to estimate D and E respectively. The treatments tested and the ultimate tensile strength (in tons per square inch) of the samples are given in the following table:

Treatment	e	ad	bde	ab	cd	ace	bc	abcde
UTS (t.s.i.)	35.1	35.4	35.3	35.4	35.2	35.4	35.2	35.6

(i) Analyse the data by analysis of variance and test the significance of the effects.
An error variance $\sigma_0^2 = 0.005$ (t.s.i.)^2 can be assumed.
(ii) Identify the sets of aliases.

ACKNOWLEDGEMENTS

Tables 1, 5, 6, 7: Condensed and adapted from the *Biometrica Tables for Statisticians* Vol. 1, with permission from the Biometrica Trustees.
Table 4: Compiled from data published by *Biometrica*, by permission of the Biometrica Trustees.
Table 6: Columns 20%, 2% and 0.1% are taken from Table III of Fisher and Yates: *Statistical Tables for Biological, Agricultural and Medical Research*, published by Oliver and Boyd Limited, Edinburgh, and by permission of the authors and publishers.
Tables 2, 3, and page 356. Reproduced from tables in BS 2564:1955 by permission of British Standards Institution, 2 Park Street, London W1.

Q

Tables

Table 1. The standard normal distribution

Proportion of the distribution lying to the right of $u = u_1$, i.e. $\dfrac{1}{\sqrt{(2\pi)}} \displaystyle\int_{u_1}^{\infty} e^{-u^2/2}\, du.$

u_1	0	1	2	3	4	5	6	7	8	9
0.0	0.5000	0.4960	0.4920	0.4880	0.4840	0.4801	0.4761	0.4721	0.4681	0.4641
0.1	0.4602	0.4562	0.4522	0.4483	0.4443	0.4404	0.4364	0.4325	0.4286	0.4247
0.2	0.4207	0.4168	0.4129	0.4090	0.4052	0.4013	0.3974	0.3936	0.3897	0.3859
0.3	0.3821	0.3783	0.3745	0.3707	0.3669	0.3632	0.3594	0.3557	0.3520	0.3483
0.4	0.3446	0.3409	0.3372	0.3336	0.3300	0.3264	0.3228	0.3192	0.3156	0.3121
0.5	0.3085	0.3050	0.3015	0.2981	0.2946	0.2912	0.2877	0.2843	0.2810	0.2776
0.6	0.2743	0.2709	0.2676	0.2643	0.2611	0.2578	0.2546	0.2514	0.2483	0.2451
0.7	0.2420	0.2389	0.2358	0.2327	0.2296	0.2266	0.2236	0.2206	0.2177	0.2148
0.8	0.2119	0.2090	0.2061	0.2033	0.2005	0.1977	0.1949	0.1922	0.1894	0.1867
0.9	0.1841	0.1814	0.1788	0.1762	0.1736	0.1711	0.1685	0.1660	0.1635	0.1611

	.00	.01	.02	.03	.04	.05	.06	.07	.08	.09
1.0	0.1587	0.1562	0.1539	0.1515	0.1492	0.1469	0.1446	0.1423	0.1401	0.1379
1.1	0.1357	0.1335	0.1314	0.1292	0.1271	0.1251	0.1230	0.1210	0.1190	0.1170
1.2	0.1151	0.1131	0.1112	0.1093	0.1075	0.1056	0.1038	0.1020	0.1003	0.0985
1.3	0.0968	0.0951	0.0934	0.0918	0.0901	0.0885	0.0869	0.0853	0.0838	0.0823
1.4	0.0808	0.0793	0.0778	0.0764	0.0749	0.0735	0.0721	0.0708	0.0694	0.0681
1.5	0.0668	0.0655	0.0643	0.0630	0.0618	0.0606	0.0594	0.0582	0.0571	0.0559
1.6	0.0548	0.0537	0.0526	0.0516	0.0505	0.0495	0.0485	0.0475	0.0465	0.0455
1.7	0.0446	0.0436	0.0427	0.0418	0.0409	0.0401	0.0392	0.0384	0.0375	0.0367
1.8	0.0359	0.0351	0.0344	0.0336	0.0329	0.0322	0.0314	0.0307	0.0301	0.0294
1.9	0.0287	0.0281	0.0274	0.0268	0.0262	0.0256	0.0250	0.0244	0.0239	0.0233
2.0	0.0228	0.0222	0.0217	0.0212	0.0207	0.0202	0.0197	0.0192	0.0188	0.0183
2.1	0.0179	0.0174	0.0170	0.0166	0.0162	0.0158	0.0154	0.0150	0.0146	0.0143
2.2	0.0139	0.0136	0.0132	0.0129	0.0125	0.0122	0.0119	0.0116	0.0113	0.0110
2.3	0.0107	0.0104	0.0102	0.00990	0.00964	0.00939	0.00914	0.00889	0.00866	0.00842
2.4	0.00820	0.00798	0.00776	0.00755	0.00734	0.00714	0.00695	0.00676	0.00657	0.00639
2.5	0.00621	0.00604	0.00587	0.00570	0.00554	0.00539	0.00523	0.00508	0.00494	0.00480
2.6	0.00466	0.00453	0.00440	0.00427	0.00415	0.00402	0.00391	0.00379	0.00368	0.00357
2.7	0.00347	0.00336	0.00326	0.00317	0.00307	0.00298	0.00289	0.00280	0.00272	0.00264
2.8	0.00256	0.00248	0.00240	0.00233	0.00226	0.00219	0.00212	0.00205	0.00199	0.00193
2.9	0.00187	0.00181	0.00175	0.00169	0.00164	0.00159	0.00154	0.00149	0.00144	0.00139

Table 1 (*continued*). Extension for higher values of the deviate

u_1	Proportion to right of u_1	u_1	Proportion to right of u_1	u_1	Proportion to right of u_1	u_1	Proportion to right of u_1
3.0	0.00135	3.5	0.000233	4.0	0.0000317	4.5	0.00000340
3.1	0.000968	3.6	0.000159	4.1	0.0000207	4.6	0.00000211
3.2	0.000687	3.7	0.000108	4.2	0.0000133	4.7	0.00000130
3.3	0.000483	3.8	0.0000723	4.3	0.00000854	4.8	0.000000793
3.4	0.000337	3.9	0.0000481	4.4	0.00000541	4.9	0.000000479

Table 2. Control limit factors for the average chart.
Limits $= \bar{\bar{x}} \pm A\bar{w}$

Sample size n	Warning factor A_1	Action factor A_2
2	1.23	1.94
3	0.67	1.05
4	0.48	0.75
5	0.38	0.59
6	0.32	0.50

Table 3. Control limit factors for the range chart.
Limits $= D\bar{w}$.

Sample size n	Lower action factor D_1	Lower warning factor D_2	Upper warning factor D_3	Upper action factor D_4
2	0.00	0.04	2.81	4.12
3	0.04	0.18	2.17	2.99
4	0.10	0.29	1.93	2.58
5	0.16	0.37	1.81	2.36
6	0.21	0.42	1.72	2.22

Table 4. Number defective quality control. Control limits from the
Poisson distribution. Values tabulated are average number of defectives
expected in the sample (λ).

Control limit	Upper action 1 in 1000	Upper warning 1 in 40	Lower warning 1 in 40	Lower action 1 in 1000
0			3.69	6.91
1	0.000	0.000	5.57	9.23
2	0.001	0.025	7.22	11.23
3	0.045	0.24	8.77	
4	0.19	0.62	10.24	
5	0.43	1.09	11.67	
6	0.74	1.62		
7	1.11	2.20		
8	1.52	2.81		
9	1.97	3.45		
10	2.45	4.12		
11	2.96	4.80		
12	3.49	5.49		
13	4.04	6.20		
14	4.61	6.92		
15	5.20	7.65		
16	5.79	8.40		
17	6.41	9.15		
18	7.03			
19	7.66			
20	8.31			
	8.96			

Table 5. *F*-distribution.

Probability level	ϕ_2	ϕ_1 (corresponding to greater mean square)											
		1	2	3	4	5	6	7	8	9	10	15	∞
0.10	1	39.9	49.5	53.6	55.8	57.2	58.2	58.9	59.4	59.9	60.2	61.2	63.3
0.05		161	199	216	225	230	234	237	239	241	242	246	254
0.01		4,052	4,999	5,403	5,625	5,764	5,859	5,928	5,982	6,022	6,056	6,157	6,366
0.10	2	8.53	9.00	9.16	9.24	9.29	9.33	9.35	9.38	9.30	9.39	9.42	9.49
0.05		18.5	19.0	19.2	19.2	19.3	19.3	19.4	19.4	19.4	19.4	19.4	19.5
0.01		98.5	99.0	99.2	99.2	99.3	99.3	99.4	99.4	99.4	99.4	99.4	99.5
0.10	3	5.54	5.46	5.39	5.34	5.31	5.28	5.27	5.25	5.24	5.23	5.20	5.13
0.05		10.1	9.55	9.28	9.12	9.01	8.94	8.89	8.85	8.81	8.79	8.70	8.53
0.01		34.1	30.8	29.5	28.7	28.2	27.9	27.7	27.5	27.3	27.2	26.9	26.1
0.10	4	4.54	4.32	4.19	4.11	4.05	4.01	3.98	3.95	3.94	3.92	3.87	3.76
0.05		7.71	6.94	6.59	6.39	6.26	6.16	6.09	6.04	6.00	5.96	5.86	5.63
0.01		21.2	18.0	16.7	16.0	15.5	15.2	15.0	14.8	14.7	14.5	14.2	13.5
0.10	5	4.06	3.78	3.62	3.52	3.45	3.40	3.37	3.34	3.32	3.30	3.24	3.10
0.05		6.61	5.79	5.41	5.19	5.05	4.95	4.88	4.82	4.77	4.74	4.62	4.36
0.01		16.3	13.3	12.1	11.4	11.0	10.7	10.5	10.3	10.2	10.1	9.72	9.02
0.10	6	3.78	3.46	3.29	3.18	3.11	3.05	3.01	2.98	2.96	2.94	2.87	2.72
0.05		5.99	5.14	4.76	4.53	4.39	4.28	4.21	4.15	4.10	4.06	3.94	3.67
0.01		13.7	10.9	9.78	9.15	8.75	8.47	8.26	8.10	7.98	7.87	7.56	6.88
0.10	7	3.59	3.26	3.07	2.96	2.88	2.83	2.78	2.75	2.72	2.70	2.63	2.47
0.05		5.59	4.74	4.35	4.12	3.97	3.87	3.79	3.73	3.68	3.64	3.51	3.23
0.01		12.2	9.55	8.45	7.85	7.46	7.19	6.99	6.84	6.72	6.62	6.31	5.65

df	α												
8	0.10	3.46	3.11	2.92	2.81	2.73	2.67	2.62	2.59	2.56	2.54	2.46	2.29
	0.05	5.32	4.46	4.07	3.84	3.69	3.58	3.50	3.44	3.39	3.35	3.22	2.93
	0.01	11.3	8.65	7.59	7.01	6.63	6.37	6.18	6.03	5.91	5.81	5.52	4.86
9	0.10	3.36	3.01	2.81	2.69	2.61	2.55	2.51	2.47	2.44	2.42	2.34	2.16
	0.05	5.12	4.26	3.86	3.63	3.48	3.37	3.29	3.23	3.18	3.14	3.01	2.71
	0.01	10.6	8.02	6.99	6.42	6.06	5.80	5.61	5.47	5.35	5.26	4.96	4.31
10	0.10	3.29	2.92	2.73	2.61	2.52	2.46	2.41	2.38	2.35	2.32	2.24	2.06
	0.05	4.96	4.10	3.71	3.48	3.33	3.22	3.14	3.07	3.02	2.98	2.85	2.54
	0.01	10.0	7.56	6.55	5.99	5.64	5.39	5.20	5.06	4.94	4.85	4.56	3.91
12	0.10	3.18	2.81	2.61	2.48	2.39	2.33	2.28	2.24	2.21	2.19	2.10	1.90
	0.05	4.75	3.89	3.49	3.26	3.11	3.00	2.91	2.85	2.80	2.75	2.62	2.30
	0.01	9.33	6.93	5.95	5.41	5.06	4.82	4.64	4.50	4.39	4.30	4.01	3.36
15	0.10	3.07	2.70	2.49	2.36	2.27	2.21	2.16	2.12	2.09	2.06	1.97	1.76
	0.05	4.54	3.68	3.29	3.06	2.90	2.79	2.71	2.64	2.59	2.54	2.40	2.07
	0.01	8.68	6.36	5.42	4.89	4.56	4.32	4.14	4.00	3.89	3.80	3.52	2.87
16	0.10	3.05	2.67	2.46	2.33	2.24	2.18	2.13	2.09	2.06	2.03	1.94	1.72
	0.05	4.49	3.63	3.24	3.01	2.85	2.74	2.66	2.59	2.54	2.49	2.35	2.01
	0.01	8.53	6.23	5.29	4.77	4.44	4.20	4.03	3.89	3.78	3.69	3.41	2.75
24	0.10	2.93	2.54	2.33	2.19	2.10	2.04	1.98	1.94	1.91	1.88	1.78	1.53
	0.05	4.26	3.40	3.01	2.78	2.62	2.51	2.42	2.36	2.30	2.25	2.11	1.73
	0.01	7.82	5.61	4.72	4.22	3.90	3.67	3.50	3.36	3.26	3.17	2.89	2.21
60	0.10	2.79	2.39	2.18	2.04	1.95	1.87	1.82	1.77	1.74	1.71	1.60	1.29
	0.05	4.00	3.15	2.76	2.53	2.37	2.25	2.17	2.10	2.04	1.99	1.84	1.39
	0.01	7.08	4.98	4.13	3.65	3.34	3.12	2.95	2.82	2.72	2.63	2.35	1.60
∞	0.10	2.71	2.30	2.08	1.94	1.85	1.77	1.72	1.67	1.63	1.60	1.49	1.00
	0.05	3.84	3.00	2.60	2.37	2.21	2.10	2.01	1.94	1.88	1.83	1.67	1.00
	0.01	6.63	4.61	3.78	3.32	3.02	2.80	2.64	2.51	2.41	2.32	2.04	1.00

Table 6. Percentage points of the t-distribution.

ϕ	50%	20%	10%	5%	2%	1%	0.5%	0.1%
1	1.00	3.08	6.31	12.7	31.8	63.7	127	637
2	0.816	1.89	2.92	4.30	6.97	9.92	14.1	31.6
3	0.765	1.64	2.35	3.18	4.54	5.84	7.45	12.9
4	0.741	1.53	2.13	2.78	3.75	4.60	5.60	8.61
5	0.727	1.48	2.01	2.57	3.37	4.03	4.77	6.87
6	0.718	1.44	1.94	2.45	3.14	3.71	4.32	5.96
7	0.711	1.42	1.89	2.36	3.00	3.50	4.03	5.41
8	0.706	1.40	1.86	2.31	2.90	3.36	3.83	5.04
9	0.703	1.38	1.83	2.26	2.82	3.25	3.69	4.78
10	0.700	1.37	1.81	2.23	2.76	3.17	3.58	4.59
11	0.697	1.36	1.80	2.20	2.72	3.11	3.50	4.44
12	0.695	1.36	1.78	2.18	2.68	3.05	3.43	4.32
13	0.694	1.35	1.77	2.16	2.65	3.01	3.37	4.22
14	0.692	1.35	1.76	2.14	2.62	2.98	3.33	4.14
15	0.691	1.34	1.75	2.13	2.60	2.95	3.29	4.07
16	0.690	1.34	1.75	2.12	2.58	2.92	3.25	4.01
17	0.689	1.33	1.74	2.11	2.57	2.90	3.22	3.96
18	0.688	1.33	1.73	2.10	2.55	2.88	3.20	3.92
19	0.688	1.33	1.73	2.09	2.54	2.86	3.17	3.88
20	0.687	1.33	1.72	2.09	2.53	2.85	3.15	3.85
21	0.686	1.32	1.72	2.08	2.52	2.83	3.14	3.82
22	0.686	1.32	1.72	2.07	2.51	2.82	3.12	3.79
23	0.685	1.32	1.71	2.07	2.50	2.81	3.10	3.77
24	0.685	1.32	1.71	2.06	2.49	2.80	3.09	3.74
25	0.684	1.32	1.71	2.06	2.49	2.79	3.08	3.72
26	0.684	1.32	1.71	2.06	2.48	2.78	3.07	3.71
27	0.684	1.31	1.70	2.05	2.47	2.77	3.06	3.69
28	0.683	1.31	1.70	2.05	2.47	2.76	3.05	3.67
29	0.683	1.31	1.70	2.05	2.46	2.76	3.04	3.66
30	0.683	1.31	1.70	2.04	2.46	2.75	3.03	3.65
40	0.681	1.30	1.68	2.02	2.42	2.70	2.97	3.55
60	0.679	1.30	1.67	2.00	2.39	2.66	2.91	3.46
120	0.677	1.29	1.66	1.98	2.36	2.62	2.86	3.37
∞	0.674	1.28	1.64	1.96	2.33	2.58	2.81	3.29

The table gives the percentage of the area under the *two* tails of the t-curve, and therefore gives the probability that t will exceed the tabular entry in *absolute* value.

Table 7. Percentage points of the χ^2-distribution.

ϕ	99%	97.5%	95%	90%	50%	10%	5%	2.5%	1%	0.1%
1	0.000	0.001	0.004	0.016	0.455	2.71	3.84	5.02	6.63	10.83
2	0.020	0.051	0.103	0.211	1.39	4.61	5.99	7.38	9.21	13.82
3	0.115	0.216	0.352	0.584	2.37	6.25	7.81	9.35	11.34	16.27
4	0.297	0.484	0.711	1.06	3.36	7.78	9.49	11.14	13.28	18.47
5	0.554	0.831	1.15	1.61	4.35	9.24	11.07	12.83	15.09	20.52
6	0.872	1.24	1.64	2.20	5.35	10.64	12.59	14.45	16.81	22.46
7	1.24	1.69	2.17	2.83	6.35	12.02	14.07	16.01	18.48	24.32
8	1.65	2.18	2.73	3.49	7.34	13.36	15.51	17.53	20.09	26.13
9	2.09	2.70	3.33	4.17	8.34	14.68	16.92	19.02	21.67	27.88
10	2.56	3.25	3.94	4.87	9.34	15.99	18.31	20.48	23.21	29.59
11	3.05	3.82	4.57	5.58	10.34	17.28	19.68	21.92	24.73	31.26
12	3.57	4.40	5.23	6.30	11.34	18.55	21.03	23.34	26.22	32.91
13	4.11	5.01	5.89	7.04	12.34	19.81	22.36	24.74	27.69	34.53
14	4.66	5.63	6.57	7.79	13.34	21.06	23.68	26.12	29.14	36.12
15	5.23	6.26	7.26	8.55	14.34	22.31	25.00	27.49	30.58	37.70
16	5.81	6.91	7.96	9.31	15.34	23.54	26.30	28.85	32.00	39.25
17	6.41	7.56	8.67	10.09	16.34	24.77	27.59	30.19	33.41	40.79
18	7.01	8.23	9.39	10.86	17.34	25.99	28.87	31.53	34.81	42.31
19	7.63	8.91	10.12	11.65	18.34	27.20	30.14	32.85	36.19	43.82
20	8.26	9.59	10.85	12.44	19.34	28.41	31.41	34.17	37.57	45.32
21	8.90	10.28	11.59	13.24	20.34	29.62	32.67	35.48	38.93	46.80
22	9.54	10.98	12.34	14.04	21.34	30.81	33.92	36.78	40.29	48.27
23	10.20	11.69	13.09	14.85	22.34	32.01	35.17	38.08	41.64	49.73
24	10.86	12.40	13.85	15.66	23.34	33.20	36.42	39.36	42.98	51.18
25	11.52	13.12	14.61	16.47	24.34	34.38	37.65	40.65	44.31	52.62
26	12.20	13.84	15.38	17.29	25.34	35.56	38.89	41.92	45.64	54.05
27	12.88	14.57	16.15	18.11	26.34	36.74	40.11	43.19	46.96	55.48
28	13.56	15.31	16.93	18.94	27.34	37.92	41.34	44.46	48.28	56.89
29	14.26	16.05	17.71	19.77	28.34	39.09	42.56	45.72	49.59	58.30
30	14.95	16.79	18.49	20.60	29.34	40.26	43.77	46.98	50.89	59.70

Answers

CHAPTER 1

1. 2.2924.   **2.** (i) $\coth(x/2)$,   (ii) $\cosh 4x + \sinh 4x \equiv \exp(4x)$.

8. $\sinh 7x = 64 \sinh^7 x + 112 \sinh^5 x + 56 \sinh^3 x + 7 \sinh x$,
$\cosh 7x = 64 \cosh^7 x - 112 \cosh^5 x + 56 \cosh^3 x - 7 \cosh x$,

$$\tanh 7x = \frac{7 \tanh x + 35 \tanh^3 x + 21 \tanh^5 x + \tanh^7 x.}{1 + 21 \tanh^2 x + 35 \tanh^4 x + 7 \tanh^6 x}$$

10. $\log_e \left(\dfrac{6 + \sqrt{(41)}}{5} \right) = 0.9083$.   **11.** $\frac{1}{6}\{2 + \sqrt{(10)}\} = 0.8603$.   **12.** 0.3215.

15. 43 mg.   **16.** 18.50 cm^3.

CHAPTER 2

1. (i) $11 - 13i$,   (ii) $\frac{1}{17}(2 + 9i)$,   (iii) $-\frac{1}{4}(1 + 5i)$.

2. (i) $\pm 7i$,   (ii) $-2 \pm 2i$,   (iii) $-1, \frac{1}{2}(1 \pm i\sqrt{7})$.

3. (i) $3\{\cos(\pi/2) - i \sin(\pi/2)\}$,   (ii) $2\{\cos(\pi/4) + i \sin(\pi/4)\}$,
(iii) $2\{\cos(2\pi/3) + i \sin(2\pi/3)\}$,   (iv) $2\{\cos(5\pi/6) - i \sin(5\pi/6)\}$,
(v) $2\{\cos(\pi/3) - i \sin(\pi/3)\}$.

4. $-2(1 + i)$.   **5.** $z + y - 3 = 0$.   **6.** Circle centre $(0, -13/3)$, radius 8/3.

7. (i) $\sin(5\theta) - i \cos(5\theta)$,   (ii) $\cos(2\theta) + i \sin(2\theta)$,   (iii) $-64i$.

8. (i) 2^{12},   (ii) -2^{18}.

9. (i) $\cos \left\{ \left(\dfrac{4k + 1}{12} \right) \pi \right\} + i \sin \left\{ \left(\dfrac{4k + 1}{12} \right) \pi \right\}$, $k = 0, 1, 2, 3, 4, 5$

(ii) $\sqrt{(2)} \cos \left\{ \left(\dfrac{8k + 1}{12} \right) \pi \right\} + i \sin \left\{ \left(\dfrac{8k + 1}{12} \right) \pi \right\}$, $k = 0, 1, 2$.

10. $(6 \tan \theta - 20 \tan^3 \theta + 6 \tan^5 \theta)/(1 - 15 \tan^2 \theta + 15 \tan^4 \theta - \tan^6 \theta)$.

11. $\frac{1}{128} \{\cos(8\theta) - 8 \cos(6\theta) + 28 \cos(4\theta) - 56 \cos(2\theta) + 35\}$.

12. (i) $\sin x \cosh y + i \cos x \sinh y$,
(ii) $2.56 + i(1.18 \pm 2k\pi)$, $k = 0, \pm 1, \pm 2, \ldots$
(iii) $\frac{1}{2}\pi(4k + 1) \pm i \log_e(3 + 2\sqrt{2})$, $k = 0, 1, 2, \ldots$

CHAPTER 3

1. (i) $3x^2 - 28x + 51$,   (ii) $\dfrac{1}{(x + 1)^2} \sqrt{\dfrac{x + 1}{4x}}$,

(iii) $\dfrac{4(x^2 - 1)}{(x + 3)^2 (x + 4)^2}$,   (iv) $\dfrac{5x^2 + x - 1}{2\{\sqrt{(x)} + 1\}}$.

2. (i) $\dfrac{\sin^2\sqrt{(x-1)}\sin 2\sqrt{(x-1)}}{\sqrt{(x-1)}}$,   (ii) $\dfrac{\cos^3 x - \sin^3 x}{(\cos x + \sin x)^2}$,

(iii) $\dfrac{-3}{\sqrt{(12x-3-9x^2)}}$,   (iv) $\dfrac{-\csc x}{\sqrt{\{\tan^2(x/2)-1\}}}$.

3. (i) $\dfrac{2x}{\sqrt{(x^4+4x^2+5)}}$,   (ii) $-\tfrac{1}{2}\csc(x/2)$,   (iii) $6(\sinh 6x - \cosh 6x)$.

4. (i) $\dfrac{37x-45}{2(2x^2+3x-11)^{3/2}}$,   (ii) $\left(\cos x\,\log_e x + \dfrac{\sin x}{x}\right)x^{\sin x}$,

(iii) $\dfrac{-2x}{x^4+1}$,   (iv) $3\sec 3x$.

11. $\dfrac{dy}{dx}=5$, $\dfrac{d^2y}{dx^2}=-10$.

12. (i) $t, \tfrac{1}{2}a$,   (ii) $-\dfrac{b}{a}\cot\theta, \ -\dfrac{b}{a^2}\csc^3\theta$,   (iii) $\tan\theta, \tfrac{1}{3}\sec^3\theta\sec 2\theta$,

(iv) $-\tan\theta, \sin\theta\sec^4\theta$.

13. Max$(1, -2)$, Min$(-3, \tfrac{2}{3})$, Asymptotes $x = 0, x = 3, y = 1$.

14. (i) Max$(-\sqrt{2}, -12\sqrt{2}-17)$, Min$(\sqrt{2}, 12\sqrt{2}-17)$, Asymptotes $x = -1$, $x = -2, y = 1$.

(ii) Max$\left(\dfrac{1}{\sqrt{2}}, -2\sqrt{2}-3\right)$, Min$\left(-\dfrac{1}{\sqrt{2}}, 2\sqrt{2}-3\right)$,

Asymptotes $x = \tfrac{1}{2}, x = 1$.   (iii) Max$(2\sqrt{3}, -3^{3/4})$, Min$(2\sqrt{3}, 3^{3/4})$, Asymptotes $x = -2, x = 2, y^2 = x$.

15. $\dfrac{2a}{\sqrt{3}}, a\sqrt{\dfrac{2}{3}}$.   **16.** 1.861.   **17.** 4.54.

CHAPTER 4

1. (i) $\dfrac{3x}{2}+\dfrac{7}{4}\log_e(2x-1)+C$,   (ii) $\log_e(4-3x)-x-\dfrac{x^2}{2}+C$,

(iii) $\dfrac{2}{3\sqrt{x}}(x^2+12x+24)+C$.

2. (i) $\log_e(3-x)-2\log_e(1+x)+C$,
(ii) $2x - 3\log_e(1-2x)+\log_e(x-2)+C$,
(iii) $\tan^{-1}x - \tfrac{1}{2}\log_e(2x+3)+C$,

(iv) $5\log_e(x-1)-\tfrac{10}{3}\log_e(3x+1)-\dfrac{1}{x-1}+C$,

(v) $\log_e x - \log_e(x+2)+\tfrac{1}{2}\log_e(2x^2+x+1)+\dfrac{1}{\sqrt{7}}\tan^{-1}\left(\dfrac{4x+1}{\sqrt{7}}\right)+C$.

3. $\tfrac{1}{6}\tan^{-1}\left(\dfrac{2x-1)}{3}\right)+C; \log_e(3-2x-x^2)+\tfrac{3}{2}\tanh^{-1}\left(\dfrac{x+1}{2}\right)+C$.

4. (i) $\sqrt{(4-x^2)} + \sin^{-1}(x/2) + C$,   (ii) $\sin^{-1}\left(\dfrac{x-1}{\sqrt{3}}\right) + C$,

(iii) $\frac{1}{3}\sinh^{-1}(3x+1) + C$,   (iv) $3\sqrt{(x^2+x-6)} - 2\cosh^{-1}\left(\dfrac{2x+1}{5}\right) + C$.

5. (i) $\frac{1}{2}\tan^2 x + \log_e \cos x + C$,   (ii) $\operatorname{cosec} x - \frac{1}{3}\operatorname{cosec}^3 x + C$,

(iii) $\frac{1}{32}(12x - 8\sin 2x - \sin 4x) + C$,   (iv) $2\tan^{-1}\left(\dfrac{\tan(x/2)-1}{2}\right) + C$.

(v) $\dfrac{1}{\sqrt{(15)}}\tanh^{-1}\left(\sqrt{\dfrac{3}{5}}\,\tan x\right) + C$,   (vi) $\dfrac{7}{10}\log_e(\sin x - 3\cos x) - \dfrac{x}{10} + C$.

6. (i) $2\sin^{-1}\left(\dfrac{x+1}{2}\right) + \dfrac{1}{2}(x+1)\sqrt{(3-2x-x^2)} + C$,

(ii) $\dfrac{1}{2}x\sqrt{(x^2-9)} - \dfrac{9}{2}\cosh^{-1}\left(\dfrac{x}{3}\right) + C$,   (iii) $\sin^{-1}\left(\dfrac{x-1}{2}\right) + C$,

(iv) $-\sinh^{-1}\left(\dfrac{2x+1}{x}\right) + C$,   (v) $\sqrt{(x^2-4)} - 2\cosh^{-1}\left(\dfrac{x}{2}\right) + C$.

7. (i) $\dfrac{x^2}{3}\sin 3x + \dfrac{2x}{9}\cos 3x - \dfrac{2}{27}\sin 3x + C$,

(ii) $(x-1)(x+3)\log_e(x-1) - \dfrac{x}{2}(x+6) + C$,

(iii) $\dfrac{1}{2}(x^2+1)\tan^{-1}x - \dfrac{x}{2} + C$,   (iv) $\dfrac{e^x}{5}(\cos 2x + 2\sin 2x)$,

(v) $x(\log_e x - 1)\log_e 2x - x\log_e x + 2x + C$,

(vi) $\frac{8}{15}$,   (vii) $\frac{2}{35}$,   (viii) $\dfrac{3\pi}{512}$.

8. $\frac{1}{5}(16 - 9\sqrt{2})$.   **9.** $\frac{1}{280}$.   **11.** $\dfrac{\pi a^4 P}{8\eta l}$.   **13.** $k_1 = 0.41, k_2 = 1.79$.

CHAPTER 5

1. (i) $\frac{64}{3}$,   (ii) $\frac{9}{2}$.   **2.** $3\pi a^2$.   **3.** $3\pi a^2$.   **4.** $3\pi; \dfrac{9\pi^2}{2}$.

5. $\dfrac{3\pi a^2}{8}; 3\pi a$.   **7.** $\pi a\sqrt{(a^2+h^2)}$.   **8.** $(\frac{5}{3}, 0)$.   **9.** $\dfrac{M}{12}(4l^2 + 3a^2)$.

10. (i) $\dfrac{3M}{20}(4h^2 + a^2)$,   (ii) $\dfrac{M}{20}(2h^2 + 3a^2)$.   **11.** $\dfrac{2M}{5}\left(\dfrac{b^5 - a^5}{b^3 - a^3}\right)$.

12. (i) $x + \dfrac{1}{3}x^3 + \dfrac{2}{15}x^5 + \cdots$ (ii) $x - \dfrac{x^3}{6} + \dfrac{3x^5}{40} - \cdots$

(iii) $-\dfrac{x^2}{2} - \dfrac{x^4}{12} - \dfrac{x^6}{45} - \cdots$.

14. (i) 0.680(5),   (ii) 0.680(4).

15. (i) 0.03192,   (ii) 0.03192,   (iii) 0.03190.   **16.** $100\pi/3$ cm^3 min^{-1}.

17. 25.643 cal.

CHAPTER 6

1. (i) 7,   (ii) 1,   (iii) 5.   **2.** $x = 2/3$.   **4.** $x = \pm\sqrt{3}, 0, 0$.

5. $\lambda = 0, \dfrac{K}{m_1 m_2}(m_1 + m_2)$.   **6.** (i) 35,   (ii) 16.

7. $\lambda = 0, \dfrac{K}{m}, \dfrac{K}{mM}(M + 2m)$.   **8.** $x = 1\frac{1}{16}, y = 2\frac{1}{21}, z = -17/24$.

9. $a = 5, b = 3/5, c = 3/2$.

10. (i) $\begin{bmatrix} 3 & 6 & -1 \\ 2 & 1 & 5 \end{bmatrix}$,   (ii) $\begin{bmatrix} 1 & -4 & -5 \\ 4 & 1 & -1 \end{bmatrix}$,   (iii) $\begin{bmatrix} 7 & 17 & 0 \\ 3 & 2 & 13 \end{bmatrix}$,

    (iv) $\begin{bmatrix} 7 & -5 \\ 23 & 17 \end{bmatrix}$,   (v) $\begin{bmatrix} 18 & 8 & -14 \\ 11 & 4 & 3 \\ 16 & 6 & 2 \end{bmatrix}$,   (vi) $[0, -10]$.

11. (i) $\mathbf{A}^{-1} = \mathbf{B}$,   (ii) $\mathbf{BA} = \mathbf{I}$,   (iii) $\mathbf{B}^{-1}\mathbf{A}^{-1}\mathbf{B}^2\mathbf{A} = \mathbf{B}$.   **12.** $\mathbf{E}_3$.

13.
$$\mathbf{A}' = \begin{bmatrix} 2 \\ 3 \\ -4 \end{bmatrix}, \quad \mathbf{B}' = \begin{bmatrix} 2 & -1 \\ 6 & 0 \\ 1 & 2 \end{bmatrix}, \quad \mathbf{AA}' = 29.$$

16.
$$\mathbf{A}^{-1} = \frac{1}{19}\begin{bmatrix} 2 & -1 & 7 \\ 5 & 7 & -11 \\ -4 & 2 & 5 \end{bmatrix}.$$
  17. $\begin{bmatrix} p'_x \\ p'_y \end{bmatrix} = \begin{bmatrix} 0 & 1 \\ -1 & 0 \end{bmatrix}\begin{bmatrix} p_x \\ p_y \end{bmatrix}$.

CHAPTER 7

1. (i) $\sqrt{(14)}$,   (ii) $\sqrt{(26)}$,   (iii) $\sqrt{(29)}$.

2. (i) $\dfrac{1}{7}(3\mathbf{i} + 6\mathbf{j} - 2\mathbf{k})$,   (ii) $\dfrac{1}{9}(-4\mathbf{i} + \mathbf{j} - 8\mathbf{k})$,   (iii) $\dfrac{1}{7}(2\mathbf{i} - 3\mathbf{j} + 6\mathbf{k})$.

3. $-14\mathbf{i} - 8\mathbf{j} + 8\mathbf{k}$.   **4.** $4\sqrt{2}; 0, -1/\sqrt{2}, 1/\sqrt{2}$.

5. $\dfrac{x - 4}{-7} = \dfrac{y - 1}{4} = \dfrac{z + 2}{4}$.   **6.** $x + 5y - 3z + 10 = 0$.

7. (i) -6,   (ii) -1,   (iii) -1.   **8.** 9.   **9.** $(13\mathbf{i} - 19\mathbf{j} + 10\mathbf{k})/3\sqrt{(70)}$.

10. (i) $21\mathbf{i} - 7\mathbf{j} + 14\mathbf{k}$,   (ii) $7\mathbf{i} + 10\mathbf{j} - 11\mathbf{k}$,   (iii) $-7\mathbf{i} - 37\mathbf{j} - 2\mathbf{k}$.

12. (i) 7,   (ii) $-17\mathbf{i} + 2\mathbf{j} - 30\mathbf{k}$.

15. (i) 2.83Å, 8.77Å, 9Å,   (ii) 4.36Å.

17. $\omega\sqrt{\dfrac{31}{3}}$; $-2\mathbf{i} + 3\mathbf{j} + 7\mathbf{k}$.

18. (i) $2t\mathbf{i} + 2\mathbf{j} + (2t + 1)\mathbf{k}$,   (ii) $8t^3 + 6t^2 + 10t + 10$,
  (iii) $(4t + 2)\mathbf{i} + (2t + 12)\mathbf{j} - (4t + 2)\mathbf{k}$.

CHAPTER 8

1. (i) $4 \tan^{-1}\left(\dfrac{x+y}{x}\right) = \log_e(2x^2 + 2xy + y^2) + C.$

(ii) $2 \log_e(y-x) + 3 \log_e(3x^2 + xy + y^2) + \dfrac{9}{5\sqrt{(11)}} \tan^{-1}\left(\dfrac{x+2y}{\sqrt{(11)}x}\right) + C.$

(iii) $\dfrac{2}{\sqrt{3}} \tanh^{-1}\left(\dfrac{y+1}{(x-2)\sqrt{3}}\right) = \log_e\{3(x-2)^2 - (y+1)^2\} + C.$

(iv) $x - 3y + 10 \log_e(x - 2y - 7) + C = 0.$
(v) $3y\sqrt{(2x+1)} = \{(2x+1)^{3/2} + C\}(x-1)^2.$
(vi) $y = (C + \tan x)\sin^3 x.$
(vii) $y^{1/4} \exp(2x) = \{\log_e(1+x) + C\}(x+1)^2.$

2. (i) $\dfrac{5!}{s^6}$, (ii) $\dfrac{1}{s+3}$, (iii) $\dfrac{4}{s^2+16}$, (iv) $\dfrac{4s+2}{4s^2+4s+17}$, (v) $\dfrac{3!}{(s-2)^4}.$

3. (i) $4\sin 2x - \cos 2x$, (ii) $\frac{1}{4}(e^{3x} - 2e^x + 5e^{-x})$, (iii) $e^x(\cos \frac{1}{2}x + 2 \sin \frac{1}{2}x)$,
(iv) $x^2 - 3xe^x + 1$, (v) $3 + x - 3\cos 3x.$

4. $\dfrac{4s}{(s^2+4)^2} ; \dfrac{3}{s^2 - 2s + 10}.$ 5. $y = 3 + x + e^x - 8 \exp(x/2).$

6. $y = 8/x^2 + 8x^2 - 4.$ 7. $x = t + 2 + 21te^t - 2e^t; y = 3e^t - 7te^t - 3.$

8. $\dfrac{2\omega s}{(s^2+\omega^2)^2} ; x = 3 \cos 2t + (2 + t/4)\sin 2t.$

9. $y = \dfrac{4}{x + A_1} + A_2.$ 10. $31.3 \text{ cm}^3.$

11. $x = a \exp(-\lambda_1 t); y = \dfrac{\lambda_1 a}{\lambda_2 - \lambda_1} \{\exp(-\lambda_1 t) - \exp(-\lambda_2 t)\}$

$z = \dfrac{a}{\lambda_2 - \lambda_1} \{\lambda_1 \exp(-\lambda_2 t) - \lambda_2 \exp(-\lambda_1 t)\} + a.$

12. 7.85 minutes.   15. $x = a\{\frac{1}{2} - \frac{1}{6} \exp(-2t) - \frac{1}{3} \exp(-t)\}.$

16. $x = A_1 \exp\left(\dfrac{n-\omega}{\mu} t\right) + A_2 \exp\left(\dfrac{-n-\omega}{\mu} t\right)$ where $n^2 = \omega^2 - k\mu.$

18. (i) $y = A_1\left(1 - \dfrac{x}{1!2} + \dfrac{x^2}{2!2.5} - \dfrac{x^3}{3!2.5.8} + \cdots\right)$

$+ A_2 x^{1/3}\left(1 - \dfrac{x}{1!4} + \dfrac{x^2}{2!4.7} - \dfrac{x^3}{3!4.7.10} + \cdots\right),$

(ii) $y = A_1 x(1 - \frac{5}{3}x^2) + A_2(1 - 6x^2 + 3x^4 - \cdots),$

(iii) $y = A_1(1 + 3x^2 + x^4 + \cdots) + A_2 x^{3/2}\left(1 + \dfrac{3x^2}{4} - \dfrac{x^4}{16} + \cdots\right).$

CHAPTER 9

1. (i) $z_x = 5x^4y + 6x^2y^2 + 3y + 1$; $z_{xy} = 5x^4 + 12x^2y + 3$; $z_{yy} = 4x^3 + 6y$.
   (ii) $z_x = \sin 2x$; $z_{xy} = 0$; $z_{yy} = -2 \cos 2y$.
   (iii) $z_x = \tan xy + y(x+y)\sec^2 xy$;
   $z_{xy} = 2 \sec^2 xy\{x + y + xy(x+y)\tan xy\}$;
   $z_{yy} = 2x \sec^2 xy\{1 + x(x+y)\tan xy\}$.

4. (i) $-\dfrac{1}{2}$ (ii) $\dfrac{\pi - 12}{18}$.   7. Max$(\frac{3}{2}, -1)$, Saddle pts $(0, 2)$, $(0, -4)$.

8. Max$(-1, 1)$, Saddle pts $\left(0, \dfrac{3}{2} \pm \dfrac{\sqrt{(13)}}{2}\right)$, $\left(-\dfrac{13}{3}, -\dfrac{2}{3}\right)$.

11. 0.097 mm.   14. $T = 2t^2 - 5t + 8$.

CHAPTER 10

1. (i) πa^2,   (ii) $\frac{9}{2}$.   2. $\frac{1}{16}(12 \log_e 3 - 10 \log_e 2 - 5)$.   3. $\pi/4$.

5. $\dfrac{\pi a^2}{8} (2 \log_e a - 1)$.   7. $\frac{1}{2}(1 - \log_e 2)$.

10. $\dfrac{\pi a^2}{2} \pm u$, where $u = a^2 \sin^{-1}(b/a) + b\sqrt{(a^2 - b^2)}$.   11. $\dfrac{3\pi}{8}$.

12. $6(3\pi - 4)$.   13. $\dfrac{35\pi}{256}$.   14. $\dfrac{3\lambda^3}{8a^3}\left\{1 - \dfrac{2a^2}{\lambda^2} + \dfrac{8a^3}{3\lambda^3} - \left(\dfrac{2a}{\lambda} + 1\right)e^{-2a/\lambda}\right\}$.

CHAPTER 11

1. $f(x) = \dfrac{2}{\pi} - \dfrac{4}{\pi}\left(\dfrac{\cos 4x}{3.5} + \dfrac{\cos 8x}{7.9} + \cdots\right) + \dfrac{4}{\pi}\left(\dfrac{\sin 4x}{3.5} + \dfrac{\sin 8x}{7.9} + \cdots\right)$.

2. $f(x) = \dfrac{1}{2} + \dfrac{4}{\pi}\left(\cos \pi x + \dfrac{\cos 3\pi x}{3} + \dfrac{\cos 5\pi x}{5} + \cdots\right)$.

3. $f(x) = \displaystyle\sum_{n=1}^{\infty} \dfrac{2n\pi}{T^2 + n^2\pi^2}\{1 - (-1)^n \exp(-T)\} \sin\left(\dfrac{n\pi x}{T}\right)$.

4. $f(x) = \dfrac{4}{\pi}\left(\dfrac{1}{2} - \dfrac{\cos 2x}{1.3} - \dfrac{\cos 4x}{3.5} - \dfrac{\cos 6x}{5.7} - \cdots\right)$.

5. $f(x) = \dfrac{4}{\pi} + \dfrac{16}{\pi^2}\displaystyle\sum_{n=1}^{\infty}\dfrac{1}{n}\sin\left(\dfrac{n\pi}{2}\right)\cos 4nx$.

6. $f(x) = \dfrac{4\pi^2}{3} - 2\left(\dfrac{\cos 2x}{1^2} + \dfrac{\cos 4x}{2^2} + \dfrac{\cos 6x}{3^2} + \cdots\right) = \dfrac{4\pi^2}{3} - 2\displaystyle\sum_{n=1}^{\infty}\dfrac{\cos 2nx}{n^2}$

7. $\theta = 9.22 - 5.48 \cos\left(\dfrac{\pi t}{6}\right) - 2.29 \sin\left(\dfrac{\pi t}{6}\right) + 0.60 \cos\left(\dfrac{\pi t}{3}\right)$

   $+ 1.64 \sin\left(\dfrac{\pi t}{3}\right) - 0.22 \cos\left(\dfrac{\pi t}{2}\right) - 0.72 \sin\left(\dfrac{\pi t}{2}\right) + \cdots$

8. $\phi = \dfrac{4}{\pi}\displaystyle\sum_{n=1}^{\infty}\dfrac{1}{2n-1}\exp\{(1-2n)y\}\sin(2n-1)x$.

CHAPTER 12

1. (a) (i) A = unit element $A^{-1} = A$, $B^{-1} = C$, $C^{-1} = B$, $D^{-1} = F$, $F^{-1} = D$.
   (ii) D = unit element $A^{-1} = B$, $B^{-1} = A$, $C^{-1} = C$, $D^{-1} = D$, $F^{-1} = F$, $G^{-1} = G$.

2. (a)

	0	2	4	6
0	0	2	4	6
2	2	4	6	0
4	4	6	0	2
6	6	0	2	4

   (b) Not a group since 4^2 is not a member of the set,

   (c)

	−1	1
−1	1	−1
1	−1	1

   (d)

	A	B	C	D
A	A	B	C	D
B	B	A	D	C
C	C	D	A	B
D	D	C	B	A

3. (a) $\{1\}$, $\{1, 4\}$, $\{1, 2, 3, 4\}$, (b) $\{D\}$, $\{D, C\}$ $\{D, F\}$, $\{D, G\}$, $\{D, A, B\}$, $\{D, A, B, C, F, G\}$.

4.

	E	A	B
E	E	A	B
A	A	B	E
B	B	E	A

   It is impossible to write the table in a different way.

5. Same group tables are obtained if

   $$B \equiv a, D \equiv b, C \equiv c, A \equiv d \quad \text{or} \quad B \equiv a, D \equiv b, C \equiv d, A \equiv c.$$

6. In 2(d) the squares of the elements are all equal to the unit element, which is not true in 2(a). Therefore the two groups are not isomorphic.

7. $A^0 = E$, $A^2 = C$, $A^3 = B$, $A^4 = F$, $A^5 = D$, i.e. cyclic group generated by A.

8. Since the group is of order 5, subgroups are of order 1 and 5, the subgroup of order 5 being the group itself which must be cyclic and therefore abelian. But $BA \neq AB$, and so the elements do not form a group. (Note: $(AB)D \neq A(BD)$).

9.

	E	i	σ_1	σ_m
E	E	i	σ_1	σ_m
i	i	E	σ_m	σ_1
σ_1	σ_1	σ_m	E	i
σ_m	σ_m	σ_1	i	E

10. E: (x, y, z); C_2: $(-x, -y, z)$; σ'_v: $(x, -y, z)$; σ''_v: $(-x, y, z)$;

	E	C_2	σ'_v	σ''_v
E	E	C_2	σ'_v	σ''_v
C_2	C_2	E	σ''_v	σ'_v
σ'_v	σ'_v	σ''_v	E	C_1
σ''_v	σ''_v	σ'_v	C_2	E

11.

			E	i	C_2	C_2'	C_2''	σ_1	σ_2	σ_3
E:	(x,y,z)	E	E	i	C_2	C_2'	C_2''	σ_1	σ_2	σ_3
i:	$(-x,-y,-z)$	i	i	E	σ_3	σ_2	σ_1	C_2''	C_2'	C_2
C_2:	$(x,-y,-z)$	C_2	C_2	σ_3	E	C_2''	C_2'	σ_2	σ_1	i
C_2':	$(-x,y,-z)$	C_2'	C_2'	σ_2	C_2''	E	C_2	σ_3	i	σ_1
C_2'':	$(-x,-y,z)$	C_2''	C_2''	σ_1	C_2'	C_2	E	i	σ_3	σ_2
σ_1:	$(x,y,-z)$	σ_1	σ_1	C_2''	σ_2	σ_3	i	E	C_2	C_2'
σ_2:	$(x,-y,z)$	σ_2	σ_2	C_2'	σ_1	i	σ_3	C_2	E	C_2''
σ_3:	$(-x,y,z)$	σ_3	σ_3	C_2	i	σ_1	σ_2	C_2'	C_2''	E

CHAPTER 13

1. (i) $\frac{1}{2}$, (ii) $\frac{1}{12}$, (iii) $\frac{1}{6}$.   **2.** $\frac{7}{12}$.   **3.** (i) 0.48, (ii) 0.52.

4. (i) $\frac{1}{16}$, (ii) $\frac{1}{17}$.   **5.** $\frac{1}{140}$.   **6.** (i) $\frac{1}{15}$, (ii) $\frac{7}{15}$.   **7.** $(\frac{4}{5})^6$.

8. $(\frac{6}{13})^3$.   **9.** 48.   **10.** (i) 336, (ii) 35, (iii) $\dfrac{(3n)!}{2}$, (iv) n.

11. $10!, \dfrac{10!}{6}$.   **12.** 336.   **13.** 24.   **14.** $1140, \frac{3}{20}$.   **15.** 700.

16. $\dfrac{11!}{48}, \dfrac{9!}{24}, \dfrac{15}{8} \times 9!$.   **17.** $\dfrac{32}{243}, \dfrac{112}{243}$.

18. (i) $\dfrac{96 \times 48!}{52!}$, (ii) $\dfrac{44}{4{,}165}$, (iii) $\dfrac{6 \times 48!}{51!}$, (iv) $\dfrac{960 \times 48!}{52!}$.

19. $\frac{3}{7}$.   **20.** 44.

CHAPTER 14

1. 1.976, 2.419.   **2.** $0.071, 0.011, 8.5 \times 10^{-6}$.   **3.** $1.0077800, 1.01 \times 10^{-5}$.

4. $2\sqrt{6}$.   **5.** 0.142, 0.70.   **6.** 1149.5, 1149.5, 1151.6.   (ii) 45.2, 240.

8. (i) 42.56%, 2.98%.   (ii) 42.56%, 10.00%.   **10.** $1{,}000.375 \text{ cm}^3, 2.58 \text{ cm}^3$.

CHAPTER 15

1. $\frac{5}{16}, \frac{57}{64}$.   **2.** 0.00032, 0.0064, 0.0512, 0.2048, 0.4096, 0.32768.

3. 4, 0.416.   **4.** 19.7, 45.6.   **5.** $\{\sqrt{(21)} - 3\}/4$.   **6.** 10.

7. $\frac{79}{192}$.   **8.** 0.082, 0.205, 0.257, 0.214, 0.134, 0.067.

9. 0.607, 32.85.   **10.** 7.65 hours, 17.97 hours.   **11.** 0.0634, 0.0025.

12. 2.303.

13. mean = 1.72, variance = 1.74.

     17.9, 30.8, 26.5, 15.2, 6.5, 2.2, 0.6, 0.1.

14. mean = 2.06.

     102.0, 210.0, 216.3, 148.5, 76.5, 31.5, 10.8, 3.2, 0.8, 0.2.

15. (i) 1.5   (ii) 1.1, 0.475.

16. (i) 0.0228, (ii) 0.1492, (iii) 0.99672, (iv) 0.6639, (v) 0.1144, (vi) 0.5948, 76.62 − 83.38.

17. (i) 151, (ii) 145.3, (iii) 48.5, (iv) 63.5.

18. (i) 7.9 days (ii) 14.1 days. **19.** 200.03 cm^3.

20. 0.851 g cm^{-3}. **21.** 0.034.

CHAPTER 16

1. 4.72, 0.08. **2.** 13.

3. average: 87.05, 88.31, 92.69, 93.95 kg cm^{-2}.
   range: 0.93, 2.16, 10.54, 13.75 kg cm^{-2}.

4. 0.04.

5. average: 82.66, 88.98, 83.79, 87.85.
   range: 0.86, 1.98, 9.69, 12.63.

6. average: 1.583, 1.612, 1.717, 1.746
   range: 0.034, 0.068, 0.280, 0.362.

7. Change in mean after sample 12.
   Limits calculated from last 8 samples: 1.640, 1.669, 1.773, 1.803.

8. 5. **9.** 0, 2, 14, 18. **10.** 3, 5.

11. Increase in the number of defectives after sample 14. Limits calculated from the last 6 samples: 8, 11.

12. 5, 7.

CHAPTER 17

1. $F(9, 9) = 2.6$; double-sided test; NS.

2. $F(9, 9) = 2.26$; single-sided test; NS.

3. (i) $u = 2.59$; single-sided test; S* (ii) 8.45 to 8.79%.

4. $F(9, 9) = 1.3$ NS; $t(\phi = 16) = 2.55$; double-sided; S.

5. (i) $t(\phi = 9) = 5.18$; single-sided; S** (ii) 0.25 to 0.63%.

6. $t(\phi = 6) = 0.7$; double-sided; NS.

7. $F(12, 10) = 1.98$; NS; $t(\phi = 22) = 1.57$ double-sided NS.

8. $\chi^2(\phi = 7) \simeq 2.5$; good fit.

9. $\chi^2(\phi = 1) = 1.55$ (with correction) NS.

10. $t(\phi = 7) = 2.42$, S.

11. $t(\phi = 18) = 1.73$, NS.

12. Yes; yes, $P(>1) = 0.44$ (use Poisson distribution).

13. Yes, $\chi^2(\phi = 1) = 3.84$ (use correction) SS test.

14. $\chi^2(\phi = 4) = 5.53$ NS.

CHAPTER 18

1. $F(3, 16) = 1.8$ NS.

2. (i) $F(2, 10) = 0.91$, (ii) $F(5, 10) = 3.17$ PS. (iii) 15.28 to 15.86%
   (iv) no interaction between analyst and batch.

3. $F_{A \times I}(6, 12) = 3.56$ S the interaction indicates that different inhibitors should be used for different types of antifreeze.

4. $F(2, 9) = 2.41$ NS.

5. $F(8, 12) = 3.50$ S, $F(3, 8) = 2.47$,
   sampling variation too great to assert difference.

6. (i) $F(2, 18) = 0.6$ NS   (ii) $F(2, 18) = 4.45$ S.

CHAPTER 19

1. CaO $= 0.099(E) + 1.32$.   2. $V = 2.032(i) - 0.223$.

3. $L = 0.0059(F) - 2.51$.   (ii) $F(3, 1) = 387$ S**.

4. (i) $M = 1.096$, $K = 19.94$.   (ii) 33.47 to 34.81.

5. (i) $T = 99.73 + 0.1077(t) - 0.001734(t)^2$ (ii) $t = 31$ min.

6. Cal. $= 8.982 - 0.001787\,(^{\circ}C) + 0.00000267\,(^{\circ}C)^2$.

CHAPTER 20

1. $n = 4$.

2. (i)

Source	d.f.	m.s.
Time	3	1329.20
Mould	2	75.59
Error	6	66.36

(ii) Strength increases with ageing time.

3. (i) Significant difference between liquids;   (ii) 0.45 to 0.47 g/denier.

4.

Source	d.f.	m.s.
Mix (M)	3	71
Speed (S)	1	37
Level (L)	2	830 S*
MS	3	26
ML	6	96
SL	2	113
MSL	6	68

5. (i) CD
   (ii) Block 1 (1), ab, acd, bcd.
        2 ac, bc, d, abd.
        3 b, a, $abcd$, cd.
        4 c, abc, ad, bd.

6. (i) $F_A(1, \infty) = 25$, S**,
       $F_B(1, \infty) = 4$, S,
       $F_D(1, \infty) = 4$, S;
   (ii) $A + BCD + CE + ABDE$
        $B + ACD + ABCE + DE$ etc.

Index